SOMA 2003

Symposium on Mediterranean Archaeology

Proceedings of the Seventh Meeting of Postgraduate Researchers at the Institute of Archaeology, University College London, 21st–23rd February 2003

Edited by

Camilla Briault
Jack Green
Anthi Kaldelis
Anna Stellatou

BAR International Series 1391
2005

Published in 2016 by
BAR Publishing, Oxford

BAR International Series 1391

SOMA 2003 - Symposium on Mediterranean Archaeology

ISBN 978 1 84171 829 3

BAR Publishing is the trading name of British Archaeological Reports (Oxford) Ltd. British Archaeological Reports was first incorporated in 1974 to publish the BAR Series, International and British. In 1992 Hadrian Books Ltd became part of the BAR group. This volume was originally published by Archaeopress in conjunction with British Archaeological Reports (Oxford) Ltd / Hadrian Books Ltd, the Series principal publisher, in 2005. This present volume is published by BAR Publishing, 2016.

Printed in England

BAR
PUBLISHING

BAR titles are available from:

BAR Publishing
122 Banbury Rd, Oxford, OX2 7BP, UK
EMAIL info@barpublishing.com
PHONE +44 (0)1865 310431
FAX +44 (0)1865 316916
www.barpublishing.com

CONTENTS

LIST OF FIGURES

LIST OF TABLES

INTRODUCTION

The Symposium on Mediterranean Archaeology (SOMA) was first established in 1997 with the chief intention to bring together students studying Mediterranean Archaeology at postgraduate level in universities across the United Kingdom. Despite the diverse character of Mediterranean archaeology, its long academic history and study by different schools of thought, the creation of SOMA lies primarily in the recognition of the role played by the Mediterranean Sea – and the landmasses that it encompasses and reaches – in the development of civilisations, cultures and societies over a wide geographical and temporal spectrum. Although the research traditions and archaeological trajectories of different Mediterranean regions are extremely diverse, there are many common threads running between them. Hence, it is more appropriate to discuss, and work towards, an integrated 'Mediterranean Archaeology' rather than maintaining separate 'Mediterranean Archaeologies' of particular regions.

It is this dialogue of 'unity in diversity' that formed the background for this SOMA conference, which took place at the Institute of Archaeology, University College London on 21st-23rd February 2003. The proceedings of that conference are presented in this volume, and broadly reflect the diversity of research areas and topics covered. Our aim was to break down traditional boundaries and address theoretical and scientific issues relating to a series of subjects under study by highlighting universal problems and attempting to identify common and useful lines of investigation. We therefore recognise the need to develop 'pan-Mediterranean' approaches to archaeology in this region, whilst acknowledging the fact that much is yet to be done to bring together common discourses in this highly regionalized and multi-disciplinary area of study.

A recent work that begins to formulate a 'pan-Mediterranean' approach within the *longue duree* of Mediterranean (pre)-history is Horden and Purcell's *The Corrupting Sea* (2000). Given the importance of this contribution to the study of Mediterranean life, history and archaeology, it was considered appropriate to invite Nicholas Purcell (St Johns College, Oxford) to present the opening lecture at SOMA 2003. Entitled *'Colonisation and Mediterranean* History', this lecture provided a detailed review of ancient and modern concepts of colonization, highlighting the social, economic and political complexities of colonization processes in Classical periods. Colonisation is also a theme examined in several other papers given by students during the course of SOMA 2003. We are indebted to Mr Purcell for this contribution, which set the tone for the conference by introducing both the long time-depth that is the preserve of archaeological approaches, and the wide geographical, historical and cultural remit of Mediterranean studies.

The second key aim of SOMA is to act as a forum for current postgraduate research in all aspects of Mediterranean archaeology, providing an opportunity for students to take part in a conference, to present papers, and to also publish their preliminary research in the conference proceedings, often for the first time. This is essential for communicating ideas between students and establishing contacts that will prove to be beneficial during the course of current and future research. Our main objective was to bring together research students of different intellectual and cultural backgrounds to obtain a more collective knowledge of the research conducted in Mediterranean archaeology, and to introduce new and alternative methods of research, and to discuss general and specific issues arising from their work in an informal setting.

By taking advantage of London's centrality, SOMA 2003 went a step further than in previous years and welcomed many students from outside British universities. Although this was not the first time that students from other countries had participated in SOMA, the response to our call for papers was exceptional. Students from fourteen countries outside the United Kingdom contributed, making this a truly international symposium. Over 125 persons in total attended the conference, with contributions from 74 participants. Not only does this reflect the wide range of academic establishments and postgraduate projects concerned with Mediterranean archaeology, but the sheer number of participants also indicates that our goal of engaging in interdisciplinary and multi-period dialogue about the Mediterranean is one shared by many.

The papers were grouped together into sessions and workshops, the themes of which included: regionalism and cultural identities, archaeological landscapes and settlement patterns, economy, trade and exchange, production and technology, religion and ritual, death and burial, destruction and conflict, art and iconography, urban and domestic spatial analysis, texts in archaeology, and archaeobotany and zooarchaeology. Cultural heritage management was another popular topic that has played an increasingly significant role in the development of archaeological sites in the Mediterranean in the last few years. The individual sessions were presented and received enthusiastically, often followed by productive and thought provoking question and answer sessions, with scope for further discussion and debate where time allowed.

Professor Ruth Whitehouse, a key researcher and lecturer in many areas of Mediterranean archaeology at the Institute of Archaeology, was asked to sum up her thoughts at the end of SOMA 2003. In this summary, there was general acknowledgment that such active postgraduate research was helping to advance Mediterranean archaeology and that the Symposium had been an overall success. Writing in the Institute of Archaeology's annual publication, Professor Whitehouse comments: "The numbers of people involved, the range of themes covered, and the enthusiasm that pervaded the symposium, left no doubt that the archaeology of the Mediterranean is as popular as it ever was and that its future is in good hands".

This publication is broadly representative of the conference proceedings. Not every speaker submitted their papers for publication, but this volume contains a large proportion of those presented at the conference. We have chosen to arrange the papers in alphabetical order by contributor's name, as some of the original conference theme are under-represented in the publication. It is hoped that those using this volume should find this the easiest way of locating papers of interest to them. As editors, we have left the content of individual papers largely at the discretion of individual authors. References, bibliographies, figures and acknowledgements are presented at the end of each paper.

Acknowledgements

There are numerous people, institutions and research groups to whom we would like to express our gratitude for all their help and support in the organisation of SOMA 2003. For financial support, we are grateful to the Palestine Exploration Fund and the Social and Cultural Dynamics Research Group (UCL). Many thanks to those who chaired sessions: Carol Bell, Jenny Bredenberg, Helen Dawson, Myrto Georgakopoulou, Claudia Glatz, Reuben Grima, Dorella Romanou and Kalliopi Vacharopoulou. We would also like to thank the volunteers from the Institute of Archaeology who helped with registration, refreshments and technical support: Diana Beatty, Rachel Fentem, Scott Haddow, Kyri Karseras, Serena Love, Lisa McNally, Maria Mina, Carrie Murray, Owen O'Donnell, Kathryn Piquette, Andrea Richards, Caroline Skelton, Lindsay Spencer and Nichola Wythe. Finally, many thanks to Professor Ruth Whitehouse and Professor Peter Ucko for their support and advice, and for encouraging us to host SOMA 2003 at the Institute of Archaeology.

The SOMA 2003 Committee:

Camilla Briault, Jack Green, Anthi Kaldelis, Anna Stellatou

Bibliography

Hordern, P. and Purcell, N. 2000. *The Corrupting Sea: A Study of Mediterranean History,* Oxford.

Whitehouse, R.D. 2000. "The Social and Cultural Dynamics Research Group", *Archaeology International 2002-2003*, London.

North Syria in the sixth century AD: coast and hinterland

Nicolas Beaudry - Université de Montréal (Canada) / Université de Paris-I Panthéon-Sorbonne (France)

Since the exploration of the Syrian hinterland by the Marquis de Vogüé (1865-67), archaeological research on Byzantine North Syria[1] has remained heavily concentrated inland, either on the provincial capitals of Syria Prima and Syria Secunda, Antioch and Apamea, or in the area of the Limestone Hills (*Figure.1.1*). The Byzantine period has received much less attention on the coast, where a mountainous landscape, a dense forest cover and a militarised border have not allowed extensive explorations like the ones carried out inland by the Marquis de Vogüé and by Howard C. Butler (1903, 1929), or aerial surveys such as that of the Syrian *limes* (Mouterde & Poidebard 1945). The early Christian remains of the Lower Orontes Valley were visited by archaeologists, but not systematically recorded, and their later reuse and modifications by Georgian and Armenian Christians often make their interpretation and dating difficult (Djobadze 1986; Lafontaine-Dosogne 1967; Mécérian 1964; Sinclair 1990). The port of Antioch, Seleukia, has seen no excavation since a survey by the Princeton University team based in Antioch turned into the excavation of a tetraconch church, and was interrupted by WWII (Campbell 1941; Stillwell 1941a). The site of al Mina, at the mouth of the River Orontes, was excavated in the 1930's (Woolley 1937, 1938) but its Byzantine finds were never published (Pamir & Nishiyama 2002: 304; Vorderstrasse 2004: 94-95). Further south, two cisterns among ploughed Byzantine remains yielded sixth-century pottery assemblages in Ibn Hani (Touma 1984), but modern Lattakia seems to forbid any significant excavation in the ancient port-city of Laodikeia. Until recently, the only other excavated and fully published Byzantine sites in coastal North Syria were the sixth-century monasteries of Saint Baarlam, on Mount Kasios (Djobadze 1986: 3-53), and of St. Symeon the Younger, on the Wondrous Mountain (Djobadze 1986: 57-113), thus the interest of a church under excavation at Ras el Bassit.

Ras el Bassit is a cape, some 50 km north of Lattakia and about halfway to Seleukia and the mouth of the River Orontes (*Figure 1.1*). It shelters a bay dominated to the north by Mount Kasios, in which lies one of the best anchorages of the Levantine coast (Sanlaville 1976). The site is best known for the excavations led by Paul Courbin from 1971 to 1984 and for their contribution to the study of the Levantine Iron Age (Courbin 1978, 1986), but Courbin's work also revealed a final period of intense activity from the late third to the sixth centuries AD. To this period belongs what he interpreted as a large three-naved basilica, one wall of which was still standing in the forest (Courbin 1972, 1973). A sondage in the north-east corner of the monument yielded two rooms in 1973, of which at least one was part of a synagogue in the late fifth to early sixth centuries, but the monument was not further investigated and it was still hardly published when Courbin died in 1994. Following a preliminary analysis of Courbin's archive (Beaudry 1998), fieldwork in Bassit resumed in 2000 to investigate what turned out to be a complex comprising an oriented three-naved church built in the dry, large ashlar masonry typical of North Syrian limestone architecture of the fifth and sixth centuries (Beaudry 2004). Work has concentrated so far on the church itself; after three missions, selective excavation has exposed most of the plan of its last phase (*Figure 1.2*) and provided insights on its elevation.

Since this church is the first urban basilical church to be excavated and recorded to this day on the North Syrian coast, comparative material must be looked for inland, mostly among the villages churches of the Limestone Hills. This paper will briefly report on the results of the ongoing excavation of this church, focusing on its architecture and liturgical installations. It will then consider whether, and to what extent, existing architectural models drawn from the North Syrian hinterland can apply to a monument like this, located on the coast. It will also point at some of its features that demonstrate different approaches to sacred space and to the staging of rituals.

The church of Ras el Bassit

The church complex of Bassit is located at the foot of the acropolis, within an enclosure wall partly identified to the west and east. It comprises rooms to the north, including the two excavated by Courbin. A room to the southwest opens to the west and supported a second floor. It is abutted by a later room opening east into what may have been a courtyard.

The church is a three-naved basilica (approx. total inner length 26.1 m; *quadratum populi* approx. 20.9 m long × 20.0 m wide). To the West, a colonnaded inner narthex forms a transversal aisle and defines, with the north and south side-aisles, a U-shaped collateral around a central nave (*Figure 1.2*). The floor of the nave is two steps down from those of the narthex and side aisles; all the floors are made of clay mortar. The west façade of the

[1] The word "Byzantine" refers, as it usually does in the Levant, to the period spanning from the reign of Constantine (first half of the third century AD) to the Arab conquest (battle of the Yarmūq, 636 AD). All dates AD / CE unless otherwise stated.

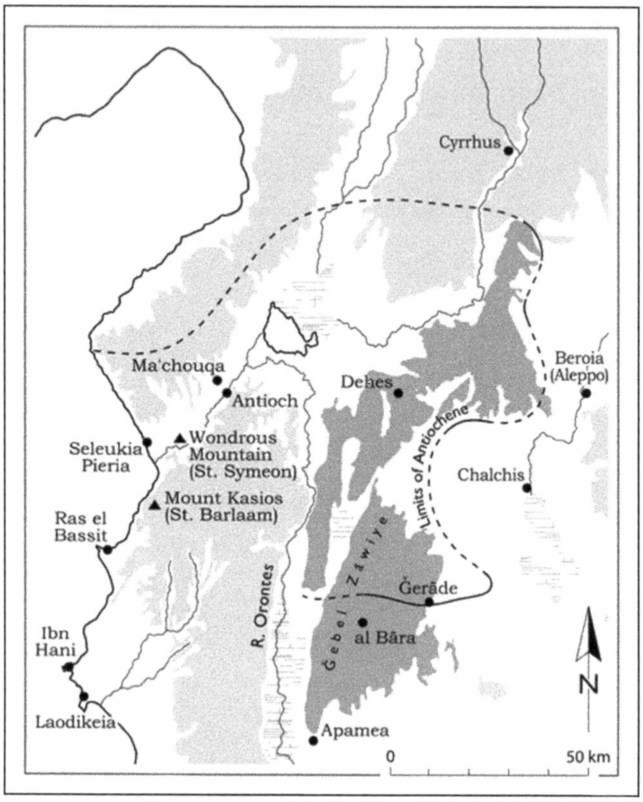

Figure 1.1. *Byzantine North Syria, the Limestone Hills (in dark grey) and the limits of Antiochene. Adapted from Seyrig 1958 and Feissel 2000.*

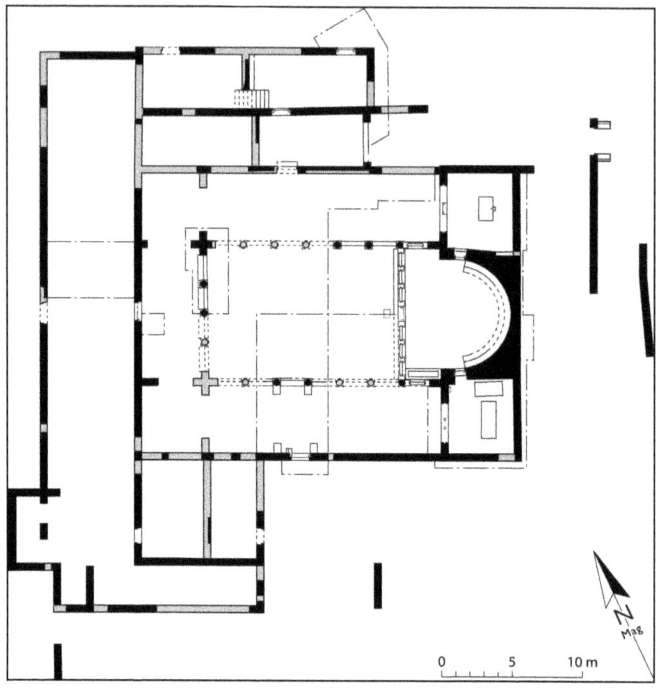

Figure 1.2. Ras el Bassit, simplified plan of the church complex, October 2002; reconstructions in grey.

church is partly preserved up to the cornice (*Figure 1.3*); a single door, whose lintel bears a cruciferous medallion, was sheltered by a porch and flanked by rectangular windows. The reverse of the façade shows the sockets of some of the beams that supported a U-shaped gallery above the narthex and side-aisles (*Figure 1.4*). In the nave itself, where the north and south colonnades fell inwards on top of each other, significant parts of the collapsed elevations could be read in the rubble, including a bay complete up to the capitals of the second order (Beaudry 2004). Architectural blocks are numbered, coordinated and catalogued, and the reconstruction and study of the architectural elevations is currently in progress.

The sanctuary occupies the last bay, and was separated from the naves by a limestone screen, partly preserved *in situ*. The apse is slightly indented and holds a two-stepped *synthronon* along its wall, partly destroyed and partly covered by a later chapel associated with burials and early thirteenth-century finds. The apse and sanctuary are at the same level (at least in their last phase), two steps higher than the nave, and were paved in *opus sectile*. A tomb follows the south chancel, and yielded human teeth and disarticulated bones. Fragments of a thick, rectangular marble table-top (*mensa*) probably belong to the main altar, but no indication was preserved of its original position.

The apse is flanked by two rooms combining ashlar and double-course masonries. Both rooms open through doors into the apse, and into the side-aisles through wide arches originally fitted with wooden screens. In the much-disturbed south room, a (robbed) tomb follows the wall of the apse, and a strong foundation of large blocks faces the arched access.

Unlike the south room, most of the north room remained sealed under its destruction layer. It yielded a virtually complete table consisting of a rectangular marble *mensa*, five colonnette legs, and a marble slab laid on a concrete bedding and surrounded with a frame (*Figure 1.5*). While legs are normally embedded into the floor or recessed into a base-plate (Duval 1994: 174-175), the *mensa*, colonnettes and base slab of Bassit were simply put on top of each other and no trace of mortar was found. This assemblage was probably unstable during earthquakes, and may thus have been one of the first local attempts at this type of table. On its eastern side, the base slab is pierced with a funnel, from which a tube leads into the sealed compartment of a reliquary (*Figure 1.6*). This compartment connects through an opening with a second, deeper compartment that remained accessible, the lid and cover-plate of which were found near by. The sealed compartment yielded two metal containers (*capsellae*) and a casket, all containing bone relics, as well as a glass flask (Beaudry 2004).

Figure 1.3. West façade, looking NE, October 2001.

Figure 1.4. Reverse of the west façade and sondage in the narthex, looking W, September 2001. Scale 2 m.

Figure 1.5 . North annex room and collapsed table, looking E, September 2001. Scale 2 m

Figure 1.6. Reliquary of the north annex room, looking W, September 2001. Scale 15 cm.

The sculpted architectural decor is relatively limited, as only the second order has profiled column bases. The side chancel plaques and parapets are simple framed panels, while only the chancel plaques facing the nave bear a simple cross. The capitals are all different: leaves and cruciferous medallions are the most common features, carved in low relief or simply outlined, but other motifs and types are also found. The decor also includes a *champlevé* marble frieze (with motifs reserved against a roughly chiselled background originally filled with coloured paste: Boyd 1982; Metzger 1980) depicting animals and frets (Beaudry 2004). The closest parallel to this frieze comes from Seleukia Pieria, and post-dates the earthquake of 528 (Boyd 2000; Stillwell 1941b). The pottery and coin evidence available so far supports a similar *post quem* in Bassit, though the sculpted decor makes a date later in the sixth century more likely (Beaudry 2004). This would be consistent with a typo-chronology of altars as defined in Jordan, where colonnette-legged tables were introduced during the second half of the sixth century (Michel 1999: 31-34, 2001: 61-68).

During this period, North Syria experienced violent earthquakes, successive visitations of the plague, a Persian invasion, and a second invasion followed by three decades of occupation (Tate 1989). The actual demographic and economic consequences of these disruptions are debatable (Biraben 1989; Durliat 1989; Kennedy 1985: 181-183; Tate 1992: 74-75; Trombley 1997, etc.), but epigraphic evidence and the results of excavations in the village of Dēhes (Sodini *et al.* 1980) both suggest that house building ceased in the Limestone Hills in the mid-sixth century, and that church building ceased in the early seventh century (Tate 1992). In Bassit, a *follis* struck by Heraclius and found under the collapsed roof sets the *post quem* for the destruction of the church to 617-619, between the short-lived Byzantine reconquest of Syria over the Persians and its loss to the Arabs.

Bassit and the North Syrian church corpus: the provincial models

Outside the capitals of the provinces of Syria Prima and Syria Secunda, Antioch and Apamea, the bulk of Byzantine archaeological data lies in the Limestone Hills, a group of mountains in the hinterland of Antioch including the Djebel Zāwiye, half of which is in Apamene (*Figure 1.1*). Until the early 20th century, these mountains were more or less deserted pasture dotted with ruins: the so-called 'Dead Cities' (Foss 1996), gradually abandoned from the seventh century onwards, thus the exceptional and often spectacular state preservation of buildings. Around 1600 villages and monasteries have been identified, of which about 100 have been studied (Peña *et al* 1987, 1990, 1999; Tate 1992; Tchalenko 1953-1958). Inscriptions date the first churches to the second half of the fourth century. Dated inscriptions also define provincial limits, since Antioch counted years in a Caesarean era (49 BC) while Apamea, Chalcis and Cyrrhus had kept a Seleukid era (312 BC) (Feissel 2000; Seyrig 1958). These limits are confirmed, with only slight variations, by the mapping of liturgical settings in churches, each province having a very strong identity regarding rituals and their staging (Donceel-Voûte 1988b, 1996, 1998: 98; Sodini 1988).

All but a few of the village churches of the Limestone Hills are three-naved basilicas (Sodini 1989; Tchalenko 1953-58, 1979-80, 1990). In the province of Antiochene, from the fourth to the sixth century, the main door moves gradually from the south side, where the church is accessed from a court, to the west façade, where an axial door allows for a more monumentalised access; in the province of Apamene, the western access comes into general use much earlier. Back on the coast, the church in Bassit is also accessed from the north and south, but its main door is to the west, facing a door in the enclosure wall (*Figure 1.2*).

The tripartite chevet is the norm in North Syria, and from the fifth century onwards the plan of the rooms flanking the apse clearly shows their specialisation (Sodini 1989). In the province of Antiochene, the north room is where the clergy are thought to have prepared for the Mass; it opens into the side aisle through a door, and usually opens into the apse as well. To the south is a martyrium, where the reliquaries are displayed; it opens on the side aisle through an arch normally closed by a screen. The reliquaries are free-standing, often taking the shape of small sarcophagi with double-sloped lids and corner acrotera; their lids are pierced to allow the introduction of a liquid, usually olive oil, which is sanctified by contact and collected through a second orifice on the face or side (Canivet 1978; Gessel 1988; Lassus 1947: 163-167). Though they can be found in other Levantine provinces, these oil reliquaries are a typical North Syrian feature; foundation relics, deposited under or within altars, are unknown in Syria, Phoenicia (Donceel-Voûte 1998: 121) and Cyprus (Papageorghiou 1985: 322-324). In the province of Apamene, the martyrium is the north annex room (Canivet 1978: 156 Figure. 1), and its access is not always as monumentalised as in Antiochene, though the plan still shows the specialisation of the rooms.

At the church of Bassit, the plan of the chevet shows no such specialisation. However, the tomb in the south annex room strongly suggests that it had martyrial functions, since privileged burials tend to gather *ad sanctos*, by the relics (Sodini 1986: 236). The foundation facing the main access into the room was probably meant for a sarcophagus or a large reliquary (to which may belong a large fragment of marble) rather than for an altar, and the tomb in the sanctuary confirms the existence of a martyrial focus to the south, as is the rule in the province of Antiochene. Churches in this province often have more than one door to the south, including a larger, more ornate one by the martyrium, to allow the pilgrims a direct access (Donceel-Voûte 1995: 195-196). Though the church in Bassit has only one door to the south, fragments of decor found in its close vicinity find no equivalent at the west and north doors.

Despite these 'Antiochean' features, the plan and elevation of the church also show features that are rare or unknown in the province of Antiochene: a *synthronon*, providing seating in the apse for the clergy; a narthex opening into the nave through a colonnade, forming a transversal western aisle; galleries above the side-aisles and narthex, allowing a vertical division of the community. No galleries are known in Antiochene. A church in Ma'chouqa, in the northern suburbs of Antioch, had a narthex and may have had a *synthronon*, but it also shows other features foreign to the models of eastern and southern Antiochene (Donceel-Voûte 1988a: 174-177). The only other confirmed *synthronon* in Antiochene is in Banqūsa, a village close to the neighbouring province of Chalcidene (Lassus 1972: 14-15). On the other hand, these features are common in the province of Apamene, and the U-shaped collateral and gallery of the church in Bassit may find their closest parallels in the churches of al Bāra (Fourdrin 1992; Lassus 1972; Tchalenko 1953-58: vol. II pl. XII).

The topography of the church in Bassit thus matches the 'Apamean' model better, save for the position of its presumed martyrium. The limits of the province of Antiochene are not defined to the south-west (*Figure 1.1*), since both Antioch and Laodikeia counted years in Caesarean eras (49 and 48 BC) (Seyrig 1958); it is thus possible that Bassit was a 'border town' between the provinces of Antiochene and Apamene. This would perhaps provide an explanation for an 'Antiochean' martyrium among more 'Apamean' features: conversely, the church of the border town of Ğerāde had, in its last phase, a martyrium to the North among typical Antiochean features (Tchalenko 1979-80: Figure. 487, 1990). Bassit may be a border town between the provinces of Antiochene and Apamene, but this would not account for the secondary altar and reliquary in the north room of its church, nor for its symmetrical chevet. Bassit also sits on the Mediterranean border of Syria, and parallels to these features must be sought outside the North Syrian Provinces.

Bassit and the Mediterranean world

The north room at Bassit does have a martyrial character, and it is equipped to satisfy a demand for secondary relics in the form of sanctified oil. However, the association of the reliquary with an altar implies a relation to the relics and a set of rituals that are foreign to Syrian liturgy. This practice, common in Constantinople, in the Aegean world and in the Western Mediterranean, and normally linked to the dedication of the church or the altar, appears to have spread throughout Palestine and Arabia during the course of the sixth century (Donceel-Voûte 1996: 328; Michel 2001: 72-81). Foundation reliquaries found *in situ* in the East church of Pella and in Khirbet es-Samra, Jordan (Michel 2001), were intended for rituals of sanctification by contact but were sealed and inaccessible and their function had thus changed. On the contrary, the one in Bassit was made specifically for the setting in which it was found, its sides and base having been left unfinished (Beaudry 2004). This shows a deliberate merging of functions, rather than the adaptation of an existing setting. Other foundation reliquaries may have allowed some other forms of sanctification by contact (e.g. in St. Basilius of Rihab and in the North church of Esbous, Jordan: Michel 2001), and the altar base of a church in Khirbet Beit Sila, near Jerusalem, may have allowed the retrieval of some sanctified liquid through the removal of the reliquary from under the altar base (Batz 2002). However, the only oil circulation systems within altar bases comparable to that of Bassit are to be found in Ravenna, Italy (Sant'Apollinare Nuovo and Museo arcivescovile: Deichmann 1969: 74-75, Figures. 112-114,118; 1974: 139, Figures. 105-108, 111).

This association of relics with an altar merges two different sets of rituals that North Syrian liturgy would have staged in two different and independant areas. These rituals are further brought together in Bassit by the transversal axis formed by the doors opening into the apse, that unify the chevet behind the screens of the sanctuary and side rooms. This axis finds parallels in some churches of Cilicia (Alahan, Dağ Pazarı, etc.: Hellenkemper 1994: Figure. 13, 25; Hill 1996, 1998), and may be paralled with the passages connecting the side-rooms behind the apse in many other Cilician churches (Korykos, Kadirli, etc.: Hellenkemper 1994: Figure. 4, 5; Hill 1996: 28-37, pl.). It may also be recognized in the passages between the main and secondary apses in the churches of Northeast Cyprus (Lamboussa, Soloi, Karpasia, etc.: Papageorghiou 1985: 301-303). It may even evoke the extension of the sanctuary to the side-aisles in some Palestinian and Jordanian churches (Donceel-Voûte 1996, Michel 2001: 54-55). It certainly implies a conception and economy of the sacred space that is different from the one expressed in the North Syrian models.

Conclusion

With its dry, large ashlar masonry, its simple volumes and its tripartite chevet, the church in Bassit does belong to the tradition of North Syrian limestone architecture. A probable martyrium to the South may relate it to the province of Antiochene, but its *synthronon*, narthex and galleries find their closest parallels in the province of Apamene.

However, these features may be found in other provinces as well, and its foundation reliquary and unified, symmetrical chevet are foreign to North Syria. They suggest further experimentation and a greater exposure of the coast to 'foreign' rituals and architectural ideas, whether from neighbouring coastal provinces like Cilicia, from Palestine, or even from the Capital and the Aegean world. Pottery assemblages have shown that the penetration of imported material culture decreased dramatically with the distance to the sea (Touma 1984); outside the big cities and military outposts, the hinterland is known to have been less receptive than the coast to Hellenisation and Orthodoxy (Mundell Mango 2000). The penetration of 'imported' rituals and architectural ideas would have similarly decreased with the distance to the sea (Krautheimer 1986: 135).

Comparison with non-Syrian data sets will certainly allow a better understanding of this church, and may allow its features to be better interpreted in relation to neighbouring regional traditions. It must also be interpreted within wider patterns of change, since its last phase belongs to the very last decades of Byzantine Syria, a period poorly represented in the North Syrian corpus, but during which the relation to relics change in Jordan and Palestine.

The church of Ras el Bassit confirms the existence of a coastal North Syria that belongs more closely than the hinterland to the Mediterranean world, a border region on the Mediterranean fringes of Syria; it already shows the importance of building a regional corpus. The site of Bassit has proved its potential for bringing fresh data and a significant contribution to this corpus, and for allowing a different perspective on sixth-century North Syria than the one offered, from the hinterland, by the Dead Cities.

Acknowledgements

I wish to thank the Directorate of Antiquities and Museums of the Syrian Arab Republic for allowing fieldwork on this monument; Prof. Jacques Y. Perreault, for entrusting me with this project and for securing the resources required; Prof. Jean-Pierre Sodini, for his invaluable support and advice. Fieldwork was funded by the Université de Montréal and staffed with students, volunteers, and local workers; I am grateful to all. This research was supported by the Social Sciences and Humanities Research Council (Canada), the Ministère des Affaires étrangères (France) and Ministère des Relations internationales (Québec), the Université de Montréal, and the Fonds pour la formation de chercheurs et l'aide à la recherche (Québec). This paper was kindly proofread by Kate Roberts and Tom Jackson and benefited from the comments of the editors; any remaining errors are my own.

Bibliography

Batz S. 2002. "The church of St. Theodore at Khirbet Beit Sila" *Israel Museum Studies in Archaeology* 1: 39-54.

Beaudry N. 1998. *Excavation at the basilica of Ras al-Basit (Syria), 1973: report from the archive.* Unpublished M.A. diss., University College London.

Beaudry N. 2004. *La basilique de Ras el Bassit, une église paléochrétienne sur la côte de la Syrie du Nord.* Unpublished Ph.D. diss., Université de Montréal & Université de Paris-I Panthéon-Sorbonne.

Biraben J.-N. 1989. "Rapport: la peste du VIe siècle dans l'Empire byzantin", in C. Morrisson & J. Lefort (eds.) *Hommes et richesses dans l'Empire byzantin* I: *IVe-VIIe siècle*. Paris: 121-125.

Boyd S. 1982. "A little-known technique of architectural sculpture: champlevé reliefs from Cyprus" *Jahrbuch der Österreichischen Byzantinistik* 32: 314-325.

Boyd S. 2000. "The relief decoration of the church building at Seleucia Pieria", in C. Kondoleon (ed.) *Antioch: the lost ancient city*. Princeton & Worcester: 220-223.

Butler H. C. 1903. *Publications of an American archaeological expedition to Syria in 1899-1900*, II: *Architecture and other arts*. New York.

Butler H. C. 1929. *Early churches in Syria, fourth to seventh centuries*. Princeton.

Campbell W. A. 1941. "The martyrion at Seleucia Pieria", in R. Stillwell (ed.) *Antioch on-the-Orontes* III: *The excavations 1937-1939*. Princeton: 35-54.

Canivet M. T. 1978. "Le reliquaire à huile de la grande église de Hūarte (Syrie)" *Syria* 55: 153-162.

Courbin P. 1972. "Rapport sur la fouille de Ras el Bassit 1971" *Annales archéologiques arabes syriennes* 22: 45-61.

Courbin P. 1973. "Ras el Bassit: rapport sur la campagne de 1972" *Annales archéologiques arabes syriennes* 23: 25-38.

Courbin P. 1978. "Une nouvelle fouille française sur la côte syrienne: a-t-on retrouvé l'antique Posideion à Ras el Bassit?" *Archéologia* 116: 48-62.

Courbin P. 1986. "Bassit" *Syria* 63: 175-220.

Deichmann F. W. 1969. *Ravenna, Haupstadt des spätantiken Abendlandes* I: *Geschichte und Monumente*. Wiesbaden.

Deichmann F. W. 1974. *Ravenna, Haupstadt des spätantiken Abendlandes* II.1: *Kommentar. Die Bauten bis zum Tode Theoderichs des Großen*. Wiesbaden.

Djobadze W. 1986. *Archaeological investigations in the region west of Antioch on-the-Orontes*. Stuttgart.

Donceel-Voûte P. 1988a. *Les pavements des églises byzantines de Syrie et du Liban: décor, archéologie et liturgie*. Louvain-la-Neuve.

Donceel-Voûte P. 1988b. "Provinces ecclésiastiques et provinces liturgiques en Syrie et Phénicie byzantines", in P.-L. Gatier, B. Helly & J.-P. Rey-Coquais (eds.) *Géographie historique au Proche-Orient (Syrie, Phénicie, Arabie, grecques, romaines, byzantines). Actes de la Table Ronde de Valbonne, 16-18 septembre 1985*. Paris: 212-217.

Donceel-Voûte P. 1995. "Le rôle des reliquaires dans les pèlerinages", in E. Dassmann, K. Thraede & J. Engemann (eds.) *Akten des XII. Internationalen Kongresses für christliche Archäologie, Bonn 22.-28 September 1991*. Münster: I, 184-205, pl. 13.

Donceel-Voûte P. 1996. "La mise en scène de la liturgie au Proche-Orient IVe-IXe s.: les 'provinces liturgiques'", in R. F. Taft (ed.) *The Christian East, its institutions & its thought: a critical reflection. Papers of the International Scholarly Congress for the 75th Anniversary of the Pontifical Institute, Rome, 30 May – 5 June 1993*. Rome: 313-338.

Donceel-Voûte P. 1998. "Le fonctionnement des lieux de

culte aux VIe-VIIe siècles: monuments, textes et images", in N. Cambi & E. Marin (eds.), *Radovi XIII. međunarodnog kongresa za starokršćansku arheologiju, Split – Poreč (25. 9. – 1. 10. 1994.) / Acta XIII congressus internationalis archaeologiae christianae, Split – Poreč (25. 9. – 1. 10. 1994).* Split & Vatican City: II, 97-156.

Durliat J. 1989. "La peste du VIe siècle. Pour un nouvel examen des sources byzantines", in C. Morrisson & J. Lefort (eds.) *Hommes et richesses dans l'Empire byzantin* I, *IVe-VIIe siècle*. Paris: 107-119.

Duval N. 1994. "L'architecture chrétienne et les pratiques liturgiques en Jordanie en rapport avec la Palestine: recherches nouvelles", in K. Painter (ed.), *'Churches built in ancient times': recent studies in early Christian archaeology*. London: 149-212.

Feissel D. 2000. "Ères locales et frontières administratives dans le Proche-Orient protobyzantin", in K. Belke, F. Hild, J. Koder & P. Soustal (eds) *Byzanz als Raum: zu Methoden und Inhalten der historischen Geographie des östlichen Mittelmeerraumes*. Vienna: 65-74.

Foss C. 1996. "Dead Cities of the Syrian hill country" *Archaeology* 9.5: 48-53.

Fourdrin J.-P. 1992. "L'église E.5 d'el Bāra" *Syria* 69: 171-210.

Gessel W. 1988. "Das Öl der Märtyrer. Zur Funktion und Interpretation der Ölsarkophage von Apamea in Syrien" *Oriens Christianus* 72: 183-202.

Hellenkemper H. 1994. "Early church architecture in Southern Asia Minor", in K. Painter (ed.) *'Churches built in ancient times': recent studies in early Christian archaeology*. London: 213-238.

Hill S. 1996. *The early Byzantine churches of Cilicia and Isauria*. Aldershot.

Hill S. 1998. "Alahan and Dağ Pazarı", in R. Matthew (ed.) *Ancient Anatolia: fifty years' work by the British Institute of Archaeology at Ankara*. London: 315-337.

Kennedy H. 1985. "The last century of Byzantine Syria: a reinterpretation" *Byzantinische Forschungen* 10: 142-183.

Krautheimer R. 1986. *Early Christian and Byzantine architecture*. Harmondsworth.

Lafontaine-Dosogne J. 1967. *Itinéraires archéologiques dans la région d'Antioche. Recherches sur le monastère et sur l'iconographie de S. Syméon Stylite le Jeune*. Brussels.

Lassus J. 1947. *Sanctuaires chrétiens de Syrie. Essai sur la genèse, la forme et l'usage liturgique des édifices du culte chrétien en Syrie du IIIe siècle à la conquête musulmane*. Paris.

Lassus J. 1972. "Églises d'Apamène" *Bulletin d'études orientales* 25: 5-36.

Mécérian J. 1964. "Expédition archéologique dans l'Antiochène occidentale" *Mélanges de l'Université Saint-Joseph* 40: 1-144.

Metzger C. 1980 "Exemples d'iconographie de mosaïque appliquée à la sculpture: à propos de deux plaques à décor « champlevé » du Musée du Louvre" *Mélanges de l'École française de Rome, Antiquité* 92: 545-561.

Michel A. 1999. "Le culte des reliques dans les églises byzantines de Jordanie" *Hortus Artium Medievalium* 5: 31-40.

Michel A. 2001. *Les églises d'époque byzantine et umayyade de la Jordanie, Ve-VIIIe siècle. Typologie architecturale et aménagements liturgiques (avec catalogue des monuments)*. Turnhout.

Mouterde R. & Poidebard A. 1945. *Le* limes *de Chalcis: organisation de la steppe en Haute Syrie romaine. Documents aériens et épigraphiques*. Paris.

Mundell Mango M. 2000. "Syria", in A. P. Kazhdan (ed.) *Oxford dictionary of Byzantium* III. New York & Oxford: 1997-2000.

Pamir H. & Nishiyama S. 2002. "The Orontes Delta Survey: archaeological investigation of ancient trade stations/settlements" *Ancient West & East* 1: 294-314.

Papageorghiou A. 1985. "L'architecture paléochrétienne de Chypre" *Corsi di cultura sull'arte ravennate e bizantina* 32: 299-324.

Peña I., Castellana P. & Fernández R. 1987. *Inventaire du Jebel Baricha. Recherches archéologiques dans la région des Villes mortes de la Syrie du Nord*. Milan.

Peña I., Castellana P. & Fernández R. 1990. *Inventaire du Jebel el-A'la. Recherches archéologiques dans la région des Villes mortes de la Syrie du Nord*. Milan.

Peña I., Castellana P. & Fernández R. 1999. *Inventaire du Jebel Wastani. Recherches archéologiques dans la région des Villes Mortes de la Syrie du Nord*. Milan.

Sanlaville P. 1976. *Aperçu géographique sur le site de Ras-el Bassit*. Unpublished report no. RCP 438. Lyon, Maison de l'Orient méditerranéen ancien.

Seyrig H. 1958. "Inscriptions grecques", in G. Tchalenko, *Villages antiques de la Syrie du Nord. Le Massif du Bélus à l'époque romaine* III. Paris: 2-62.

Sinclair T. A. 1990. *Eastern Turkey: an architectural and archaeological survey*, IV. London.

Sodini J.-P. 1986. "Les 'tombes privilégiées' dans l'Orient chrétien (à l'exception du diocèse d'Égypte)", in Y. Duval & J.-Ch. Picard (ed.) *L'inhumation privilégiée du IVe au VIIIe s. en Occident*. Paris: 233-243.

Sodini J.-P. 1988. "Géographie historique et liturgie: l'opposition entre Antiochène et Apamène", in H. Ahrweilwer (ed.) *Géographie historique du monde méditerranéen*. Paris: 201-206.

Sodini J.-P. 1989. "Les églises de Syrie du Nord", in J.-M. Dentzer & W. Orthmann (ed.) *Archéologie et histoire de la Syrie* II: *La Syrie de l'époque achéménide à l'avènement de l'Islam*. Saarbrücken: 347-372.

Sodini J.-P., Tate G., Bavant S., Biscop J.-L. & Orssaud D. 1980. "Déhès (Syrie du Nord), campagnes I-III (1976-1978): recherches sur l'habitat rural" *Syria* 75: 1-301.

Stillwell R. 1941a. "Outline of the campaigns", in R. Stillwell (ed.) *Antioch on-the-Orontes* III: *The excavations 1937-1939*. Princeton: 1-34.

Stillwell R. 1941b. "Catalogue of sculpture", in R. Stillwell (ed.), *Antioch on-the-Orontes* III: *The excavations 1937-1939*. Princeton: 116-134.

Tate G. 1989. "La Syrie à l'époque byzantine: essai de synthèse", in J.-M. Dentzer & W. Orthmann (ed.), *Archéologie et histoire de la Syrie* II: *La Syrie de l'époque achéménide à l'avènement de l'Islam*. Saarbrücken: 97-116.

Tate G. 1992. *Les campagnes de la Syrie du Nord du IIe au VIIe siècle: un exemple d'expansion démographique et économique à la fin de l'antiquité* I. Paris.

Tchalenko G. 1953-58. *Villages antiques de Syrie du Nord. Le Massif du Bélus à l'époque romaine*. Paris.

Tchalenko G. 1979-80. *Églises de village de la Syrie du Nord*. Paris.

Tchalenko G. 1990. *Églises syriennes à bêma*. Paris.

Touma M. 1984. *La céramique byzantine de la Syrie du Nord du IVe au VIe siècle*. Unpublished Ph.D. diss., Université de Paris-I Panthéon-Sorbonne.

Trombley F. R. 1997. "War and society in rural Syria c. 502-613 A.D.: observations on the epigraphy" *Byzantine and Modern Greek Studies* 21: 154-209.

de Vogüé C. J. M. 1865-67. *Syrie centrale, architecture civile et religieuse du Ier au VIIe s.* Paris.

Vorderstrasse T. 2004 "The romanization and christianization of the Antiochene region: the material evidence from three sites", in I. Sandwell & J. Huskinson (eds.) *Culture and society in later Roman Antioch. Papers from a colloquium, London, 15th December 2001*. Oxford: 86-101.

Woolley C. L. 1937. "Excavations near Antioch in 1936" *The Antiquaries Journal* 17: 1-15, pl. I-XIV.

Woolley C. L. 1938. "Excavations at al Mina, Sueidia, I: the archaeological report" *Journal of Hellenic Studies* 38: 1-30.

Intra-regional variation in long distance trading relationships on the northern Levantine coast – the key to site survival?

Carol Bell – University College London

The northern Levant coast is blessed with many excavated sites and, to this wealth of archaeological data are added the rich textual finds of Ugarit, many of which concern trade. Despite this richness, few attempts have been made to synthesise patterns of trade across the LBA/Iron Age transition on a regional scale. Why Ugarit was never rebuilt after being destroyed at the end of the LBA is a question of regional importance. Up to its destruction early in the 12th century BC, this was *the* major port of trade between the Hittite and Egyptian empires. One of the most prosperous cities in the Eastern Mediterranean literally ceased to trade, while other centres were either reconstructed rapidly or were not destroyed at all.

Ugarit has the largest number of published Mycenaean ceramic artefacts in the Levant. As well as Mycenaean wares, a considerable quantity of Cypriote wares have also been found at Ugarit, making these two categories of wares suitable candidates for the study of the consumption of 'imported goods' across a broad social scale. This paper considers the imported ceramic evidence from northern Levantine coastal sites and interrogates these data for intra-regional variation.

The excavators of Ugarit have suggested that the sheer quantity of material from the Argolid found there suggests that it was a key destination for Aegean trade: a "destination de choix pour le commerce égéen" (Yon *et al* 2000: 18). However, the existence of a *direct* trading relationship between the Aegean and Ugarit is not borne out in the rich textual records of Ugarit.

Although Ugarit's texts attest commercial relationships with many areas, there is not a single record of an Aegean merchant trading in Ugarit, which Singer does not see as a simple oversight (Singer 1999: 676). In contrast, contact with Cyprus, 160 km away, is well documented (particularly at the very end of the LBA).

The analysis that follows attempts to quantify the density of Mycenaean and Cypriote finds (defined as finds per 100 square metres of excavation) such that valid comparison can be made between similar contexts at Ugarit and other sites. Only after 'normalising' the data in this way can intra-regional variation in consumption of these vessels on the northern Levant coast, and what that might mean for LBA trading relationships, begin to emerge from the mass of data.

The published excavations of Ugarit, Tell Sukas and Sarepta form the data set for this analysis. Like Ugarit, Tell Sukas was also one of the sites destroyed at the end of the LBA, but was rebuilt with little delay. Sarepta, located in what became Phoenicia in the Iron Age (between Tyre and Sidon), was entirely spared destruction, and its stratigraphy proceeds smoothly into the Iron Age without break.

First, similar contexts with good excavation publications were identified in the different sites for comparative quantitative analysis of Mycenaean wares after normalising the data for the different aerial extents of the excavations. The relative abundance of Cypriote and Mycenaean wares at the different sites was also reviewed, where possible, as well as the trend over time of percentages of imported wares within the total pottery assemblages.

Turning first to Mycenaean pottery, this class of find is suitable for a regional-scale study such as this as it is present in meaningful quantities along the length of the Syro-Palestinian coast, from Ras el Bassit in the north to Gaza in the south.

Figure 2.1 shows the comparative percentages of Mycenaean pottery shapes found at the Northern Levantine coastal sites. One thing immediately leaps out: the high percentage of stirrup jars that dominate the assemblage at Sarepta. At Ras Shamra and Minet el-Beida they represent 20% of the published assemblage but at Sarepta, more than 50% of the vessels were stirrup jars, while the situation at Tell Sukas lies between the two extremes of Ugarit and Sarepta.

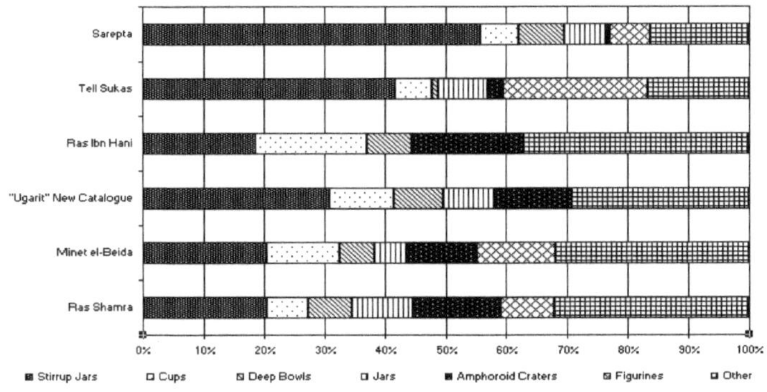

Figure 2.1: Comparative Mycenaean Shapes

The published excavation records were consulted to ascribe all the finds with known find locations into four context categories; domestic, funerary, palatial and ritual. The purpose of this exercise was to identify individual excavation areas on the different sites of similar context type in order to make cross-site comparisons of like contexts; for example to compare domestic contexts from one site with those of another, and not with, say, a funerary context at the second site where the presence and quantity of imported vessels might be motivated by different factors and practices.

After considering the quantity and quality of published excavation data available, my search for comparable contexts homed in on the domestic contexts of Sarepta, Tell Sukas and the tell of Ras Shamra at Ugarit.

In selecting a well-documented excavation area at Ras Shamra for detailed comparison with domestic contexts at Sarepta and Tell Sukas, the choice is narrowed down to those excavated after 1970. Of the 24 hectares or so of the tell area some 6 hectares have been excavated. Only 3,000 square metres, however, represent virgin areas excavated since 1970. The Centre de la Ville area was excavated between 1978 and 1994. *Ras Shamra-Ougarit III* (Yon *et al* 1987) appeared in 1987 and details some of the results from the excavation of the Centre de la Ville area. This location gave Marguerite Yon and her team the chance to excavate a virgin area, high on the tell and at its geographic centre, using modern methods.

The excavation yielded remains of a number of houses (which were not elite mansions found elsewhere on the tell), a cult place (the Temple aux Rhytons, so called because of the large number of rhyta found there, including 11 Mycenaean ones) and tombs integral to the houses as well as evidence of industrial activities such as olive oil manufacture. Evidence of olive oil production has been found both within a household context and in what seems to be a purpose-built building operating on an industrial scale of manufacture (Yon *et al* 1987: 197). No pottery from tombs was included in the publication. Overall, this was a domestic context, in which ordinary citizens lived and worked.

Figure 2.2 shows the distribution of shapes and styles of Mycenaean wares in this residential area. The absolute numbers of Mycenaean main shapes is charted across the Late Helladic style categories. The intent of showing the data in this form is to give an overall impression of the spread of styles and diversity of shapes recovered from this domestic context, rather than to try to communicate large quantities of data. The breadth of the range of different Mycenaean vessel types ordinary people had access to at Ugarit is striking. A point to note from Fig. 2 is that stirrup jars form a significant part of this assemblage, but do not dominate it.

On closer inspection of *Ras Shamra-Ougarit III*, it is clear that not *all* ceramic finds have been published.

With the exception of the figurines, for which a complete catalogue appears for both domestic and imported examples, lists of "materiel répresentatif" have been presented from the houses. The high proportion of figurines in the chart could well be overstating their importance in the overall repertoire, therefore. Assuming that the published corpus is a fair reflection of the total record for the *imported* wares, *Ras Shamra-Ougarit III* confirms the wide availability of Mycenaean (as well as Cypriote) ceramics to the inhabitants of an ordinary domestic quarter of Ras Shamra. The density of published Mycenaean pottery finds calculated per unit area of excavation published in *Ras Shamra-Ougarit III* is 6 pottery finds per 100 sq m of excavation (67 finds/1200 sq m of exposure).

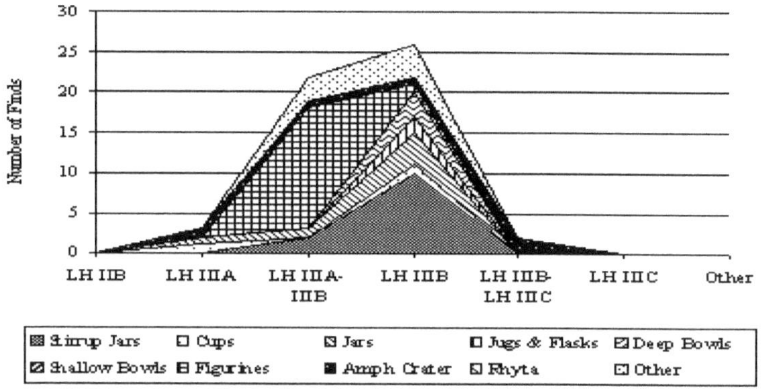

Figure 2.2: Centre de la Ville - Excluding Temple aux Rhytons, Source: Leonard 1994; Yon et al 1987.

Turning to Tell Sukas, final reports on the Danish excavations (1958-1963) on this coastal site, south of Jableh, are still being published, and ten volumes have appeared so far (between 1980 and 1996). Tell Sukas lies at the western end of a fertile plain some 26 kms south of modern Latakia. The tell settlement is ideally positioned from the point of view of engaging in maritime activities, being on a headland between two harbours.

On the tell site, excavations reached the LBA and the evidence shows that Tell Sukas was partially destroyed by fire at the end of the LBA (Level J has a destruction layer that the excavators dated to approximately 1170 BC). However they believe that the devastation was neither as widespread nor complete as that of Ugarit and found no significant delay in reconstruction and occupation on a similar scale to the LBA. Work on the tell revealed the remains of several domestic buildings of varying sizes.

The analysis that follows concentrates on the excavation results from the domestic buildings at the crest of the tell and compares these with Centre de la Ville. Based on my calculations, an area of 2,370 square metres was excavated at Tell Sukas, close to twice that of Centre de la Ville.

The principal publication covering the Mycenaean pottery finds from the tell is *Sukas II* (Ploug 1973) which, unfortunately, does not cover Cypriote wares as these appear to have been outside the original brief of the excavation.

Figure 2.3 shows the distribution of finds on the tell site at Tell Sukas. The Mycenaean pottery this part of the site is not abundant and only 61 examples have been published. Like Ugarit, the assemblage has a broad range of shapes and stirrup jars do not dominate. The large number of figurines found in domestic contexts also resembles the situation at Centre de la Ville. The overall density of finds is low, being only 61 objects in an area of 2,370 sq m – or 3 finds per 100 sq m excavated. This confirms the relative, as well as absolute, validity of the statement made by the excavators that Mycenaean wares are not numerous at Tell Sukas.

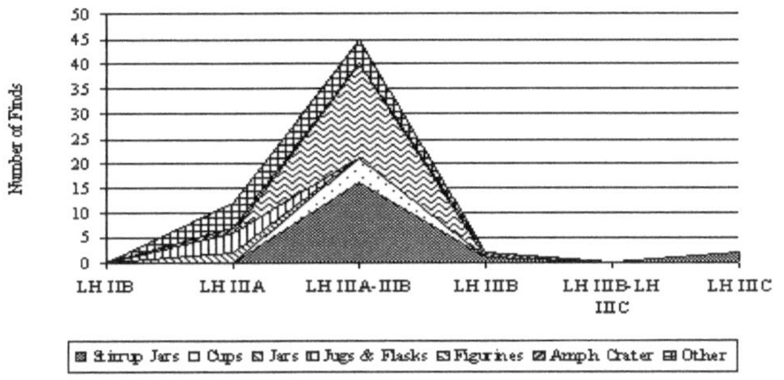

Figure 2.3: Tell Sukas - Shape and Style Overview – Tell Site, Source: Leonard. 1994; Riis 1970; Ploug 1973.

The site of Sarepta lies on a low mound on the Lebanese coast 50 kms south of Beirut near the modern village of Sarafand, located roughly equidistant between Sidon and Tyre (being south of Sidon and north of Tyre). The site was excavated between 1969 and 1974 by James Pritchard of the University of Pennsylvania with the specific goal of documenting the stratigraphy of a Phoenician urban site in their homeland and identifying a Phoenician city known from texts. More than 50 ancient sources refer to Sarepta, including 14th century BC Ugaritic texts and Egyptian papyri.

No site in Lebanon had produced well-stratified evidence of occupation between 1200 and 600 BC prior to this and, thirty years on, Sarepta remains the only coastal city to be extensively excavated and published in Lebanon. A late Roman port (Area I) and the tell itself (Area II) were excavated.

Area II yielded remains of LBA and Iron Age occupation from two soundings. Sounding X revealed a continuous sequence of occupation from the 13th – 6th/5th centuries BC. The area excavated was an industrial quarter (with the exception of a small shrine). 22 kilns were found, attesting a specialised pottery industry in addition to purple dye and olive oil manufacture and other industrial activities. There was no sign of destruction in any of the strata excavated, indicating that Sarepta did not experience the same fate as Ugarit and Tell Sukas in the early 12th century BC. Sounding Y, located approximately 100 m SE of Sounding X, in contrast, was a mainly residential area, containing what Pritchard described as "modest homes" (Pritchard 1978: 74). The Sarepta expedition meticulously recorded *all* pottery finds and the final excavation reports provide a complete picture of the assemblages of the two soundings that penetrate the LBA/Iron Age transition.

Area II Y at Sarepta seems to be a similarly ordinary domestic quarter to Centre de la Ville at Ras Shamra during the LBA. This area lies at the highest part of the mound in a position not dissimilar to that of Centre de la Ville on the tell of Ras Shamra or the excavated areas on the Tell at Sukas. The scale of the excavation is an order of magnitude smaller, however, with only 100 sq m exposed.

Figure 2.4 shows a completely different pattern of shapes and styles, with stirrup jars dominant and the presence of Mycenaean IIIC:1 bowls (in stratum G - LBA/Iron Age transition) – shown as LH IIIC on the graph. The number of pottery finds included in this graph is only 20, but as the area of excavation is small, this converts to a density of 20 finds per 100 sq m of excavation (contrasted with only 6 finds per 100 sq m at Centre de la Ville).

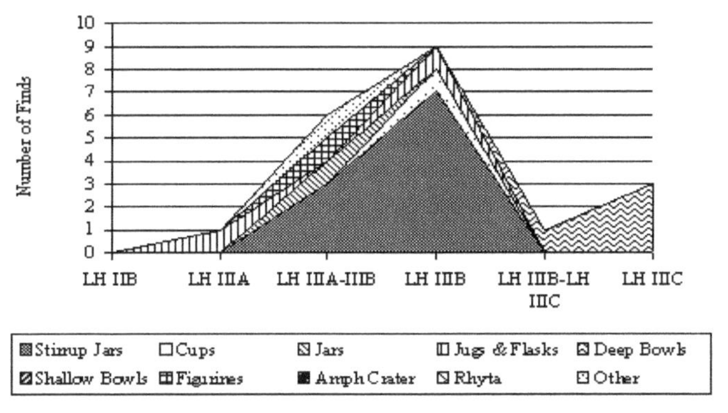

Figure 2.4: Sarepta Area II-Y Finds (Domestic Context) Source: Leonard 1974; Anderson 1988.

Because the sample size is small, but also because excavations at Centre de la Ville at Ugarit revealed industrial functions (including one industrial building), a comparison has also been made with Sarepta area II-X. Like Centre de la Ville, the 800 square metre area of Area II-X also contained a structure attributed to a religious function, a shrine according to the excavators, and 18 finds of diverse shapes were recovered from this part of the site.

Figure.2.5 shows a similar profile of shapes and styles to that seen in the smaller Area II-Y, with stirrup jars dominating. 108 objects were recovered here in an excavation of 800 sq m. Consequently, the density of Mycenaean finds is 14 finds per hundred sq m of excavation. This is not as high as in Area II-Y, but well over twice that of Centre de la Ville (based on *Ras Shamra-Ougarit III*).

Turning now to the relative importance of Cypriote wares, unfortunately, no LBA/Iron Age Cypriote wares from the tell excavations at Tell Sukas were published in *Sukas II*

(Ploug 1973), which concentrates on Aegean imports only from this period. The following discussion will, therefore, focus on Centre de la Ville and the two soundings at Sarepta.

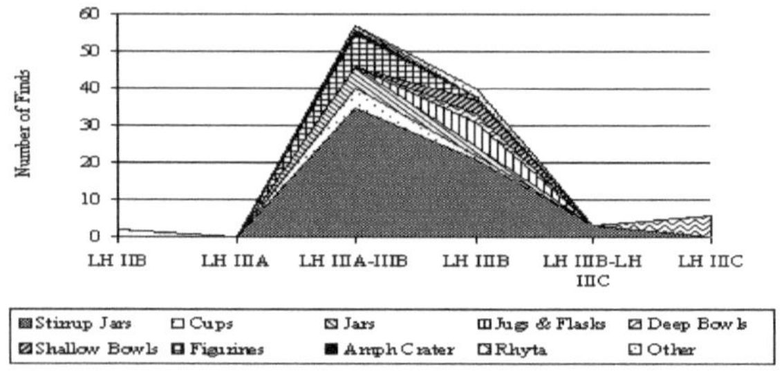

Figure 2.5: Sarepta Area II-X Finds
Source: Leonard 1994; Koehl 1985.

Table 2.1 summarises the ceramic assemblages published in *Ras Shamra-Ougarit III* (Yon *et al* 1987: 11-27) for three houses. Pottery finds have been divided into three groups, namely Local, Cypriote and Mycenaean. Obviously, local wares are likely to be severely under-represented in this published sample and therefore have not been used to calculate any percentages of imports within the total assemblage.

Cypriote wares appear to slightly outnumber Mycenaean ones in this representative sample. Given both the proximity of Cyprus, 100 km away by sea, and the textual evidence of close trading contacts between the two areas, this is probably what one should expect – especially in an ordinary residential quarter such as this one where the inhabitants would be less likely to have access to exotica than those inhabiting palatial areas or large mansions.

The very recent publication of *Ras Shamra-Ougarit XIV* (Yon & Arnaud 2002) catalogues the finds from another house, the Maison au Sud du Temple aux Rhytons, which is located at the southern end of the excavation and partially within the 1200 sq m area published in *Ras Shamra-Ougarit III*. This adds 33 unpublished Mycenaean vessels found in a domestic context to the corpus, one amphoroid crater having been previously published in 1990. The number of Cypriot finds are more than double that of the Mycenaean finds, which also suggests a greater availability of Cypriote wares to the inhabitants of this ordinary domestic quarter. The addition of these data to the calculation of find densities of Mycenaean wares increases the density to 8 finds per 100 sq m of excavation area – still considerably lower than that at Sarepta.

Turning to Sarepta, complete quantitative ceramic data exist in the final reports and an interesting picture emerges, as can be seen in *Table 2.1*.

In sounding Y, during period H, there is a marked increase in Cypriote imports, while Mycenaean wares make their first appearance at this time. In stratum G, on the other hand, there is a complete absence of Base Ring ware, and White Slip ware is also almost completely absent, while the Mycenaean count rises substantially, and exceeds the total number of Cypriote wares. Over the whole period, the percentage of imports declines substantially. In Sounding X, Stratum II marked the peak of imports and the frequency of Cypriote LBA ceramics dropped over the first three strata, while the proportion of Mycenaean greatly increased after c. 1350 BC – according to the chronology of the excavators at Sarepta. Mycenaean sherds outnumbered Cypriote ones throughout the period considered.

Discussion

This analysis presented in this paper highlights some significant differences in the find densities of Mycenaean wares at the different sites in parts of the sites where ordinary people lived and worked. It also identifies differences in the relative amounts of Mycenaean and Cypriote wares.

Figure 2.6 shows the density of Mycenaean finds found in each of the four areas expressed as finds/100 sq m of excavation exposure and takes into account the recently released data in Ras Shamra-Ougarit XIV. The results are clear: both Sarepta Area II soundings have a substantially higher density of Mycenaean wares than does either Centre de la Ville or Tell Sukas. Perhaps the lesser amount of Mycenaean ware at Tell Sukas compared with Ugarit can be explained by the likelihood that this site was on the periphery of Ugarit's territory and one might expect the centre of that world to have more by way of imports than its fringes.

House	*Local Wares*	*Mycenaean Wares*	*Cypriote Wares*
Maison A	50	7	6
Maison B	>53	4	5
Maison E	33	3	6
Total	>136	14	17

Table 2.1: Centre de la Ville Cypriote Wares Published in Ras Shamra-Ougarit III
Source: Yon et al 1987: 11-27

Unquestionably, in my view, more shipping would have unloaded 'international' cargoes at Minet el-Beida and Ras Ibn Hani than at either of Tell Sukas' harbours. However, the reverse trend between Centre de la Ville and Sarepta cannot be explained in such simple terms. The difference is not a case of a percentage point or two but is large enough to suggest that there may be a substantially different consumption pattern, and flow of trade, in imported ceramics at Sarepta, especially when taken together with the relative amounts of Cypriote and Mycenaean wares.

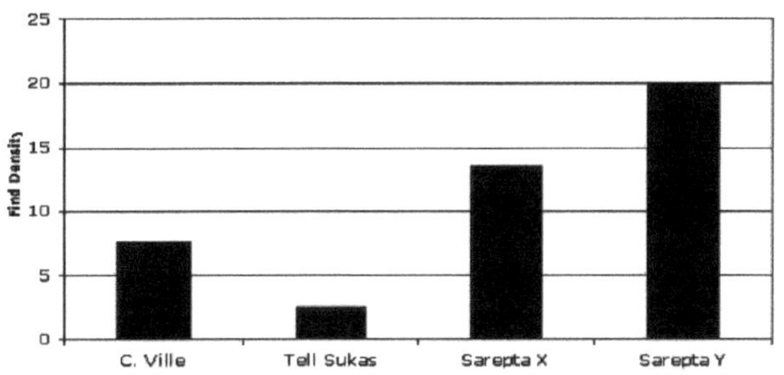

Figure 2.6: Find Densities of Mycenaean Wares (finds per 100 sq m of excavation)

The impression created by the absolute numbers of Mycenaean finds published from the different sites with regard to the importance of their commerce with the Aegean can, therefore, be misleading. Absolute find numbers are high at Ugarit, but once the data are corrected for aerial extent of excavations in comparable contexts, it seems that Sareptans possibly had a greater appetite for these wares and Koehl speculates:

> " ... *taste for foreign pottery at Sarepta was for Mycenaean and the merchants who handled the trade, be they Phoenicians, Cypriotes or Mycenaean colonists on Cyprus, sought out Mycenaean pottery where it was available.*" (Koehl 1985: 147)

Certainly Sareptans possessed Mycenaean wares in greater quantities in areas where ordinary people lived and worked when the data are expressed per unit area excavated.

Another factor that requires discussion is the dominance of stirrup jars in the Sarepta assemblage. Stirrup jars are usually held to be containers for the transportation of liquids, with the smaller varieties believed to be vessels for valuable – perhaps, sometimes, perfumed – oils. Koehl has made the suggestion that Sarepta may have been a "distribution centre" through which Mycenaean vessels passed inland, to sites like Kamid el-Loz, Tell 'Ain Sherrif and Tell el Ghassil, whose Mycenaean assemblages are also dominated by stirrup jars (the rest being flasks and other sorts of jars mainly) (Koehl 1985: 144). The gist of this argument is that the more specialised open forms, such as amphoroid craters, remain at port, while stirrup jars go inland for contents and Koehl suggests a theoretical route from Sarepta inland in a northerly direction along the Litani River.

At Ugarit, on the other hand, perhaps van Wijngaarden (1999) is correct when he suggests that stirrup jars may have represented luxury versions of everyday items (that is, oil containers) within the households of Ugarit across the range of social classes. Certainly, no large concentrations of them have been found that could represent trade goods awaiting transhipment.

Preliminary Conclusions

This paper presents evidence that the quantities of Mycenaean and Cypriote wares in the published archaeological record of Sarepta are significantly different from that of the two other sites further north. Not only is the Sareptan assemblage dominated by stirrup jars, but also Mycenaean finds are twice as abundant per unit of area of excavation than in a comparable context at Ras Shamra. The evidence published so far suggests that Cypriote wares were more abundant than Mycenaean imports at Ras Shamra while the picture is reversed at Sarepta, which is fully published.

Hankey, as early as 1967 (Hankey 1967, 1971), suggested the involvement of Cypriote middlemen in distributing Mycenaean wares to the Levant. In broad terms, my results for Ras Shamra fit this hypothesis, and would also fit in numerically with Renfrew's (1975) ideas about down the line exchange.

The ceramic evidence from Ras Shamra demonstrates that inhabitants of Centre de la Ville had more Cypriote than Mycenaean wares. Also, the absence of Aegean merchants in the vast textual archive of Ugarit, and the abundance of Cypriot references mentioned earlier, supports the involvement of Cyprus in this trade.

Koehl's suggestion (Koehl 1985: 144) that Sarepta might have been a distribution centre for Mycenaean wares itself, particularly for the stirrup jars (and presumably their contents), to LBA settlements further inland could be used to argue that a more direct relationship existed between Sareptan consumers (or middle-men) and Argive manufacturers than existed at Ugarit or Tell Sukas.

The general picture of maritime trade at the end of the LBA in the Eastern Mediterranean has moved away from notions of an Aegean, or any other, inter-regional thalassocracy and it is now considered that there were many active trading fleets. Under such conditions, long distance trade probably reached new highs in the last century or so of the LBA, both in terms of overall volume and in that of the range of items traded (as evidenced by the diversity of the cargo of the Uluburun and Cape Gelidonya wrecks).

Stratum	*% Imports*	*Myc. Count*	*Cyp. Count*
II-Y H	4.30	3	51
II-Y G1	0.52	17	9
II-Y G2	0.47	8	6
II-Y F	0.25	3	4
II-X II	1.43	10	7
II-X III	1.04	29	6
II-X IV	0.87	7	4
II-X V	0.64	10	9

Table 2.2: Cypriote Wares: Sarepta Area II
Source: Anderson 1988: Table 2; Koehl 1985: Tables 2A/B

Despite notionally being vassals of Great Powers, be they under the overlordship of Egypt (Sarepta), or the Hittites (Ugarit and Tell Sukas), the cities of the Northern Levant Coast seem to have enjoyed reasonable autonomy to select their own material culture. In such a decentralised world that was receptive to free trade in ceramics, it is conceivable that Sarepta had evolved a different suite of trading relationships to import its preferred group of ceramics (and their contents) from Ras Shamra and Tell Sukas. A glance at a map of the Eastern Mediterranean reminds one that ships from the Aegean would have to sail past, or partly around, Cyprus to Ugarit and Tell Sukas, while this is not so for Sarepta, assuming that by the end of the LBA, more direct shipping routes were possible. The higher density of Mycenaean finds at Sarepta, together with the fall-off in Cypriote wares (once Mycenaean wares have been introduced) suggests that a different strategy was being employed there (compared with the sites further north) to obtain imported wares.

To return to the events at the close of the Bronze Age, it is intriguing that the site that escapes destruction (perpetrated or at least catalysed by invaders that are linked with a Mycenaean cultural background elsewhere in the Eastern Mediterranean) is the one that has the greatest density of Mycenaean pottery finds when comparable contexts are examined. It is far too early to draw any definitive conclusions from this analysis of imported ceramics. However, to return to Marguerite Yon's words, among the sites considered here, in the direct sense of the phrase, perhaps Sarepta was a *"... destination de choix pour le commerce égéen"*. What is certain is that the stratigraphy of Sarepta shows no evidence of destruction at the end of the Late Bronze Age.

Bibliography

Anderson W.P. 1988. *Sarepta I The Late Bronze and Iron Age Strata of Area II, Y.* Beirut.

Hankey V. 1967. "Mycenaean Pottery in the Middle East" *Annual of the British School of Archaeology at Athens* 62: 107-147.

Hankey V. 1971. "Mycenaean trade with the south-eastern Mediterranean" *Mélanges de l'Université Saint-Joseph* 46: 9-30.

Khalifeh I.A. 1988. *Sarepta II The Late Bronze and Iron Age Periods of Area II X.* Beirut.

Koehl R.L. 1985. *Sarepta III The Imported Bronze and Iron Age Wares from Area II, X.* Beirut.

Leonard A. 1994. *An Index to the Late Bronze Age Aegean Pottery from Syria-Palestine.* Jonsered.

Ploug G. 1973. *Sukas II: The Aegean, Corinthian and Eastern Greek Pottery and Teracottas.* Copenhagen.

Pritchard J.B. 1978. *Recovering Sarepta, a Phoenician City.* Princeton.

Renfrew C. 1975. "Trade as action at a distance", in J. Sabloff & C.C. Lamberg-Karlovsky (eds.) *Ancient Civilization and Trade* Tucson: 3-60.

Riis P.J. 1970 . *Sukas I: The North-East Sanctuary and the First Settling of Greeks in Syria and Palestine.* Copenhagen.

Singer I. 1999. "A Political History of Ugarit", in G.E. Watson, G.E. & N. Wyatt (eds.) *Handbook of Ugaritic Studies*. Leiden.

Wijngaarden G.-J. van 1999. "An archaeological approach to the concept of value – Mycenaean pottery at Ugarit, Syria" *Archaeological Dialogues* 1999-1.

Yon M. *et al* 1987. *RAS SHAMRA - OUGARIT III: Le centre de la ville, 38-44e campagnes (1978-1984).* Paris.

Yon M., Karageorghis, V. & Hirschfeld, N. 2000. *RAS SHAMRA - OUGARIT XIII: Céramiques mycéniennes d'Ougarit.* Nicosie et Paris.

Yon M. & D. Arnaud (eds.) 2002. *RAS SHAMRA - OUGARIT XIV: Etudes Ougaritiques, 1. Travaux 1985-1995.* Paris.

The south Italic fighting technique

Michael T. Burns - University College London

Introduction

This paper discusses the fighting technique practised by the Italic peoples of southern Italy during the 4th and early 3rd centuries BC, from the regions the Romans knew as Campania, Lucania, Apulia and Samnium. There are a number of references in ancient sources which assert that the Romans had adopted their method of fighting from the south Italic people known as the Samnites. Sallust claims that

> *"our ancestors . . . were never too proud to take over a sound institution from another country. They borrowed most of their armour and weapons from the Samnites"* (Sallust *Cat.*51.37-38).

The *Ineditum Vaticanum* is even more specific, stating

> *"the Samnite oblong shield was not part of our national equipment, nor did we have javelins, but fought with round shields and spears . . . But when we found ourselves at war with the Samnites, we armed ourselves with their oblong shields and javelins . . ."* (*Ineditum Vaticanum,* trans. Cornell 1995: 170).

The Samnites, as described by the Romans, were characteristic of most south Italic peoples, who all fought in a broadly similar fashion.

I was inspired to investigate south Italic fighting methods by Connolly's 'The Roman fighting technique deduced from armour and weaponry' (1991: 358-363). Connolly believed it was possible to reconstruct Roman fighting methods of the early Imperial period from the armour and weapons used, especially when it could be shown this equipment had been adapted to meet the threat posed by the weapons and fighting styles of their enemies. This approach stressed the importance of the relationship between defensive and offensive equipment and treated their functional aspects as integral components of a single fighting system, or technique. Analysing the equipment in this manner created a greater awareness of the technical features exhibited in armour and weaponry, and a deeper understanding of what these changes and modifications in equipment entailed in actual usage. This paper intends to shed light on the south Italic fighting material in a similar manner.

A wealth of military equipment and iconographic representations, depicting its use, has been recovered from tombs in southern Italy, especially at the sites of Paestum, Capua and Ruvo. Before addressing the south Italic evidence, however, it is important to lay out in plain language exactly what the accoutrements of war were intended to do, at their most basic and functional level. To understand the mechanics of the fighting method we must first examine the killing potential of the weapons and the defensive capacity of the armour. The relationship between these two variables is intrinsically linked to the type of fighting technique practised.

Part I. The killing potential of weaponry and the defensive capacity of armour

Weaponry causes damage to the human body in three main ways: through puncturing, lacerating and crushing wounds. Most fatalities from punctures and incising wounds occur from massive blood loss. What many people fail to realise is that wounds, even fatal ones, inflicted with swords, spears, arrows, and axes seldom incapacitate an individual immediately. It is often the case that in the heat of combat, pumped up with adrenaline, the human body can sustain an incredible amount of pain and injury before passing out from blood loss, or expiring (Lurz 2000). Of these three types of wounds punctures are by far the most efficient way to kill someone. It requires a penetration of only 2 centimetres to inflict a lethal wound to a vital area of the human body, with a modest expenditure of energy. It is also far more difficult to stop the flow of blood from a penetrating wound than a laceration. The vulnerability of the human body to certain types of wounds was well understood in the ancient world. This is clearly exhibited in numerous south Italic tomb paintings of duels in which wounds are depicted often in graphic detail. Celsus (V.28) states that when "the wound has been made by a sharp weapon, a transverse wound is the more favourable". Lacerations, or transverse wounds, rarely penetrated the body to a lethal depth, although their capacity to injure or disable an enemy should not be discounted.

Even today these points are stressed to modern soldiers, who are trained in bayonet drills to wound or disarm an enemy by slashing manoeuvres, or to incapacitate them with a butt stroke. It is then they are told "kill without mercy" by thrusting repeatedly (U.S. Army F.M. 23-25, 1953). The analogy between modern techniques of killing with hand weapons and those used in ancient warfare is relevant as far as human physiology remains as vulnerable to the effects of these wounds, as they were in ancient times. They also illustrate the importance that has been placed on developing an established technique to wound and then dispatch an enemy. This methodology is clearly evident in south Italic representational evidence, and shows that the ancients were well aware of the capabilities of their arms and their effects on the human body.

When considering ways of protecting the human body we must think in terms of a defensive system; this includes not only armour but agility as well. The primary purpose of armour is to protect the human body from the full effects of weaponry. This does not mean it was expected to make one impervious to weapons, but rather to give the warrior an acceptable amount of protection, that would allow him to manoeuvre and inflict casualties on the enemy. The degree to which a warrior was armoured was largely dependent on the types of weaponry he expected to encounter on the battlefield, and what resources were available to him. Sallust (*Bell. Jug.*105.2) for example, states that during the Jugurthine war in the 2nd century BC *"a cohort of Paelignians [were] equipped with light armour, which allowed them to march at a good pace and yet protected them as well as heavier armour would have done against the light missiles used by the enemy".* It is the balance which maximises the killing potential of the weaponry, while optimising the protective capacity of armour, that was a important consideration in ancient warfare. One that led to the continual, albeit gradual, development in the evolution of weaponry and armour, and the tactical methods used to employ them. As will be shown, south Italic methods of warfare were predicated by the relationship between the offensive and defensive capabilities of the warrior.

Part II. The weaponry of southern Italy

The most common weapon found in tombs and paintings from southern Italy during the 4th century BC was the javelin or throwing spear. A wide variety of types and sizes were used, but overall they fall into two main categories: wide-bladed and narrow-pointed (Small 2000: 223). An excellent example of these two types of spears comes from Paestum, tomb 2/1957 of the Gaudo necropolis, and is dated to the middle of the 4th century BC (*Figure 3.1*, after Pontrandolfo & Rouveret 1992: 381-2). Wide bladed spears and javelins were designed to inflict larger wounds resulting in a greater amount of tissue damage and immediate blood loss. Warriors who were not armoured were especially vulnerable to these weapons as they had the potential to hit vital organs and sever arteries. Narrow pointed weapons were much more specialised than other dual-purpose spears and javelins. Their primary purpose was to be armour piercing, concentrating the impact on a smaller area, which gave the weapon a greater degree of penetrating power. These types of armour or shield-piercing throwing weapons were heavy and often had a long iron shank (Connolly 2000: 43). This gave the warrior the ability to penetrate the shield of an enemy and then carry on through to enter his body. The deep punctures that could be inflicted by these weapons would be much more difficult and dangerous to extract as the wound often closes up around the weapon. Appian describes a sea battle in which a commander was wounded by a similar type weapon *"in the thigh by a barbed Spanish javelin of solid iron, which was impossible to extract quickly. Menacrates thus became unable to fight..."* (Appian V.82).

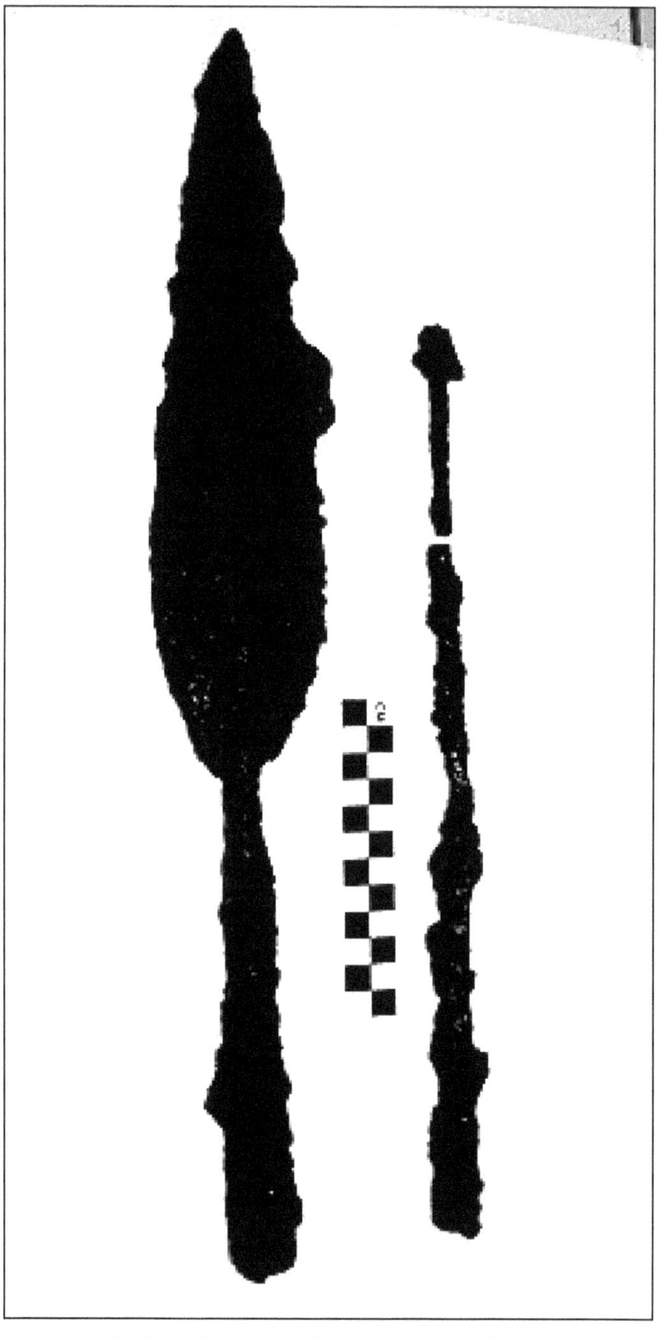

Figure 3.1. Wide bladed and narrow pointed spears from Tomb 2/1957 Gaudo necropolis Paestum, 360-340 BC.

An important feature often overlooked by those studying javelins and throwing spears is the *amentum*, or throwing thong. The amentum was a leather thong about 18 centimetres long, which was wound or tied around the javelin so as to create a loop, at or near the centre of the shaft. The javelin was cradled in the palm and held in place by the third and fourth fingers. The first two fingers were inserted into the loop of the amentum.

This gave the javelin a rotary motion when thrown, which increased its velocity and improved both accuracy and penetrating power. The potential of these weapons is graphically illustrated in the duelling scenes from many south Italian tomb frescos. The javelin is often shown completely impaling the arms and legs of adversaries, with the amentum hanging from the shaft of the weapon. Ennius (III.168), writing in the middle of the 3rd century, makes references to troops throwing 'loop-handled lances' (*hastae ansatae*). Diodorus (XVI.63.82) repeatedly makes reference to south Italic peoples, such as the Lucanians and Bruttians, whose enemies were *"shot down with javelins"*. The killing potential of this weapon is evident from the description of Alexander of Epirus' death *"when a Lucanian exile cast a javelin which transfixed him"* at long range (Livy VIII.24). A tremendous amount of power would have been necessary to 'transfix' a man moving at long range, moreover Alexander was probably armoured, thus requiring even greater velocity upon impact to penetrate.

Part III. The armour of southern Italy

The shield was always regarded as the warrior's primary form of defensive equipment. In southern Italy the shield most frequently depicted on vases and paintings was the round *aspis*, or hoplite shield, which had been adopted from the Greeks (Snodgrass 1999: 75). The *aspis* however, was gradually eclipsed by the oblong shaped *scutum*, which was made of plywood and leather (Polybius VI.23). The Romans claimed they had adopted this type of shield and the use of javelins from the Samnites (Sallust *Cat.* 51.37-38). The *scutum* was much better adapted for defence against missile weapons and still sturdy enough to be used in close combat. Another important item of defensive equipment in the south Italic panoply was the helmet. A variety of open-faced helmets, such as the Pilos, Apulo-Corinthian, Samno-Attic and Montefortino types were used in southern Italy. These usually had cheek pieces attached to provide additional protection without obscuring vision. The advantage of open-faced helmets is that they allowed the warrior an unrestricted field of vision and did not hinder hearing. Being able to look and listen were essential to warriors fighting in open order formations, which the use of javelins necessitated. The design of all these helmet types made use of rounded surfaces, which could help deflect the full impact of a thrown weapon (cf. Burns 2003: 11-12).

The shield and helmet were sometimes supplemented by body armour. The most common types of body armour used during the 4th Century BC were the triple-disc and rectangular anatomical cuirasses. These cuirasses were a type of harness, which composed of a breast and back-plate, around 24-30 cm square, held in place by two side and shoulder plates, all of which was made of bronze (Connolly 1986: 117-118). This type of armour was an essentialist form of protection designed to cover only the most vital areas of the human torso, the heart and thoracic blood vessels. Polybius (VI.23) echoes this concern when he states that Roman soldiers of the 2nd century wore a similar breastplate *"which is placed in front of the heart, and called a heart protector* (*kardiophylakes*)". It should also be stressed that the protection afforded by armour is as much psychological as physical. If a soldier believes he is adequately protected he need not be totally encompassed in armour. In this case mobility is just as important, especially when using missile weapons. Livy (XXXIV. 39) comments that it was *"a run before hurling . . . which is what gives the greatest velocity"* to the javelin. Modern experiments with 100 pound pull bows and arrows with a 553-grain iron head, could only occasionally penetrate bronze plate of 2mm thickness to a potentially mortal depth of .75 inches at a distance of ten yards (Gabriel & Metz 1991: xx; Wheeler 2001: 178). The thickness of most south Italic bronze armour averages 1mm or less, which is comparable to that found in Greek armour (Jarva 1995: 141). If we consider that most hand thrown javelins, even those fitted with the *amentum*, would have a much lower velocity than the arrows from a hundred pound bow, it is clear that Italic armour would be an effective form of protection.

Part IV. The south Italic fighting technique

By examining the actual equipment of the south Italic warrior it is possible to get an idea about the killing potential of the weaponry and the defensive capacity of the armour. But the capabilities of the equipment cannot explain the actual fighting techniques that were practised. It is fortunate then that a large amount of pictorial evidence survives from tomb and vase paintings and provides an invaluable insight into how the accoutrements of war were actually used. Tomb 12/1969 from the Andriuolo necropolis, 370-360 BC, depicts two warriors who have each thrown a javelin with amentum before engaging each other with thrusting spears (Figure 3.2, after Pontrandolfo & Rouveret 1992: 202-205). The warrior on the right having hit his opponent in the head with a javelin finishes him off with a thrust to the abdomen. But even in triumph we see that the victor has been wounded by a javelin, which penetrates his shield and then his shoulder. A painting from tomb 7 in the Gaudo necropolis, 370-360 BC, depicts an almost identical scene with the added detail that each warrior has thrown two javelins, instead of one, before closing in hand to hand combat (Pontrandolfo & Rouveret 1992: 251-253). The number of javelins carried by Andriuolo necropolis Paestum, 370-360 BC.warriors may have varied in actual practice, tomb 28 from the Andriuolo necropolis, 330-320 BC, shows a warrior equipped with five javelins! (Pontrandolfo & Rouveret 1992: 156-159). But the most commonly depicted amount appears to have been two javelins.

Figure 3.2. South Italic Fighting technique depicted in painting from Tomb 12/1969, Andriuolo necropolis Paestum, 370-360 BC.

The paintings correspond well with the description of fighting given by Ennius (III.160-161) in the middle of the 3rd century BC, when he states *"after they were tired out from standing and spattering each other with loop-handled lances (hastis ansatis), they engaged with javelins on all sides"*. This is one of the earliest descriptions of Italic combat and by a writer who had served not as a Roman, but as an allied soldier from southern Italy. What is clearly evident from Ennius' account and these paintings is that an exchange of javelin fire occurs before warriors engage in close combat. This technique gave the warrior the potential to kill, or more probably, wound an enemy from afar. With specialised throwing weapons and modifications, like the *amentum*, the possibility to disable an opponent was greatly increased. In the rush to hand to hand fighting which followed the warrior would have had a distinct advantage over an enemy who was already injured.

Conclusion

It has been the purpose of my paper to go beyond the usual description and discussion of weaponry and bring into focus a very important aspect of warfare that is often neglected. The actual mechanics of how these weapons could injure or kill a soldier and the methods by which these were practised. Armour and weaponry are interrelated, in that improvement in one often results in a corresponding advancement in the other. Understanding the relationship between the capabilities of these two factors is crucial in any study of fighting methods. The evidence from Southern Italy shows that a very specific fighting technique was practised based on a clear understanding of what were the killing potential of their weapons and the defensive capacity of their armour.

Bibliography

Burns M.T. 1999. *The Military Equipment and Methods of Warfare of the Samnites and other Oscan Speaking Peoples*. Unpublished Masters Thesis, University of Massachusetts, Boston.

Burns M.T. 2002. "The homogenisation of Italian military equipment under the Roman republic" in A. Merryweather and J. Prag (eds.), *"Romanization?"*

(Proceedings of a postgraduate conference, Institute of Classical Studies, London, 2 November 2002.). *Digressus supplement vol.I:* http://www.digressus.org (last accessed 10th June 2004).

Connolly P. 1981. *Greece and Rome at War*. London.

Connolly P. 1986. "Notes on the Development of Breastplates in Southern Italy" in J. Swaddling (ed.) *Italian Iron Age Artefacts in the British Museum*. British Museum Press: London.

Connolly P. 1991. "The Roman Fighting Technique Deduced from Armour and Weaponry", in B. Dobson and V.A. Maxfield (eds.) *Roman Frontier Studies 1989: proceedings of the XVth International Congress of Roman Frontier Studies.* Exeter: University of Exeter Press.

Connolly P. 2000. "The reconstruction and use of Roman weaponry in the second century BC" *Journal of Roman Military Equipment Studies* 11: 43-46.

Cornell T. 1995. *The Beginnings of Rome.* London.

Gabriel R.A. & Metz K.S. 1991. *From Sumer to Rome: The Military Capabilities of Ancient Armies.* Westport.

Jarva E. 1995. *Archaiologia on Archaic Greek Body Armour*. Studia Archaeologica Septentrionalia. Rovaniemi.

Lurz, F. 2002. *The Dubious Quick Kill.*

Pontrandolfo A. & Rouveret. A. 1992. *Le Tombe Dipinte di Paestum.* Modena.

Richardson T. 1998. "Ballistic testing of historical weapons" *Royal Armouries Yearbook.* Vol.3. Leeds: 50-52.

Small A. 2000. "The Use of Javelins in Central and South Italy in the 4th Century BC" in E. Herring, R. Whitehouse & J. Wilkins (eds.). *Ancient Italy in its Mediterranean Setting: Studies in honour of Ellen Macnamara.* London. 221-234.

Snodgrass A. 1999. *Arms and Armour of the Greeks.* Baltimore.

U.S. Army. 1953. *Combatives Field Manual.* FM 23-35. Department of Defense: Ft. Benning, Ga.

Wheeler E. L. 2001. "Firepower: missile weapons and the 'Face of Battle'" in *Roman Military Studies 5.* Krakow.

The Necropolis of Capestrano: New Excavations and Finds

Monica Capodicasa - Università G. d'Annunzio Chieti, Italy

Capestrano is a medieval village in the province of L'Aquila (Italy), which extends across a territory including the Gran Sasso National Park and the Laga Mountains. The village looks out towards the Tirino Valley. On this fertile plateau and at the source of the Tirino River, is the extensive necropolis of Capestrano (see *Figure 4.1*). The necropolis was discovered in 1934 by a local farmer, and the small village has become renowned above all for the "statue of the Warrior" discovered there (*Figure 4.2*). The information available for this burial area is somewhat limited. However, what we have at our disposal indicates that it was an important necropolis in the territory of the *Vestini* tribe, in central Abruzzo. The *Vestini* were part of the Italic population named the *Safin* of central Adriatic Italy. The area under discussion extends from Collelongo to Ofena and includes Capo d'Acqua and Capestrano.

This necropolis was used for local burials for approximately eight hundred years, between the seventh and first Centuries BC. No clear evidence of a contemporary settlement is yet known, however the necropolis was positioned on a river terrace in the vicinity of a Neolithic settlement, as attested by pottery finds at the site. In 1992, approximately two meters from the edge of the field, remains of a Neolithic structure c. 3-4 metres in diameter, ceramic fragments, oil presses and flint artefacts were also discovered.

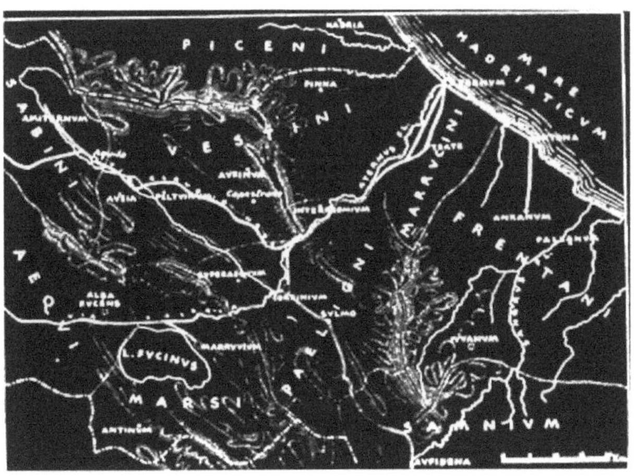

Figure 4.1. Map of the Abruzzo region showing tribal territories in the 6th century BC. (After Moretti 1936).

The "statue of the Warrior" dates to the sixth century BC (Moretti 1936: 110), and has many similarities with other statues found in Abruzzo belonging to the same period. Similar statues, although of an inferior artistic quality, have been found in Atessa, Bellante, Guardiagrele, Collelongo and Loreto Aprutino. They have parallels in Etruscan sculpture, for example from

Figure 4.2. "Statue of the Warrior" from Capestrano (Moretti 1936)

Casale Marittima, formerly Caere in ancient times (*Figure 4.3*, Colonna 2001: 105). The "statue of the Warrior" represents an original interpretation of Greek-Etruscan versions, and is also unique on account of its inscription, transliterated as: "*ma kuprì koram opsùt anini rakinevìi pom(p---)ì*". From this inscription we know the name of the artist, Aninis, and the name and social standing of the person who commissioned it, King Nevio *Pompuledio* (La Regina 1986: 7-10). For archaeologists the inscription is extremely important for understanding the role of warriors in central Italy in the seventh and sixth centuries BC and the social organisation of the *Safin* people.

Only a few metres from the findspot of the "statue of the Warrior" in the necropolis, a unique bust of a woman was also discovered (*Figure 4.4*), dating to the same period. Like the "statue of the Warrior", she has her arms crossed and her hands open to indicate the symbols of prestige. The jewels she wears are also symbols wealth and power; and although there are no parallels, she must be related to the "statue of the Warrior" and his milieu.

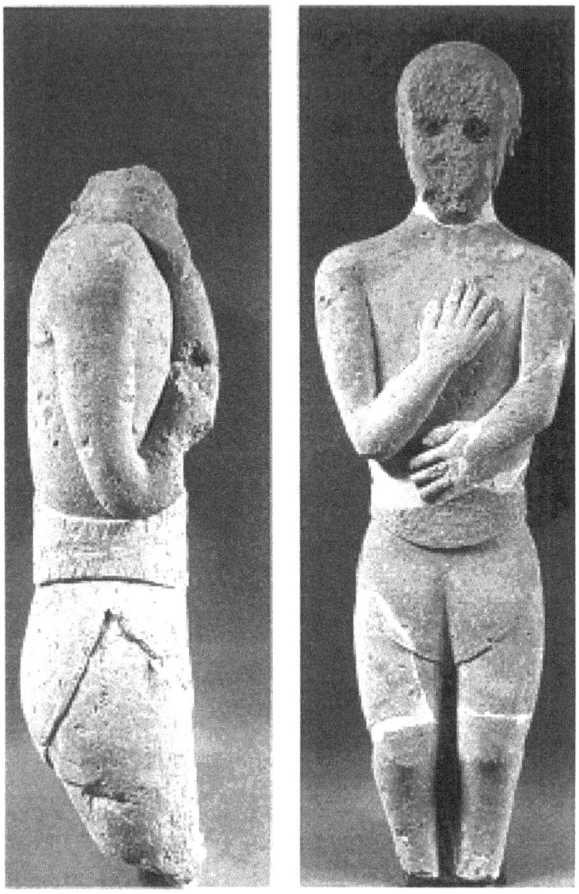

Figure 4.3. Etruscan sculpture from Casale Marittima (ancient Caere). (Collona 2001)

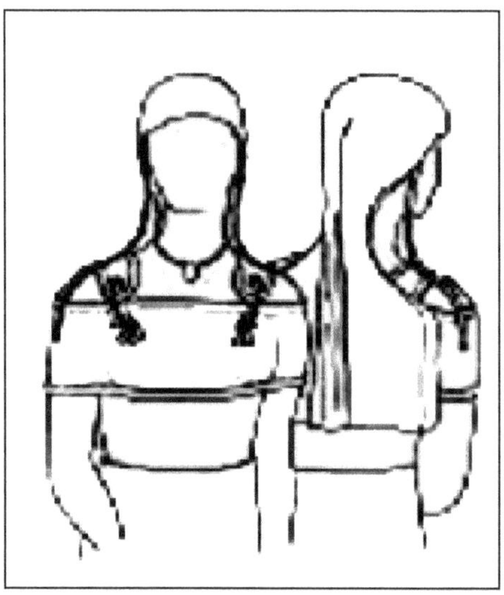

Figure 4.4. Reconstruction of female bust from Capestrano (Moretti 1936)

I believe that it is now possible to identify the burial location of the individual represented in the "statue of the Warrior". Tomb no.3 is not only close to the findspot of the "statue of the Warrior" (marked with a star in *Figure 4.5*), but is also different to the other burials in its orientation, position and size, as well as the quantity and distinguishing features of its funerary equipment. The rectangular space adjacent to tomb no.3 is also evident (*Figure 4.6*). It contained two spears, a sword with a cross-handle, a bronze basin (lebetis) used as a lid, a razor and footwear. Also present were the following objects: a stand, a semi-circular iron blade and an iron sword (Moretti 1936: 95-96). The tomb is shorter and wider than other tombs nearby; namely tomb nos.12, 13, 14, 15 and 21, positioned in a semi-circle surrounding tomb no.3. Moreover, these burials all have rich and elaborate funerary equipment compared with others in the same necropolis. The arrangement of the tombs could suggest that the warrior was placed at the centre of the burial area, in the highest position, both physically and symbolically (Papi 2000: 141-142). There may have been a tumulus for tomb no.3, with the "statue of the Warrior" placed in a central position. However, this hypothesis is unlikely, as there is insufficient space for such a reconstruction.

Another tomb in the Capestrano necropolis, close to tomb no.3 discussed above, also has an additional adjacent space. This is tomb no.8, which contained a well-preserved skeleton, but no other finds (Moretti 1934:97). The adjacent space has the same lengthwise dimensions as tomb no.8, but is only 50cm wide.

Regarding tombs thought to belong to women, tomb no.2 is one of the richest. Despite the lack of osteological analysis, it is considered to be a female burial principally because of the accompanying jewellery (*Figure 4.7*), which included three bronze fibulae, a necklace, pendants hanging from one of the fibulae, and two bronze fingerings. The jewellery ensemble is similar to that painted on the female bust mentioned above (*Figure. 4.4*), with the fibulae joined together by a small decorative chain. The necklace was found on the body and consisted of yellow glass paste pendants with blue "eyes" alternating with pieces of Italian amber. The burial is particularly interesting as it contains bronze objects related to wine serving (*Figure 4.8*) which are usually found in the tombs of males. These objects include a ladle, a strainer, a simpulum (a libation cup) and a stamnos (a small vase) (Moretti 1936: 96).

Helmets and cuirasses were not found at Capestrano, probably because they were made from perishable materials such as wood or leather, or because they were not considered as symbols of prestige like swords and daggers; a similar situation is attested in Campovalano cemetery (d'Ercole 1996a: 118). The huge helmet represented on the "statue of the Warrior" probably gave him great prestige and is the most exaggerated feature of the statue. However, it seems that at the time of burial, the helmet was not part of the warrior's funerary ensemble (Cianfarini 1969: 17).

Animal remains are also attested in the Capestrano necropolis. During the 1965 excavations, piglet remains were found between the legs of a young individual (Zanco 1967: 327), perhaps a votive offering, or possibly an animal beloved to the child buried. In Abruzzo, similar burials are attested at Corropoli, where a woman was buried with a dog (d'Ercole1996b: 149 In Sulmona, a skeleton of a small dog

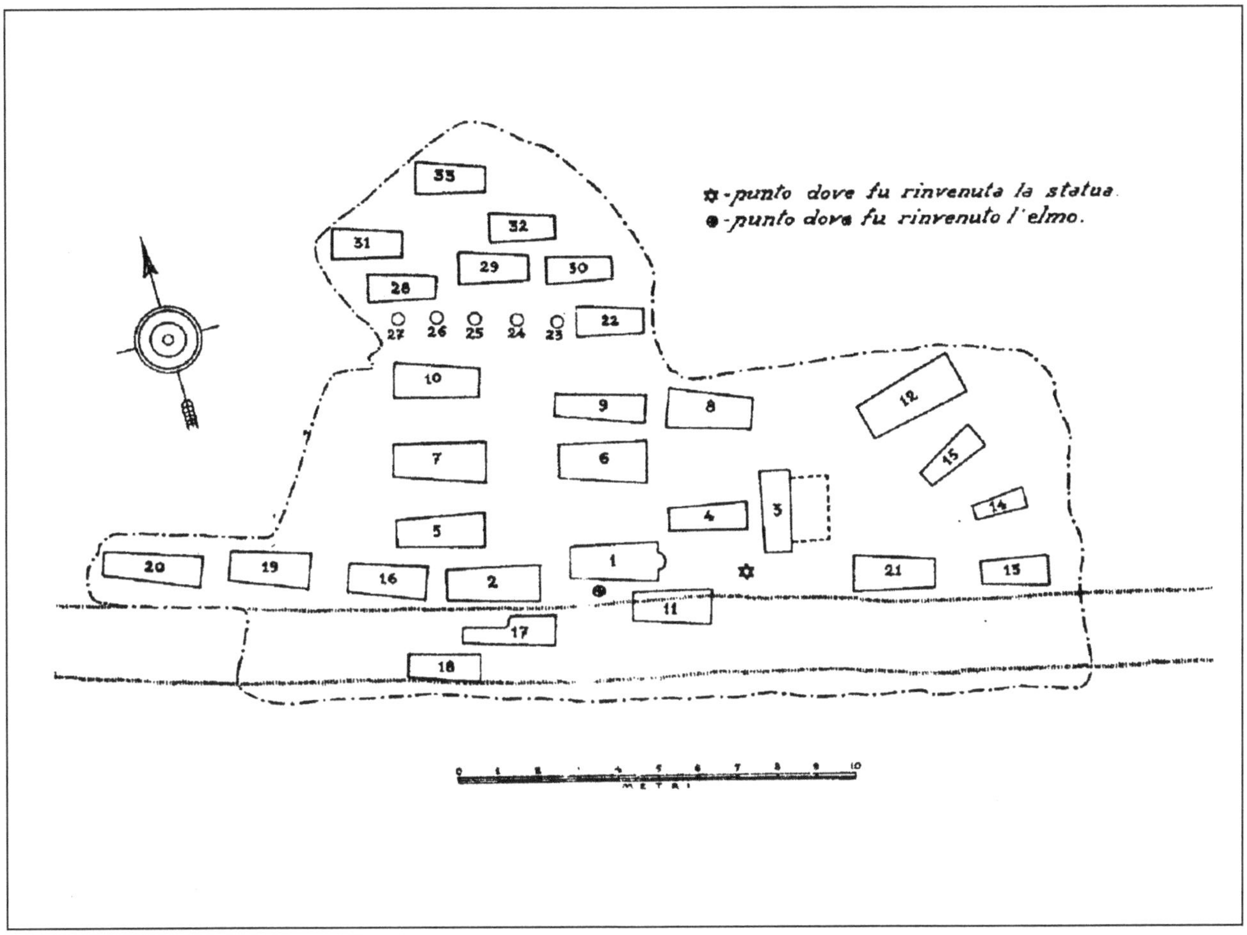

Figure 4.5. The Capestrano necropolis: burials excavated in 1934. (Unpublished archives of the local Archaeological Soprintendenza of Abruzzo, Chieti).

(d'Ercole1996b: 149). In Sulmona, a skeleton of a small dog was found adjacent to a burial urn (Zanco 1967: 327, no. 4).

Hard to explain are two tombs excavated in 1934 at Capestrano. Two individuals in a supine position had their faces covered by tiles; perhaps used to protect the face. The corpses themselves were covered with stones. These were probably child burials, attested in this cemetery and others, such as Fossa and Bazzano in the territory of the Vestini tribe.

During excavations at Capestrano in 1992, the remains of a female burial were found. The sex was determined through osteological analysis and the accompanying funerary equipment. It consisted of two bracelets in semicircular sections joined by a small clasp, two iron fibulae, two bronze plaques making up part of a belt decorated with a series of animals (*Figure 4.9*), a vessel in local bucchero (*Figure 4.10*), and a huge impasto jar. These are typical Iron Age objects associated with the *Picene* area, as documented by Lollini (1976: 107-196). Furthermore, they are typical of the middle Adriatic area in the Abruzzo region. The decorated plaques are unique within the necropolis at Capestrano. Another plaque, but without decoration, was found in 1934 and came from tomb no.13. Only two ends survive today, which were made from bronze and had nine clasps. Belts decorated with mythological animals date from the end of the seventh century to the second half of the sixth century BC (Zanco 1974: 57-61). No other belts of this type are known from the sixth century BC onward.

Figure 4.6. Tomb no. 3 from Capestrano (Unpublished archives of the local Archaeological Soprintendenza of Abruzzo, Chieti).

Figure 4.7. Jewellery from tomb no.2. (Moretti 1936)

Figure 4.8. Bronze wine-serving vessels from tomb no.2 (Moretti 1936)

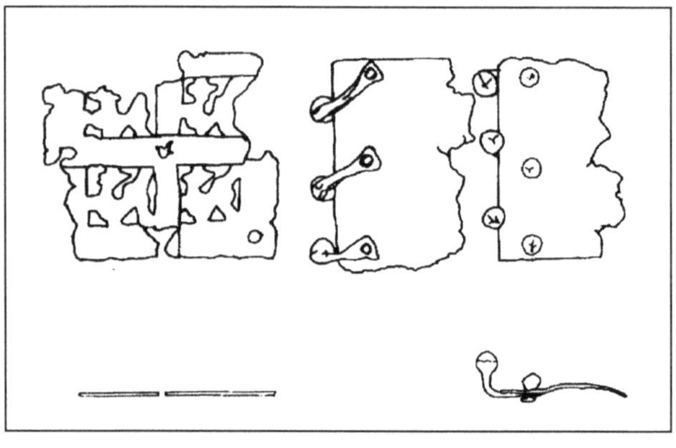

Figure 4.9. Bronze plaques found in 1992 (Illustrated by the author)

The local bucchero vessel (a small amphora – *Figure 4.10*) associated with the female burial was decorated with a geometric zigzag pattern, and is also typical of the Picene area in the sixth century BC (Landolfi 2001: 73-76). It is the only one of its type found in Capestrano. Perhaps this was a burial of a high status woman belonging to a distinct ethnic group (d'Ercole 1996a: 183). This theory is supported by similar examples known from burials at Campovalano, a site also found in Abruzzo wihtin the *Picene* tribal territory. Like Campovalano, Capestrano has a large number of bronze vessels that are not attested in other cemeteries of the *Vestini*, such as Bazzano e Fossa.

Also unique to the Capestrano necropolis are the engraved fragments of a funerary bed made of bone. The remains are few and not well preserved. They are not artistically important but they show that the use of the Capestrano necropolis continued during the Hellenistic period (3rd-1st Centuries BC). The tombs that date to the Hellenistic or late Hellenistic period are fewer in number than those in the sixth and mid-fifth Centuries BC. Among them are two chamber tombs with funerary equipment including bowls and black glazed dishes, *balsamaria* and oil lamps. One of the chambers contained a Corinthian vase. Some of these objects have two letters inscribed on them in particular: M and O.

Burials of the Archaic period (6th-4th Centuries BC) at Capestrano were in simple graves. The body was placed on its back with the arms along its side, with the head oriented towards the East. This burial position was typical in areas occupied by the *Vestini* and *Equi* tribes. There are no examples of wooden coffins or stones to cover the body. Only two burials show traces of stone coverings, which were only used to cover the heads. We could hypothesize that children were also present in the cemetery, such as the two burials discovered in 1973, the only child burials known from Capestrano in this period. Grave tumuli were not popular in the necropolis at Capestrano, however they were used more widely in Abruzzo at the time. There is perhaps only one instance of a grave mound at Capestrano. It belongs to tomb no.1 and was discovered in 1965. The corpse was placed on its back, and lowered into an earth grave lined with irregular stone slabs. It was subsequently covered entirely with stones (Zanco 1967: 320).

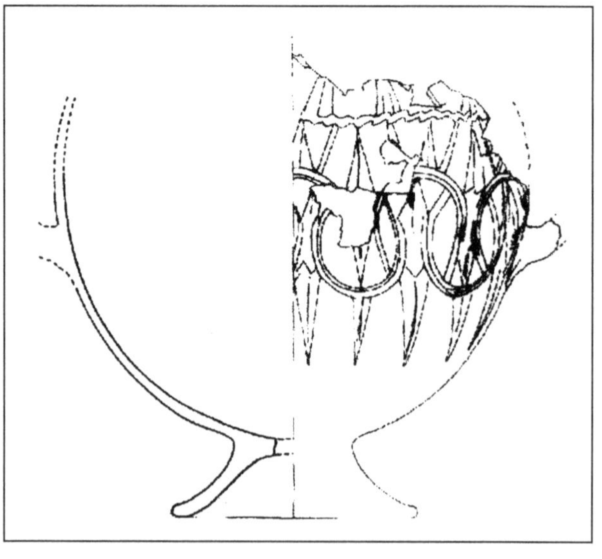

Figure 4.10. Bucchero amphora found in 1992 (Illustrated by the author)

Information that we have at our disposal concerning the Hellenistic and late Hellenistic phase (3rd Century BC to 31 BC) is not very clear. It is uncertain whether burials in chamber tombs occurred in this period. A diary of the excavations at the site reports the finding of paving and lateral walls for a wooden enclosure; only hook-shaped nails now exist to support the identification of such a wooden enclosure.

Cremation urns were also found at Capestrano. In burials of the sixth Century BC and fourth and third Centuries BC, urns were placed along the road at the site. Even chamber tombs have their corridor paths oriented towards the road, and are dated to the Roman period. It therefore appears that the necropolis was in use over many centuries.

Perhaps the road running through the necropolis could have provided access to a nearby settlement. Numerous remains suggest that there were a variety of settlements in the hillsides over the centuries. One hypothesis suggests that the nearest settlement to Capestrano was Collelongo, rather than Ofena (Zanco 1967: 329). This would mean that the necropolis was quite far from the settlement. Research conducted in 1962 by Prof. Silvio Ferri from the University of Pisa (Italy), suggests that such a road linking the necropolis to a settlement may have existed in antiquity. At the time of Professor Ferri's research, local farmers spoke of a "funerary road". This information is reported only in documents preserved in the local Soprintendenza (Ferri 1939). The distance from the necropolis to the settlement area was approximately one kilometre. The archaeological study of the necropolis at Capestrano has now been interrupted for a long time, and only recently has Dr. V. d'Ercole of the local Soprintendenza begun to re-excavate the site.

Acknowledgments

I would like to thank on this occasion Dr.V.d'Ercole for giving me permission to study the cemetery of Capestrano and Dr. S. Cosentino and G. Mieli for their help and advice.

Bibliography

Colonna G. 2001. "La scultura in pietra". In G. Colonna (ed.) *Eroi e Regine, Piceni popoli d'Europa*. Roma: 104-110.

Cianfarani V. 1969. *Antiche civiltà d'Abruzzo*. Roma.

d'Ercole V. 1996a. "La necropoli di Campovalano". *Documenti dell'Abruzzo teramano* IV (1). Teramo: 167 – 192.

d'Ercole V. 1996b. "Rassegna Paletnologica", *Le Valli della Vibrata e del Salinello. Preistoria e Archeologia*. Pescara: 133 – 151.

d'Ercole V. 2000. "Il Guerriero e la Necropoli di Capestrano" In *Principi europei dell'Età del Ferro* (various authors), National Archaeological Museum of Chieti, Museum Exhibition Catalogue. Chieti. Pp. 43-46.

Ferri, S. 1939. Unpublished archives of the local Archaeological Soprintendenza of Abruzzo, Chieti.

Landolfi M. 2001. "Forme ideologiche e costume funerario". In G. Colonna (ed.) *Eroi e Regine, Piceni popoli d'Europa*. Roma: 73 – 76.

La Regina A. 1986. "I Vestini in epoca arcaica". In E. Mattiocco (ed.), *Centri fortificati vestini*, Sulmona: 7 – 10.

Lollini D.G. 1976. "La civiltà Picena" In *Popoli e Civiltà dell'Italia Antica* V. Roma: 107 - 196.

Moretti G. 1936. "Il Guerriero Italico e la necropoli di Capestrano". *Bullettino di Paletnologia Italiana*. Roma: 94 – 114.

Papi R. 2000. "Continuità e trasformazione dell'ideologia militare nei territori sabellici medioadriatici". *Studi sull'Italia dei Sanniti*. Milano: 104 – 110.

Zanco O. 1974. *Bronzi arcaici da Campovalano*. Roma.

Zanco O. 1967. "Capestrano. Su alcune tombe italiche rinvenute in località Capo d'Acqua". *Notizie e Scavi* XXI. Roma: 320 – 330.

Corn-mummies come to light

Maria Costanza Centrone - University of Swansea, Wales

Introduction

As part of my PhD research on Egyptian corn-mummies, I have recently concentrated on the symbolic meaning lying behind these artefacts that were presumably manufactured during the Osirian Mysteries in the month of Khoiak.[1]

The items we are dealing with are commonly called 'corn-mummies' although the term 'corn' might create confusion. In fact, they are packages, ca. 35–50cm long, made of earth and grain (emmer wheat or barley), wrapped in resin-soaked linen and, at times, covered by a linen shroud and tied up with a linen band. They were moulded or modelled into a mummiform shape and provided with wax mask comprising face, divine beard and royal crown and, occasionally, with an erect wax phallus (i.e. ithyphallic). Royal sceptres, scarabs and uraei made in wax were sometimes placed over the body. Usually corn-mummies have been found enclosed in wooden falcon-headed coffins, although the specimens from Wady Qubbanet el-Qirud (or 'Valley of the Tombs of the Monkeys', one of the south-west valleys of the Theban hills) were apparently simply buried in the sand (Lortet and Gaillard 1907-1909, II: 247–248, IV: 209; Lortet 1906: 43–46).

Inscriptions on the coffin lids, addressing Osiris as 'lord of Westerners', the royal sceptres, divine beard and the *atef* crown with which the mummies were often provided, all symbols of the sovereignty of the god over the Netherworld, led to the conclusion that these figures are representations of Osiris.

As for the chronological period these objects belong to, this appears to be one of the most difficult matters to investigate: what we have at our disposal is limited to stylistic evidence provided by the coffins corn-mummies are placed into, and the masks they wear. A plausible chronological range would date the artefacts to the period between the Third Intermediate Period (XXII Dynasty, c.950 BC) and the Early Ptolemaic Period (c. 300-200 BC).

In this paper I will discuss some examples of corn-mummies, focusing on the materials employed in their manufacture, rather than on the decorative and stylistic features by which they are characterised. I intend to demonstrate that the substances used in the process of making the figurines have symbolic connotations for the ritual and the artefacts themselves.

Corn-mummies

The first specimen to be examined is in the British Museum (EA 60746) (*Figure 5.1*). It consists of a rounded ithyphallic Osirian mummy, and is 37cm long, 10cm wide, and wrapped in bitumen-soaked bandages. Traces of straw are visible on the surface, along with remnants of plaster. It is provided with a bulbous headdress (presumably representing a white crown), is wrapped in bandages and then covered with black bitumen. A green wax mask is placed on the face. Its original context has not been recorded, but a clear resemblance to other provenanced specimens allows us to classify the item as coming from Wady Qubbanet el-Qirud.

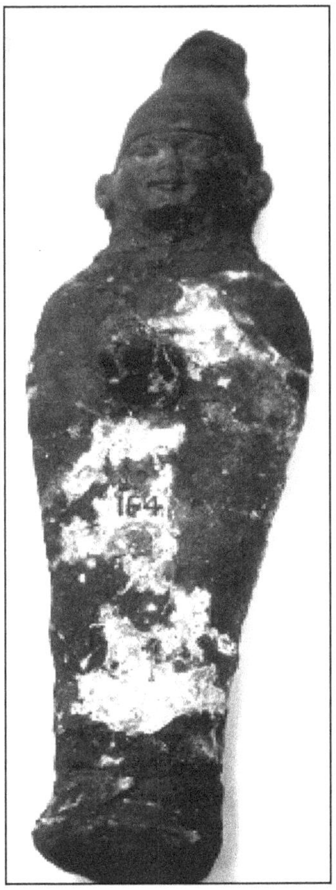

Figure 5.1. Corn mummy in British Museum, EA 60746. Photo taken by the author, Courtesy of the Department of Ancient Egypt and Sudan, The British Museum.

[1] The issue has been largely investigated in my Ph.D. thesis, currently in its final writing up session. The Khoiak Festival was the annual celebration of life, regarded as the renewal of nature, the restitution of the universal Order or, most specifically, as the resurrection of Osiris, the dead king who continues to rule after death, exercising the supreme power in the world. He had the authority of maintaining the World Order; he embodied the essence of life and his renewal.

Budapest Museum of Fine Arts has recently brought to my notice a specimen of an Osirian mummy kept in their collection (n. 68.2) (*Figure 5.2*). In view of its fragile condition, the object has never been fully examined, nor subject to scientific analysis (X-ray and Infra-Red Spectroscopy, Gas Chromatography/Mass Spectrometry and such like). It has a wrapping composed of linen bandages alternating with thin layers of resin. The outer coating is seemingly made of a linen shroud of which only a few traces are still visible. No mask is placed over the head, which is in tubular form, as is the probable white crown.

Figure 5.2: Corn-mummy in the Museum of Fine Arts in Budapest, No. 68.2. © Egyptian Collection of the Museum of Fine Arts, Budapest.

The Field Museum of Natural History in Chicago displays in the Egyptian Hall an Osirian figurine (inv. n. 174050), which is 45.5cm long, 12.7cm wide. Its provenance is, on the record, unknown (*Figure 5.3*). Nevertheless, I would link this specimen with the class of items from Tehne el-Gebel, in view of the remarkable resemblance to the mummies excavated by Lortet in Tehne (Lortet and Gaillard 1907–1909, IV: 210–213). In additional support of this interpretation, a note attached to the relevant museum files refers to a letter mentioning Tehne el-Gebel, dated November 1909 and addressed to Mr. Ayer, who collected the item in Egypt at some point before 1908.

The figure seems to be wrapped in resin-soaked linen and covered with a thick layer of wax. The museum note, dated October 1918 by the then curator of the collection, refers to a layer of stucco covering the mummiform packages in order to whiten them. A green wax mask, comprising a face and *atef* crown, was placed on the mummy, to which a divine beard, feathered-crown, horns and uraeus were applied. The mummy is ithyphallic. Positioned on the chest, a fist, in wax, held royal sceptres also made from wax. A wax scarab and 4 uraei, bearing the sun disc, were placed in the coffin along with the corn-mummy. According to the museum file, the figure was found along with four smaller packages (inv. nos. 31190–31192, 31610), approximately 15.8 cm in length, and each provided with a wax mask. I presume that these packages represented the four Sons of Horus, in view of the other Tehne specimens (*Figure 5.4*). The note also describes a roughly oval mummiform object covered with a layer of wax, 6.6 cm in length. A photograph of the object is not available and the note is imprecise. I would, however view this oval object to be the wax scarab, often placed below the head of other mummies from Tehne.

The last example of a corn-mummy I would like to describe is in Cairo Museum (JE 36539) (*Figures 5.5–5.6*). Although it has been recorded as coming from Tehne (Lefebvre 1903: 229–231), it shows stylistic features that differ from the Chicago specimen. In fact, it has a black coffin of sycamore wood, c. 61.5cm in length, c. 22.7 cm in width, with a gilded falcon's head and a blue tripartite wig. The decoration is in white paint; the space between the front lappets of the wig is occupied by a falcon's head bearing a sun disc rising from a winged heart (or a pot?). This is similar to the vignette depicted on another Cairo specimen (JE 36540, v. Lefebvre 1903: 229–231; Kessler 1981: 267, Pl. IX, 4; Raven 1982: 22). The central column of inscriptions, flanked by two rows of protective genii, reads: *'words spoken by Osiris, foremost of the West, the great god, lord of Bẖ (t); by Osiris, lord of Mr-nfr(t), the great god, lord of R3-st3'.* The inscription, by addressing Osiris as *'lord of Bẖ (t)'* and *'lord of Mr-nfr(t)'*, two places that have been identified as localities of the Tehne area (Gauthier 1925–1931, II: 4–5, III: 51–52; Kessler 1981: 269; Vandier 1961: 47–49), confirms the origin of the item.

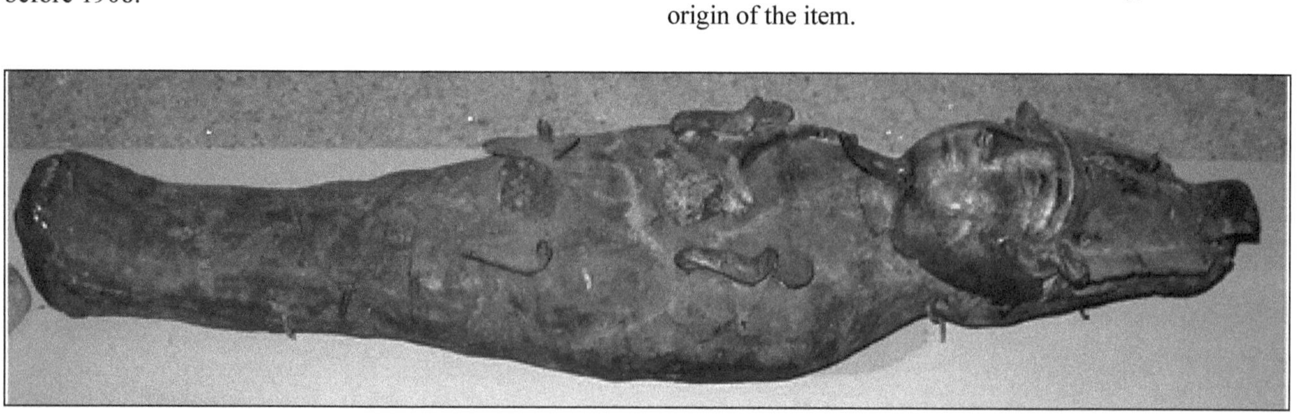

Figure 5.3: Corn-mummy at the Field Museum of Natural History, Chicago, Inv. No. 174050. © Collections of The Field Museum, 174050.

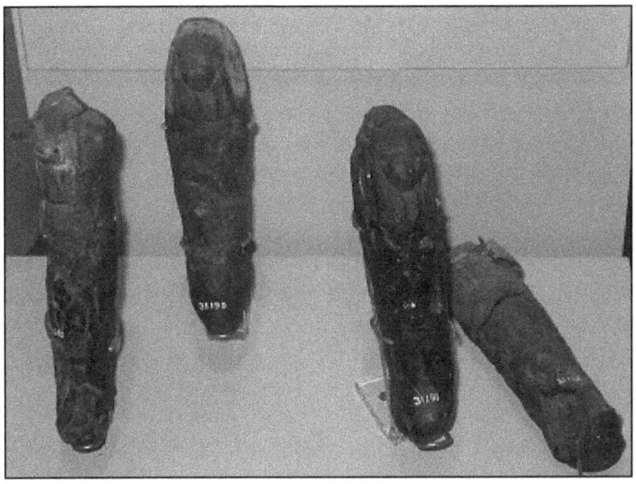

Figure 5.4: Four Sons of Horus, found near the corn-mummy, Field Museum of Natural History, Chicago.© Collections of The Field Museum, 31190–92, 31610.

The coffin case contains an ithyphallic Osirian mummy, c. 50cm length, c. 13cm in width, wrapped in black bitumen-soaked bandages and covered by a linen shroud. It is provided with a wax mask comprising a face, an *ʿtef* crown with uraeus and feathers in gilded wax. Traces of a wax divine beard are still visible. On the chest a fist in wax holds the royal sceptres in place. The coffin also contains a wrapped ball, placed near the neck, and a scarab in wax. Also present in the coffin are four mummiform packages wrapped in bandages and provided with wax masks. These are positioned symmetrically in relation to the mummy, with two at the head and two at the feet of the mummy.

Materials

I will not consider the presence of grain among the substances used to produce corn-mummies, nor will I analyse the symbolic connotations this product had in ancient Egyptian civilisation and specifically in the Osirian context. This matter deserves greater a degree of analysis than space permits me within this paper. I will therefore limit my investigation to other components, namely the materials themselves.

For instance, resin and bitumen were essential ingredients of the mixture from which the corn-mummies, the mummiform packages of the Sons of Horus, the scarabs and all the other objects placed in the coffins were made. These materials were also employed to wet the bandages the mummiform packages were wrapped in, to give the mummies, especially those buried without a coffin, a solid shape.

In fact, since most of the items have not been subject to modern chemical tests, in order to determine the nature of the components, we have to rely almost exclusively on the impression at sight and the museums' catalogues files which were, in most cases, compiled at the beginning of last century and never updated.[2]

In the nineteenth and early twentieth century the terms 'bitumen' and 'resin' were used indiscriminately to mean the substance employed in the embalming process but indeed, they are very different substances. Bitumen is a viscous mixture of hydrocarbons (*Concise Oxford English Dictionary*, 2002: 140) while resin is an organic substance exuded by plants and tress (*Concise Oxford English Dictionary* 2002: 1218). Lucas (1962: 348–353) demonstrated that often resin was used by the ancient Egyptians in the preservation of the corpse, rather than bitumen. In fact, the blackened appearance might have suggested to early scholars that the mummy had been treated with bitumen when most probably it is resin that was originally black or had become blackened with age. The presence of fatty acids may have caused the resin to become black, in addition to any heating to which the material was probably subjected, to render it sufficiently liquid to be poured over the body or to soak the wrapping bandages (Serpico 2000: 465).

Figure 5.5 (left): Coffin containing corn-mummy in Cairo Museum, JdE 36539. Photo taken by the author, courtesy of the Egyptian Museum, Cairo.
Figure 5.6 (right): Corn-mummy in Cairo Museum, JdE 36539. Photo taken by the author, courtesy of the Egyptian Museum, Cairo.

The only specimens of corn-mummies subject to meticulous scientific analysis of the materials, are those

[2] Tests carried out on materials thought to be bitumen or resin due to the blackened appearance, have sometimes turned out to be blackened paraffin wax from the time of excavations.

in the Archaeological Museum in Krakow, Poland (Klosowska 1997: 17–24). In these cases the tests showed that the bitumen-like substance which formed a thick layer covering the mummies, in fact contains no bitumen, but rather a mixture of different adhesives such as vegetable gum, animal glue and calofonium-like resin. Scientific analyses have also been carried out on other specimens, such as in the Museo Civico of Padua (inv. n.143, v. Zampieri 1981: 12–14.) and in the Brotmuseum in Ulm, Germany (inv. n. 0-755, v. Sokiranski, Adrario, Zwaka and Pirsig 2003: 137-142). However, in those cases little attention has been given to the resin-like substances employed to produce the mummies.

I am going to argue that resin or any other material that could give the same appearance as resin was not only intended to give the mixture of earth and grain the mummy was composed of a compact structure, or merely to fix together the bandages over the package. In fact, resin as well as resin-like substances were considered by the ancient Egyptians not only for their physical properties and technical potency but also for their symbolic connotations (Raven 1989: 7–21).

Resin was well known as a preservative and embalming agent, considering its viscosity and excellent sealing qualities. For these reasons it was deemed to provide a guarantee of the afterlife (Raven 1989: p.18). Burials were occasionally covered with great quantities of resin, presumably to ensure the renewal of the deceased.

Because of its gumminess and malleability, it was a suitable material also for the manufacture of jewellery, amulets and was generally used in the plastic arts. These objects are related to the magical arts of creating and destroying at will; consequently, resin itself was considered to have the same supernatural and divine properties (Raven 1989: p.15). Numerous funerary amulets were made of resin, such as the figurines representing the Four Sons of Horus in resin found in the XXII Dynasty in the cemetery in the Ramesseum (Petrie 1937: 26, nos. 600-603), the resin scarab found in the XXII Dynasty necropolis at Lahun (Petrie, Brunton & Murray 1923: Pl. LV A.2), and the heart scarab made of resin discovered in Tutankhamun's tomb (Carter 1927: 83, Pls. XXV, XXVI A).

Another material widely utilised in this context is wax. It was used to shape the Osirian masks placed over the corn-mummies and those applied on the small mummiform packages representing the characteristic facial traits of the four Sons of Horus. Wax was also employed to produce component parts, such as the phallus, fists, royal sceptres, scarabs, and uraei-wearing sun discs, occasionally found over the grain packages.

Wax was characterised by a wide range of technical applications. For example, it was employed for sealing, fixing, painting, coating, and also in medicine, pharmacology and the plastic arts (Büll 1977; Raven 1983). Its magical properties were well known. In fact, its imperviousness to air, water, moisture and microbes made wax highly suitable for embalming; consequently this kind of invulnerability was regarded as a guarantee of the afterlife and survival. Secondly, when in contact with the heat, mysteriously (to a non-scientific observer) it started becoming soft, melting, changing shape and miraculously giving birth to something new. Finally, in the plastic arts wax was considered as one of the best materials, thanks to its malleability that allowed the human sculptor to become master of supernatural powers, to create and destroy at his will in an amazing way (Raven 1983: 28–32).

As a symbol of divine energy, wax obviously played a very important role in Egyptian magic (Barta 1972: 1233–1237). It was employed for manufacturing figurines and wax objects so that benefit could be gained from their divine protective influence. The same intent lay behind the practice of placing, along with the mummified body, scarabs, facial masks or any other amulets that could ensure the resurrection of the deceased (Varga 1964: 6–12).

A symbolic meaning also lay behind the use of sycamore wood for the Osirian falcon-headed coffins containing the corn-mummies. According to chapter 109 of the *Book of the Dead* (Faulkner 1985: 102) twin sycamores (*nehet*) of turquoise were at the eastern gate of heaven from which the sun god Ra rises every day (Wilkinson 1994: 90). Moreover, the sycamore tree was considered as a manifestation of the goddesses Nut, Isis and Hathor (Wilkinson 1994: 90; Buhl 1947: 80–97) who offered nourishment to the deceased. That is why sycamore trees were often planted near tombs, amulets in the shape of leaves of this tree were manufactured, and why coffins were made from sycamore wood.

Conclusions

In this paper my aim was to demonstrate that the materials employed to manufacture the corn-mummies and the coffins they were often buried in, contribute symbolic meanings to the artefacts and to the Osirian ritual they were presumably involved with. The resin in which the Osiris body (in the shape of the corn-mummy) was wrapped, assured the renewal and regeneration of the god. The sycamore wood coffin Osiris was placed in protected and nourished the god during his journey to the Underworld. The wax objects placed over the corn-mummies, thanks to their powerful revivifying qualities, undoubtedly played a role in ensuring the symbolic rebirth of Osiris.

Bibliography

Barta W. 1972. "Materialmagie und-symbolik", in W. Helck & W. Westendorf (eds.) *Lexikon der Ägyptologie* III. Wiesbaden: 1233–1237.

Buhl M.L. 1947. "The Goddesses of the Egyptian Tree Cult" *Journal of Near East Studies* 6: 80–97.

Büll R. 1977. *Das grosse Buch vom Wachs: Geschichte, Kultur, Technik.* London/München.

Carter H. 1927. *The tomb of Tut,Ankh.Amen* II. London.

Faulkner R.O. (ed.) 1985. *The ancient Egyptian Book of the Dead.* London.

Fuchs R. 1972. "Wachs", in W. Helck & W. Westendorf (eds.) *Lexikon der Ägyptologie* VI Wiesbaden: 1090-1094.

Gauthier H. 1925-31. *Dictionnaire des noms géographiques contenus dans les textes hiéroglyphiques,* II, III & V. Cairo.

Kessler D. 1981. *Historische Topographie der Region zwischen Mallawi und Samalut.* Tübinger Atlas des Vorderen Orients 30. Wiesbaden: 253-290.

Klosowska M.J.A. 1997. "The conservation and technical examination of three corn-mummies at the Archaeological Museum in Cracow" *Materialy Archeologiczne* 30: 17-23.

Lefebvre M. G. 1903. "Sarcophages égyptiens trouvés dans une nécropole gréco-romaine à Tehneh" *Annales du service des antiquités de l'Égypte* 4: 229-231.

Lortet L. & Gaillard C. 1907–1909. *La faune momifiée de l'ancienne Egypt.* Vols. II & IV. Lyon.

Lucas A. 1962. *Ancient Egyptian Materials and Industries.* London.

Petrie W. M. F. 1937. *The funeral furniture of Egypt*, 26, nos. 600-603. British school of archaeology in Egypt and Egyptian research account, 59. London.

Petrie W. M. F. Brunton G., Murray M.A. 1923. *Lahun.* II. British school of archaeology in Egypt and Egyptian research account, 26. London.

Raven M. J. 1982. "Corn-mummies" *Oudheidkundige Mededelingen iut het Rijksmuseum van Oudheden te Leiden* 63: 7–34.

Raven M. J. 1983. "Wax in Egyptian magic and symbolism", *Oudheidkundige Mededelingen iut het Rijksmuseum van Oudheden te Leiden* 64: 7–47.

Raven M. J. 1989. "Resin in Egyptian Magic and Symbolism" *Oudheidkundige Mededelingen iut het Rijksmuseum van Oudheden te Leiden* 70: 7–21.

Serpico M. 2000. "Resins, Amber and Bitumen" in P.T. Nicholson & I. Shaw (eds.) *Ancient Egyptian Materials and Technology.* Cambridge/New York: 430-474.

Serpico M. & White R. 2000. "Oil, Fat and Wax" in P.T. Nicholson & I. Shaw (eds.) *Ancient Egyptian Materials and Technology.* Cambridge & New York.

Sokiranski R., Adrario C., Zwaka P. & Pirsig W. 2003. "Der spriesende Osiris" *Antike Welt* 2: 137–142.

Vandier J. 1962. *Papyrus Jumilhac.* Paris.

Varga E. 1964. "Les monuments d'une coutume funéraire égyptienne de Basse Époque" *Bulletin du Musée Hongrois des Beaux-Arts* 25: 3–17.

Wilkinson R. H. 1994. *Symbol and Magic in Egyptian art*, London.

Zampieri, C. 1981. "Cento opere restaurate del Museo Civico di Padova" *Bollettino del Museo Civico di Padova* 70: 12–16.

The Tomb S1 of Cyrene: from the Hellenistic phase to Christian re-use

Luca Cherstich – University of Oxford

Cyrene, in Libya, has one of the best-preserved and largest cemeteries of the Greco-Roman world. Although many of its tombs have been looted they can still provide precious information on the local ancient society (Cherstich 2002). This paper is one of the products of a survey in the southern necropolis, which is part of a larger project of the Italian Archaeological Mission in Cyrene's necropolis.

This paper examines the example of tomb S1, a monument already known, but never fully investigated (Cassels 1955: 34; Stucchi 1975: 165n1; Tomlinson 1967: 250-251, 254). A study of this tomb provides a context relevant to the Late Hellenistic (2nd -1st century BC) Cyrenean burial practices and to the general picture of the Ptolemaic Mediterranean, especially in North Africa. Attention will be focused on the 'loculi' in this tomb. These are long, corridor-like, rock-cut recesses, usually with a rectangular plan and a roughly rectangular section, and divided into two or three levels. At Cyrene they are accessed directly from the façades of the tombs or via internal chambers (as shown in *Figure 6.5*). Although loculi are quite widespread in the Hellenistic Mediterranean, the phenomenon is not yet well investigated. Their diffusion in territories ruled or influenced by the Ptolemies is noteworthy, but this fact could simply be a coincidence and must be examined further.

Position

The ancient road to Balagrae, starting from the Southern Gate of Cyrene, goes through an important religious area with continuity of use from Archaic to Imperial times (Stucchi 1975: 252, 260, 261, Fig. 406). The position of tomb S1, the first tomb on the right side after the last temenos, was quite prestigious: not only it was the first tomb seen by travellers from Cyrene, but due to its monumental columns it certainly appeared visually to be a continuation of the nearby temples. The family owner of the tomb was probably amongst the most prestigious or richest of Late Hellenistic Cyrene.

Tomb exterior

S1 is a false-façade tomb (Tomlinson 1967), a class that seems typical of south-eastern cemeteries of Cyrene. In these flat areas it was impossible to cut tall façades like those in the steep north-western cemeteries. Therefore large isodomic façades were constructed as monumental buildings with nothing behind them. The entrance to the real sepulchre was situated beneath the façade: either on the same side or entirely independent and situated off-centre behind the façade, which is the case for tomb S1.

The tomb has a huge courtyard (25m x 30m), delimited on the side near the road by 7 simple rock-cut graves (*Figure 6.1*), which were looted and therefore impossible to date. It is noteworthy that these people were buried outside the tomb proper. They could have been related to the principal family, but perhaps belong to a lower status. Alternatively, if these graves are later than the principal tomb, these people may have been buried outside the tomb, owing to a change in burial customs.

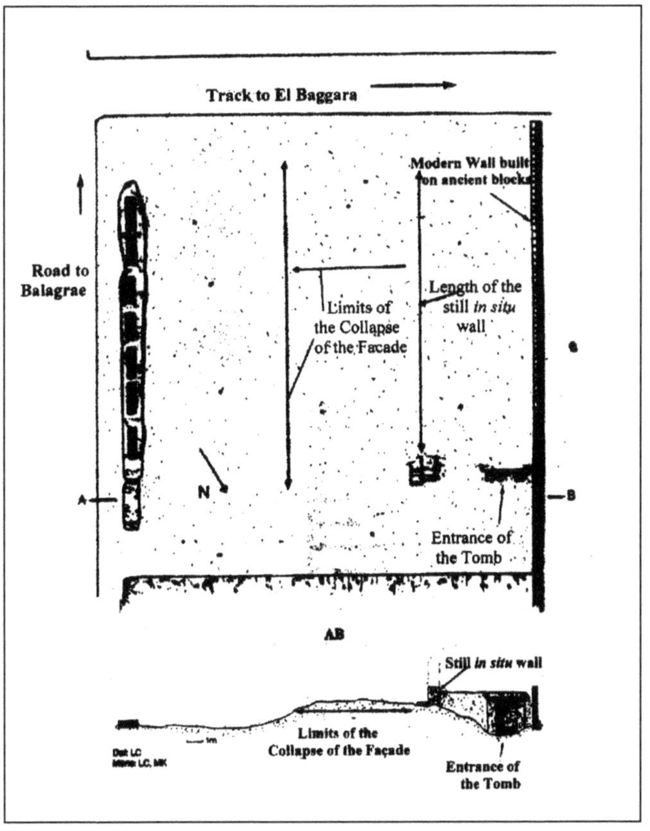

Figure 6.1 Exterior of the tomb

On the other side of the courtyard there is a low, half-buried wall about 20 m long. This is what remains of the lower part of the façade. The rest is collapsed and now fills the courtyard. The courtyard is nowadays more buried than in 1956, the time of Tomlinson's investigations (Tomlinson 1967: pl.46 a-b). He was able to recognize features not visible today: an anta-wall with an engaged-column. From what is visible, one can only argue that the façade was a colonnade of 6 (or 8) columns between two half-columns. Although excavation is needed to better understand its plan and elevation, it is quite clear that this was a highly monumental façade, with dimensions challenging contemporary temples.

On top of the façade there stood a typical Cyrenaican

funerary statue (*Figure 6.2*), roughly dated to the Hellenistic period (Beschi 1972: 251-261, Type L). The inscription on its base (*Figure 6.3*) is reminiscent of inscription no. 5169 in the Corpus Iscriptionum Graecarum (Boeck *et al* 1827-1828), reported in the early 19th century by the French explorer Pacho. Inscription no. 5169 has not been rediscovered since that time, and one wonders if was actually a fabrication by Pacho, who perhaps borrowed and elaborated upon the fragmentary funerary statue base inscription.

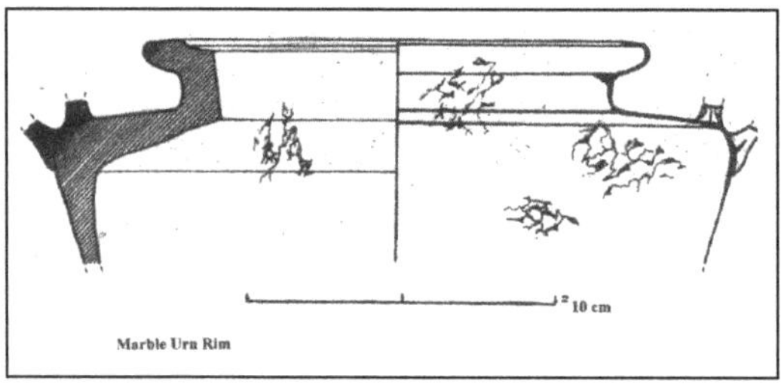

Figure 6.2: marble urn rim

Considering the statue and the complicated mouldings of a gheison block (*Figure 6.4*), the tomb should be dated no earlier than the Late Hellenistic period (2nd-1st Century BC). This date is compatible with the mixture of Ionic (Ionic fluting on column drums) and Doric (mutules and frieze) elements in the architectural blocks visible today (Bacchielli 1980).

Tomb interior

Down some now buried steps, a dromos leads to a long chamber, from which 12 loculi can be entered (*Figure 6.5*). The chamber was originally designed to contain 20 loculi, but the last 8 remained uncut. At the end of the chamber there is a basin. At its base a small channel connects the basin to a pit.

In the vicinity a slab was found (*Figure 6.5*) whose dimensions fit the basin's sides. The distance between the two holes in the slab is equal to the distance between the two ends of the small channel in the basin. This suggests that the slab acted as a lid for the basin. Through the holes some liquid offerings could have been poured into the basin and then flowed through the small channel into the pit. The pit could therefore be a *bothros* for offerings to the dead or underworld divinities. For the first time this kind of slab is found close to its original context in Cyrene, and it is now known that it was probably used as a table for offerings.

The offering procedure is associated with the rear part of the tomb. From other Cyrenaican tombs we know that the chamber was usually completely cut prior to the loculi, which were subsequently cut only when there was need for them. For this reason, the loculi are only found in the front half of the tomb S1 interior. This means that in S1 the offering system had already been designed and incorporated in the original phase of tomb construction. In fact, at the time of construction some rock was deliberately set aside for cutting the basin.

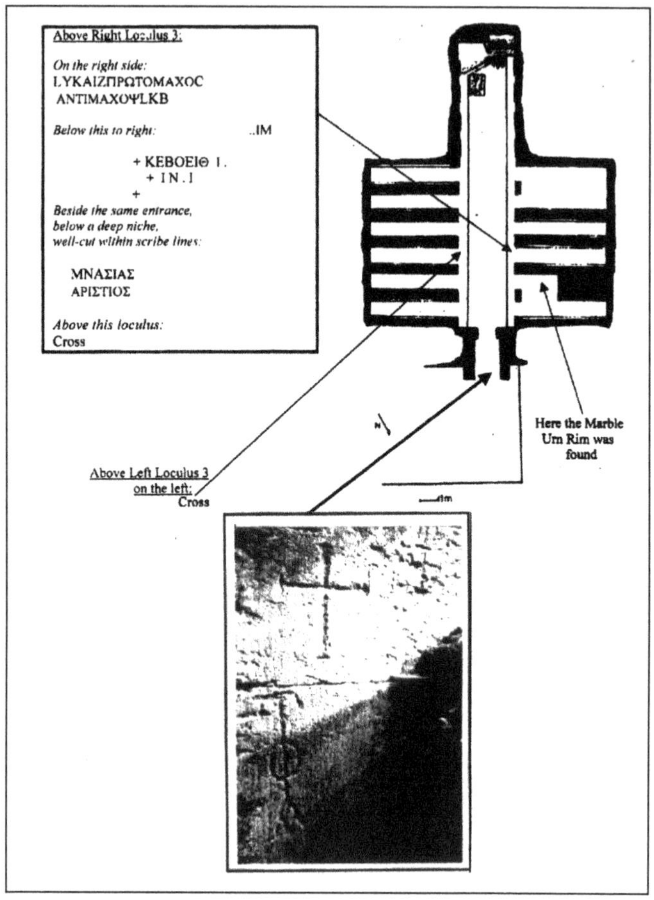

Figure 6.3: the tomb interior

In Hellenistic Cyrenaica, inhumation was the predominant burial rite, and loculi the predominant method for burying dead in rock-cut tombs. In Cyrene, the cutting of loculi did not continue after Early Imperial times (1st century A.D.) and this means that all the S1 loculi must be roughly dated to a 200-250 years span (probably 2nd Century BC- 1st Century AD).

Tomb S1 was looted and therefore no human remains are present in the loculi. However, if one considers the number of loculi, and the fact that some loculi have two levels, it becomes clear that S1 could have stored more than 40 corpses. It is uncertain whether all these dead were members of a large clan or family, or rather individuals united by other social links.

Many tombs in Cyrene have numerous loculi organised around rectangular chambers. What makes S1 different is the fact that because it was subterranean, it had to be reached via a stepped dromos. This situation is quite atypical in Cyrene. Other tombs have interiors at the same level to the exteriors because they were cut into the vertical cliffs of the valleys.

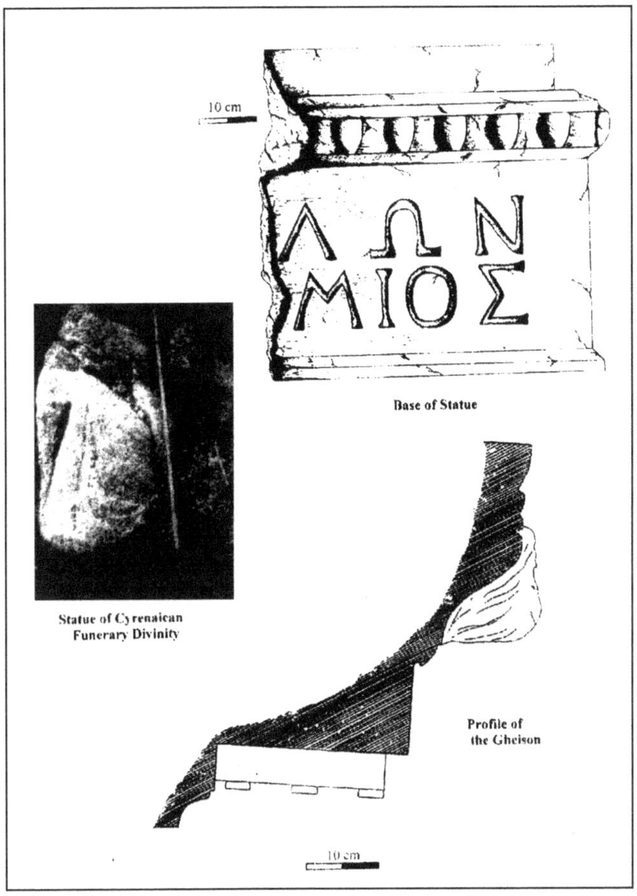

Figure 6.4: Various features from the tomb exterior

The flat landscape of the Southern Necropolis could have influenced the subterranean siting of S1 but it could have also been the result of a conscious choice. For this reason, it could be worthwhile to search elsewhere for tombs with interiors similar to S1.

In the Kydonia necropolis (Crete), there is a tomb smaller than S1 but featuring a similar plan (Μαρκουλακη *et al.* 1982: 15-117). It is accessed through a stepped dromos and it has a rectangular chamber from which various loculi can be accessed. Similar to those at Cyrene, these loculi have two levels and the names of the dead inscribed above the entrances. A Cyrenaean coin dated to the 4th-3rd Century BC was found in this tomb.

Because the relationships between Crete and Cyrene are well attested (for example van Enteferre 1948: 120-126), it is quite plausible that Cyrenean customs influenced those who commissioned the construction of the Kydonia tomb. On the other hand one cannot suppose that Crete influenced Cyrene. In Cyrenan tombs loculi are not only numerous but they appear at date (4th Century BC) which is long before Crete's loculi. Cyrenean practices could have influenced Kydonia or, more probably, the Cyrenean influence came through a medium: Alexandria.

More interesting is the link between S1 and Alexandria. The Shatby Chambrettes A and C (Adriani 1952: 3-6, Fig. 9, 10, 12) are both underground tombs, with loculi organised around rectangular chambers and reached through stepped dromoi. These tombs are dated to the late 4th – early 3rd Centuries BC, certainly predating tomb S1. Although later tombs in Alexandria tend to have more complicated plans, it is noteworthy that the first Alexandrian tombs are very similar to Cyrenaican examples. They probably were the first tombs in Alexandria to use loculi.

Alexandria was founded in 331 BC and the Greco-Macedonian elites certainly had their own burial customs (e.g. tumuli; Andronikos 1994: 98-99 fig.55), at least at the beginning of the Hellenistic period. None of the Macedonian-style tumulus tombs are currently known in Alexandria. The only probable example of a Macedonian-style tomb is the so-called 'Alabaster tomb' (Venit 2002: 8-9).

Loculi are the most commonly used burial device known in Alexandria but their inspiration is unknown. Both the Levant and Pre-Ptolemaic Egypt could have been sources of inspiration, although the evidence for Alexandrian-type loculi is not so clear (Venit 2002: 16). Whatever the origin, loculi became the official burial-type in Hellenistic Alexandria, and probably become more common elsewhere thereafter, often with local interpretations. In Petra, for example, the tombs with Alexandrian-type façades have loculi which are often cut in very high positions in the walls, unlike any Alexandrian or Cyrenean tomb (McKenzie 1990:113-4). In any case they are still loculi and the position itself could be a sign of local interpretations. The idea of loculi could have come to the tombs of Petra from Alexandria in conjunction with the "Baroque architecture" so well studied by McKenzie.

At Cyrene some loculi tombs can be dated to the 4th century BC due to architectural features in their façades that have comparanda in well-dated public buildings (Stucchi 1975: 47-52 for Buildings, Fig. 59, 60 for Tombs). No dated loculi in Cyrene predate the 4th Century BC. However one must also consider that tomb W28 contains a burial installation that seems to be a rudimentary antecedent of a loculus, and is dated to the 5th Century BC (Thorn forthcoming: IV.1.2).

There is a stronger possibility that loculi appeared in Cyrene before those in Alexandria. Considering the proximity of the two places and the strong presence of Cyreneans in Alexandria, it is also possible that Cyrenean loculi inspired the Alexandrian ones. In any case, because the shape of a loculus is so simple, one cannot exclude the possibility that the feature was created in many different places at around the same time.

Certainly what seems to be more Alexandrian is the stepped dromos. Alexandria's landscape is more flat than the rocky landscape of Cyrene, and therefore subterranean tombs were more common: Alexandrian tombs were usually cut under the level of the terrain, Cyrenean ones were cut in the steep walls of the valleys. For this reason, the dromos in S1 could be an Alexandrian fashion and the fact that the tomb was built during the Late Hellenistic period suggests a Ptolemaic influence.

The same identical scheme of the Shatby tombs (stepped dromoi, rectangular chambers with loculi) can be found in Cyprus (Gardner *et al* 1888: 267, fig. 2) and Berenike (Cyrenaica) (Dent *et al* 1983: fig. 3; Ghazal 1983: fig. 2). In Cyprus stepped dromoi were already typical but the use of loculi could testify to Ptolemaic customs, whereas the tombs in Berenike could possibly reflect both local Cyrenaican traditions and Ptolemaic fashions. Both sites were part of the Ptolemaic empire and were perhaps responsive to these influences for this reason. All this evidence could place the underground interior of tomb S1 into a much broader and more complex picture of Alexandrian–Cyrenaican burial customs, that may have influenced various places also under Ptolemaic rule.

Outside Cyrene, tombs have real façades or do not have any façade at all. For the moment, false-façades seem typical only of Cyrene. The façade of S1 is not well preserved but it shows similarities with tomb S4 nearby that certainly reflects Alexandrian influences.[1] The position, size and elaboration of sculptural and architectural elements are all suggestive that the tomb owners were high ranking. Furthermore, the tomb probably demonstrates signs of the most up-to-date (Ptolemaic) fashions.

Second Phase: Around 2nd Century AD

At some point in its history, the interior of tomb S1 was modified. Two pairs of loculi were transformed into two chambers (*Figure 6.3*). The fact that this was done rather than cutting new loculi, for which space was available, demonstrates changing burial customs.

In the Early and middle Imperial period (1st-3rd century AD), numerous tombs in Cyrene had their loculi transformed into wider chambers. This change is also detectable in Ptolemaic Egypt and Idumaea (the area between the Negev desert and the Dead Sea) (Venit 2002: fig. 26, 143, 151, 152). In Cyrenaica, in Hadrianopolis (near Euhesperides) there are also tombs that in 2nd century AD were originally cut with these wide chambers (Jones & Little 1971: fig.4).

During the Roman Empire, Greek North Africa saw a change of burial customs that required larger spaces than the old loculi. These rooms could have been for larger coffins, or to contain the marble sarcophagi that from the 2nd Century AD onwards were becoming more widespread in the Mediterranean.

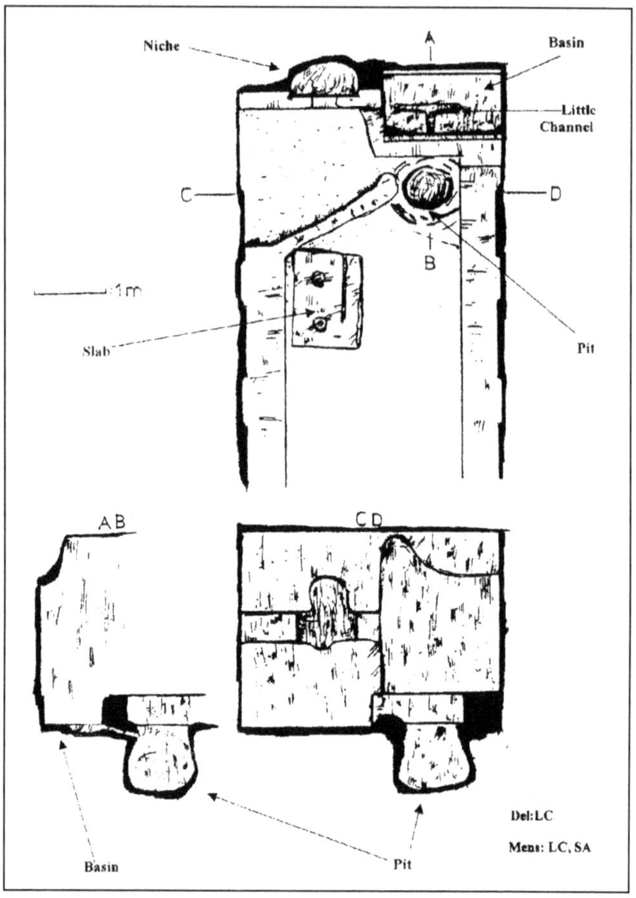

Figure 6.5: Various features from the tomb interior: the votive system

Paradoxically, in tomb S1 the only burial that could be dated to the Roman Imperial period is one that requires less space: a cremation. In the second, right-hand entrance, I personally discovered a marble urn rim, the first ever found in Cyrenaica (*Figure 6.2*).

Numerous stone urns have been found in Tripolitania (Di Vita-Evrard *et al* 1996: 89-120, Tav. XXXIX-XLVI; Mabruk 1987: Fig. 4; Mahmoud 1977: fig. 4). These urns are not identical, but share the same dimensions and general shape as our example. All the Tripolitanian examples are roughly dated to the 2nd Century AD, suggesting a similar date for the piece in tomb S1. Cyrenaica has no marble quarries and so this object was certainly imported from elsewhere. Egypt could be a possibility, but further analysis is required.

Finding a cremation in Imperial Cyrene is unusual, not only because inhumation was becoming widespread throughout the Empire at this time (Morris 1992: 52-69), but also because Cyrenaica, in addition to many other Greek regions, had throughout its history a preference for inhumation. Many explanations could be proposed for

[1] Tomb S4 will be published among the papers of the Conference of Cyrenaican Archaeology held in Chieti (Italy) on 24-26 November 2003.

the S1 urn. It could be the burial of a Cyrenean who died abroad, or a non-local individual (a Tripolitanian?) with different burial traditions who was accepted into the family. Furthermore the rarity of installations for cremations in the necropolis (Di Valerio *et al*: this volume) could indicate that in S1 it marked something else, perhaps status or fashion? These hypotheses cannot be verified at present, as this is the only evidence for a marble urn ever found in Cyrene. The possibility that the urn was never used for a cremation must also be considered.

Third Phase: Late Antique period (?)

In the dromos of tomb S1, on the right side (the most visible part of the tomb for someone approaching from the road) a Monogram, a Latin Cross and possibly a smaller cross were inscribed (*Figure 6.3*, inset). The similarity of all these carved signs could indicate that they are contemporary. The presence of the cross suggests a late date, probably the 4th Century AD or later. The Monogram has no parallel and remains a mystery.

Other inscriptions are above the entrance of the third loculus on the right. Noteworthy is the interpretation that Dr. J. Reynolds has suggested for one of them: the formula *"Lord help ... "* followed by the illegible name of the dead (Reynolds *et al* 2003: 177 - but their location is wrong). The fact that this inscription is Christian, is assured by the large central cross in its vicinity. It is shallowly cut above the loculus, an element also repeated above the third loculus on the left. It is possible that these inscriptions are related and contemporary with those in the dromos. Their presence above the loculi demonstrates for the first time in Cyrene a re-use in Late Antique times (possibly 4th-6th century AD?) of an older Hellenistic burial feature without any additional modification.

All this evidence suggests that a Christian, important enough to have his own monogram, re-used the sumptuous old Greek tomb. It is difficult to say whether during its four centuries of use, the same family used the tomb.

Conclusions

The tomb S1, although looted, shows an evolution of burial customs in Cyrene over roughly six centuries (2nd Century BC - 4th Century AD). The potential for roughly dating this sequence could help further studies of Cyrene's necropolis, like the survey project in the necropolis by the Italian Archaeological Mission of Chieti University (see Di Valerio *et al*: this volume).

Furthermore the changes visible in S1 are comparable with other tombs in North Africa, Crete and Cyprus, demonstrating how much Cyrenaican tombs can potentially tell us. Cyrenaica has still not received the same attention that other parts of the Greek World have had to date, and the study of its tombs could be a key to a better understanding of what we have lost in other important cemeteries such as Alexandria.

Acknowledgements

Thanks to Debora Lagatta, J. and D. Thorn, J. Ehrenwerth, R. Tomlinson, J. Reynolds, E. Fabbricotti

Bibliography

Adriani A. 1952. *Annuarie du Musee Greco-Romain III (1940-1950)*. Alexandria.

Andronikos M. 1994. *Vergina. The royal tombs.* Athens.

Bacchielli L. 1980. "La tomba delle Cariatidi ed il decorativismo nell'architettura tardo-ellenistica di Cirene" *Quaderni di Archeologia della Libia* 11: 11-34.

Beschi L. 1972. "Divinita` funerarie Cirenaiche" *Annuario delle Scuola Archeologica di Atene e delle Missioni Italiane in Oriente* XLVII-XLVIII: 133-341.

Boeckh A., Franz, J., Curtius E., Rohl H. & Kirchhoff A. 1827-1828. *Corpus Iscriptium Graecarum.* Berolini.

Cassels J. 1955. "The cemeteries of Cyrene" *Papers of the British School at Rome* XXIII: 1-43.

Cherstich L. 2002. "Surveying the Cemeteries of Ancient Cyrene. Preliminaries of a Topographic Approach", in G. Muskett *et al* (eds.) *SOMA 2001: Symposium on Mediterranean Archaeology*, Oxford: 20-27.

Dent J.S., Lloyd J.A., Riley J.A. 1983. "Some Hellenistic and Early Roman Tombs from Benghazi" *Libya Antiqua XIII-XIV*: 131-212.

Di Vita-Evrard G., Fontana S., Mellegni F., Munzi M., Musso L. 1996. "L`Ipogeo dei Flavi a Leptis presso Gasr Gelda" *Libya Antiqua N.S. II*: 85-133.

Gardner E.A., Hogarth D.G., James M.R., Elsey Smith R. 1888. "Excavations in Cyprus 1887-8" *Journal of Hellenic Studies IX*: 264-270.

Ghazal A. 1983. "Recent discoveries or Roman rock-cut tombs at Sidi Hussein-Benghazi" *Libya Antiqua XIII-XIV*: 219-234.

Jones G.D.B. & Little J.H. 1971. "Hadrianopolis" *Libya Antiqua VIII*: 53-67.

Mabrouk G. 1987. "L`Ipogeo ed i materiali rinvenuti" *Libya Antiqua XV-XVI*: 45-50.

Mahmoud S.A. 1977. "Neo-Punic tombs near Leptis Magna" *Society for Libyan Studies Annual Report 1976-77*: 27-40.

McKenzie J. 1990. *The Architecture of Petra.* Oxford.

Morris I. 1992. *Death ritual and social structure in Classical Antiquity.* Cambridge.

Reynolds J., Ward-Perkins J.B. & Goodchild R.G. 2003. *Christian monuments of Cyrenaica.* Hertford.

Stucchi S.1975. *Architettura Cirenaica.* Roma.

Thorn J.C. forthcoming. *The Necropolis of Cyrene: two hundreds years of exploration*

Tomlinson R.A. 1967. "False-Façade tombs at Cyrene"

Papers of the British School at Athens LXII: 241-256.

van Enteferre H. 1948. *La Crete et le Monde Grec de Platon ò Polybe*. Paris.

Venit M.S. 2002. *Monumental tombs of Ancient Alexandria*. Cambridge.

Μαρκουλακη Σ. & Νινιου–Κινδελη Β. 1982. "Ελληνι στικος Λαξευτος χανιων Ανακαφη Οικοπεδου Μαθιουλακη" *Archaiologikon Deltion XXXVII*, Athens: 1-118.

Lilith across the ages

Virginie Danrey – Arché Orient, Lyon

Monstrous creatures have been a recurring theme in the history of humanity. They are defined by new departures from the norm and can be constructed in an infinite variety of ways. However, if one classifies the different types of fantasy beings found in the Mediterranean area, one notices that only a limited number of monsters have been copied throughout history. Many monsters stemming from the Mesopotamian tradition have come down through centuries and have come to the European continent through the medium of imagery and texts. The aim of this article is to show the different routes of diffusion taken by these figures, and to analyse the ways in which these symbols migrate from one culture to another. This will be illustrated by the famous figure of Lilith, which spread in the whole Mediterranean basin and apparently also came from the Mesopotamian tradition.

Lilith appears in an apocalyptic poem from the book of Isaiah composed during the 8th century BC, which describes the end of the kingdom of Edom and the world's return to the state of first chaos. Lilith is described as a feminine demonic entity, quoted amongst wild animals and evil beings:

> *"Wildcats shall meet with hyenas,*
> *goat-demons shall call to each other;*
> *there too Lilith shall repose,*
> *and find a place to rest.*
> *There shall the owl nest*
> *and lay and hatch and brood in its shadow"*
> *(Isaïah 34*, 14*).*

The talmudic texts contain various references to Lilith which give some descriptions of her appearance and her personality. In *Nidda*, *"Lilith is a demoness with a human appearance except that she has wings" (Nidda* 24b*)*. Erubin adds that *"Lilith has long hair" (Erubin* 100b*)*. Anyhow, Lilith is generally associated with a nocturnal bird-demon, both seductive and ravenous.

The Babylonian Lilitû

An old-Babylonian terracotta from the Louvre, dating from around 1700 BC, curiously reminds us of the criteria presented above (see *Figure 7.1*). We can see a winged goddess wearing a horned mitre, with bird talons instead of feet. She is accompanied by two horned animals, which may be ibexes. A similar figure is represented on a terracotta found in Babylon dating from the tenth century BC, today conserved in the British Museum (*Figure 7.2*), but this time, two lions and two owls – nocturnal birds *par excellence* – are set on each side.

The identification of this figure can be discussed: she is frequently associated with Ishtar who shares many similarities with the Jewish Lilith, such as her sovereignty, her sexual aspects, and her link with the sky and the underworld, amongst others. Nevertheless, the bird talons are not usually attributes of Ishtar. In the Mesopotamian tradition, there are other lists of demons which mention *Lilû* and his partner *Lilitû*, sometimes called *Ardat Lili*. These two names come from the Sumerian "LIL", whose Akkadian transcription *lulu* and *lulti* refers to the notion of "lasciviousness". The male demon *Lilû* seduced women during their sleep, whereas *Lilitû* was described as an unsatisfied virgin, a nocturnal ravisher who attacked married men and their homes. The latter was sometimes associated with the wind spirit (from the sumerian lil "wind").

A prophylactic inscription engraved on a plaque from Arslan Tash dating from the seventh and sixth century BC mentions the demoness Lili, *"that flies in a dark chamber"*.[1] Further, the Hebraic dictionary identifies the name of Lilith with the male and female owls, connecting it with the Hebraic root *laïl* meaning "the night", which we find in old-chaldean through *lilia*. This root is moreover frequently associated with *lou'a*, which means the swallowing/absorption. Mesopotamian texts do not give a precise physical description of the demonesses Lilitû and Lili. However, the onomastic link between them and their characters bring them close to the Jewish Lilith.

The Jewish Lilith

Lilith appears many times in the two fundamental books of the Jewish tradition, the Talmud and Zohar, in which she bears many different names (*Erubin* 18b). She is linked to the rank of demons who are considered to be God's envoys, agents of calamities and illnesses. Lilith lives longer than men, but she is bound to die on judgment day (*Zohar* 1.55a). In the Hebraic tradition, Lilith is essentially described as a nocturnal entity, who devours children and seduces men (Kabbala, *Zohar* 2:267b, 3:76b-77a ; *Zohar, Sitrei Torah* 1:147b-148b; Talmud, *Shab.* 151b.), with whom she had evil and often criminal dealings. Likewise, Lilith attacked legitimate wives so that she could reach the conjugal bed (*Zohar* 3:19a ; *Bacharach, 'Emeq haMelekh* 19c).

Lilith has been developed in the popular imagination. She was particularly feared by women in labour, whose children she devoured:

> *"The evil Lilith,*
> *Who causes the hearts of men to go astray*
> *and appears in the dream of the night*

[1] Rosenthal F. *in* Pritchard 1969: 658 ; Patai 1978: 222.

and in the vision of the day,
Who burns and casts down with nightmare,
attacks and kills children,
boys and girls..."[2]

Incantations were said in order to send Lilith away from homes, and amulets were hung in the rooms of women in labour and babies:

"This is the spell for a demon and spirits and Satan
and Lilith in order to banish them from the entire house.
Cut off the king of the demons,
the greater ruler of the liliths.
I adjure you, whether you are male or female,
I adjure you [...]
Go, leave and depart from the house!"[3]

Lilith in the Jewish-Christian tradition

The strictly kabbalistic tradition does not give way to mythical constructions and their plastic representations. Apart from some written mentions of Lilith, no art has survived. However, she became very famous in the Judeo-Christian tradition, where she was designated as the first wife of Adam:

"There is a female,
a spirit of all spirits,
her name is Lilith,
and she was at first with Adam" (Zohar 3:19*)*[4]

The New Testament does not speak expressly of Lilith. Indeed, she seems to arrive late in the Jewish mystic tradition. The most ancient trails of this first Eve come from the third century BC in Rabbi Yehouda Bar Rabbi's writings (Rabba's Genesis 18:4; 17:7) even if she is not clearly called Lilith.

The two epics of creation have been connected in the Alphabet of Ben Sir, also called *Midrach*, which is a compilation of proverbs dating from the tenth and eleventh centuries, according to which woman was created twice (Ben Sira's Alphabet 23a-b). The first woman Lilith was made equal to man. Refusing to submit to him, she invoked the name of the 'Ineffable' who endowed her with wings, in order to fly away from the Garden of Eden. She received a severe punishment for this: each day she would see one hundred of their children dying. The three angels that went to find her could not convince her to come back, but having mercy, gave her a special power over new-borns. God created a second woman, Eve, from Adam's rib. Lilith went with the master of fallen angels, Samael, with whom she had many demonic children. Because she would not forget Adam and make up for her daily loss of children, Lilith blamed Eve, identifying herself with the tempter snake:

"The Serpent, the Woman of Harlotry, incited and seduced Eve [...]" (Bacharach: 'Emeq haMelekh 23c-d)[5]

The various aspects of the Lilithian heritage

In this way, from the Middle Ages, the iconography of the antique Lilith increased: she is represented with bird talons and long tail to one side of the temptation tree in the *Original Sin* of Hugo van der Goes[6] and in other manuscripts of the 15th century (*Figure 7.3*). In full degeneration, her image is of a hideous creature, very different from the beautiful and seductive Lilith known until then. Beside the representations inherited from the antique bird-woman appear numerous representations of Lilith under the aspect of a snake-woman coiling around the temptation tree (*Figure 7.4*): in the manuscript *Les très riches heures*[7] *du Duc de Berry*, in Bosch's *Paradise*, and in many other examples of Late Medieval and Renaissance art.

Lilith was not only endowed with a new appearance; her symbolism was also enriched in every way. So far known as queen of demons, her throne was set either in Rome, or in the kingdom of Saba, or in Egypt, or in the sea depths. The Hebraic dictionary draws a link between Lilith and the mermaids. Lilith lived a long time in European popular beliefs. Her story has been developed and distorted by external traditions. For instance, some pious readings were made to counter Lilith during the evening before the circumcision of a child, to send her away. Lilith was so feared that, until the sixteenth century in central Europe, children who smiled during their sleep were woken up, for fear that they were playing with Lilith and that she would take them away. Therefore, Lilith was sometimes associated with the dark moon. The erudite tradition of the sixteenth century compares the Jewish Lilith to the Arabic Ghula[8], a nocturnal demoness who sucks the blood of men and children, or designates her as a marrow-sucking strige.[9] Later on, Lilith would also be identified with a centaur with a donkey body.

2 Extract from a Lilith Exorcised on a Persian Bowl conserved to the Semitic Museum of Harvard University (Patai 1978: 228). Cf. also Patai 1978: 225 and Segal 2000.

3 Extract from an aramaic inscription on a bowl dating from the 6th century conserved to the Pennsylvannia University (Patai 1978: 226).

4 Cf. also *Zohar* 1:34b.

5 Cf. also *Zohar, Sitrei Torah* 1:148a-b.

6 The *Original Sin,* realised between 1467 and 1468, is conserved to the Kunsthistorisches Museum of Vienna

7 Cf. the temptation, fall and expulsion scenes from *Les très riches heures du Duc de Berry,* realised between 1411 and 1416 by the brothers Limbourg; the Lutwin's temptation dating from the 15th century ; the temptation scene from the cathedral Notre Dame (Paris); the Bosch's *Paradise*, painted around 1510, today conserved in the Prado Museum (Madrid) ; the *Fall of man* painted around 1570 by Titian, and the famous Michelangelo's temptation and fall scenes from the Sistine Chapel (Rome).

8 Cf. Gesenius' *Thesaurus*; Rosenmuller 1793 (commentary of the book of Isaiah).

9 A nocturnal and evil spirit corresponding to a metamorphosed live or dead person. Cf. Robertson 1680.

Figures 7.1-7.2. Old-Babylonian Terra-cotta, Musée du Louvre AO 6501 & Terra-Cotta from Babylon (10th century B.C.), British Museum / Collection Col. Norman Colville.

Figures 7.3-7.4. The Spread of the Heritage in Lilith's representations
(detail from illumination of the 15th century B.C. & Snake-woman from Bosch's Paradise).

'Lilithian' figures?

Looking at the various aspects of this lilithian heritage, we can nevertheless distinguish some stable "lilithian" features, especially her hybrid and nocturnal character, her devouring conduct resulting from an unsatiated maternal desire, and above all her extraordinary power of seduction. From this statement, a relationship could be established between Lilith and different mythological figures stemming from other cultures of the Mediterranean basin.

Harpies, for instance, are represented sometimes as winged women, and sometimes as birds with women's heads. In ancient Greece, the three daughters of Electra and Thaumas are ravishers with strong bird talons, which enabled them to take hold of children and to extract the souls from the bodies of dead people in order to bring them to Hades.[10] We recognize again the harpies in the Islamic world with the figures of murgh-i âdamî, 'anqa, zaghsâr and bahrî. The Lamies, sometimes represented as birds with a woman's head,[11] are the Hellenic descendents of Lamme or Lamia, often mentioned alongside Lilitû. In the Idylls's Theocritus, Lamia is designated as one of the first mistresses of Zeus. Jealous, Zeus's new wife, Hera, killed all of Lamia's children and deprived her of her sleep. Desperate, Lamia went to hide in a cavern, envying the happy mothers whose children she kidnapped.

In the same way, the *Empusa* terrified women and children, whose flesh they ate. The former seduced dozing men and sucked up their vital strength until they died. In Greek mythology, the Minyad sisters (Leucippe, Arsippe and Alcithoe, the daughters of king Minyas from Orchomenos in Boeotia), denying the divinity of Dionysos and refusing to take part in his cult, were metamorphosed and were endowed with bat's wings.[12] Finally, the Greek Erinyes, equivalent of the Roman Furies, are winged genies with snake hair, living in the heart of darkness. In charge of keeping the established order, they track their victims whom they drive to madness.[13] The attributes and the history of each of these mythical figures evoke untiringly those of Lilith.

On the whole, all these "lilithian" figures are connected to a multi-millenary tradition which stems from the Mesopotamian heritage. The Sumerian-Akkadian bestiary had a great impact on the whole of the Near East and throughout the Mediterranean world thanks to the influence of Greek civilization and of biblical scriptures. Classical authors also played an important role in the transmission of the oriental heritage, as did also trade and artistic exchanges, or military campaigns. The routes of diffusion of those symbols can be explained by the mingling of the different cultures of the Mediterranean basin. We have studied the figure of the bird–woman, but it would be possible to illustrate the same idea with other Mesopotamian monstrous creatures, such as the *hydra*, the winged bull and many others.

Bibliography

Patai R. 1978. *The Hebrew Goddess.* New York.

Pritchard J. B. 1969. *Ancient Near Eastern Texts relating to the Old Testament.* Princeton.

Robertson. 1680. *Thesaurus linguae sanctae.* London.

Rosenmuller. 1793. *Scholia Jesiaie Vaticina.* Leipzig.

Segal J. B. 2000. *Catalogue of the Aramaic and Mandaic incantation Bowls in the British Museum.* London.

10 Cf. for example the statue of an harpy into tears from the National Archaeological Museum of Athens, dating from the 4th century B. C.

11 Cf. for example the Lamia on the Stamos' vase dating from the 6th century B. C.

12 Cf. for example the fresco of the Casa dei Vettii from Pompeii.

13 Cf. for example *Oreste and the Erinyes* painted by Gustave Moreau in 1891.

Cycles of island colonisation in the prehistoric Mediterranean

Helen Dawson – University College London

Understanding processes of island colonisation and abandonment in the prehistoric Mediterranean is the remit of my current research. The geographical scope is pan-Mediterranean, and chronologically it encompasses prehistory from the earliest times that human presence on the islands is documented to its establishment (roughly from the end of the Pleistocene to the Iron Age). By using a comparative approach, I aim to highlight and investigate patterns and anomalies, to question established geographical boundaries and chronological restrictions, and, ultimately, to think of alternative models to offer as a critique of prevailing theoretical paradigms. This paper presents a series of observations, drawn from my ongoing analysis of the colonisation data, which question the prevailing view that island colonisation was a linear process linked to the gradual filling of available arable land during the Neolithic. While this is in part an explanation, in order to understand various phases of colonisation correctly, we should first consider them as a whole, as opposed to adopting a single model to explain different processes. By viewing the data in their entirety, it becomes clear that island colonisation was a much more irregular phenomenon than previously seen (even during the Neolithic). The causes underlying both the patterns and the anomalies in the islands' record will be addressed in a separate paper (in preparation).

Geographical and environmental data provide the main starting point for much analysis of colonisation and abandonment patterns. Island biogeographers argued that geographical parameters (mainly size and distance) were responsible for why organisms select islands for colonisation (MacArthur and Wilson 1967). Cherry, who applied island biogeography to the Mediterranean islands, concluded that 'patterns in *human* island colonisation displayed a great deal of *noise*' (1990: 199 emphasis added). By this he meant anomalies, or "colonisation events not following general models", that he attributed mainly to gaps in the archaeological data. Numerous island-based projects over the past ten years have produced new data, but these anomalies remain: the rigid application of biogeography to cultural transmission is a likely culprit for this 'noise'. Instead, a more sophisticated approach is necessary, not just to account for other important geographical variables (such as island configuration and the geographical distribution of resources), but also for culture-specific perceptions of the environment and of what constitutes a demographically sustainable community.

The bulk of the data I consider comes from published island survey and excavation projects. New regional syntheses have become available since Cherry's seminal article on island colonisation (1981), including studies of the central Adriatic islands (Gaffney *et al.* 1997, 2000; Bass 1998) and of the Aegean islands (Broodbank 1999, 2000). Overall, the data have had different effects on our understanding of island cultural dynamics: in some cases, our views on island colonisation are broadly confirmed (e.g. Malta and Crete), whilst – elsewhere – they are changing incrementally (e.g. the Aegean islands). In the case of Cyprus and the Balearic islands, research has had quite opposite effects on chronology: Cyprus now provides the earliest known instance of human occupation of a true island in the Mediterranean, by a group of hunter-gatherers in the 11th millennium cal. BC (Simmons 1989, 1991, 1998, 1999; Peltenburg *et al.* 2000, 2001, 2002). Further research in the Spanish islands has resulted in the definition of a later colonisation horizon than previously thought (Ramis and Alcover 2001, Ramis *et al.* 2002, *contra* Waldren 1982, 1992), highlighting the fact that further research does not always result in earlier dates.

At a broader level, the new data indicate the need for new distinctions between different phases and types of island-human interaction, and how this human activity varies spatially and temporally. Human activity on islands can take place at a variety of levels, and can be labelled in various ways, e.g. visitation, utilisation, integration, colonisation, establishment, abandonment, and re-colonisation. Guerrero claimed that "these episodes, stages or phases are regularly to be found in *every* colonizing process, and *never* in any other order" (2001: 140, emphasis added). The islands' archaeological record, however, contradicts this rigid view of colonisation. These phases do not necessarily follow each other in a linear trajectory: an island can be repeatedly visited without this resulting in its colonisation or permanent settlement, while it may become integrated in an exchange network. Abandonment clearly involves prerequisite phases, though their order remains to be investigated. Instead, these activities are often seen as preparatory to colonisation or permanent settlement, which, especially for the Neolithic, is often seen as the ultimate aim of human endeavour in an island setting (Evans 1977; Guerrero 2001). Under this paradigm, colonisation is explained with a variety of migration theories, usually based on some variation of Ammerman and Cavalli Sforza's 'Wave of Advance Model' (1973, 1979). Van Andel and Runnels (1995), who adapted this model, envisaged colonisation as a punctuated set of events, rather than a continuous process, but still emphasised the idea of a long-term trajectory guided by some form of economic pioneering. Recently, Zilhão has also maintained that colonisation entailed the 'leapfrogging' of communities (2000: 170), which he associated with a form of 'pioneer ethic' in the spread of the Neolithic across the western Mediterranean (2000: 173).

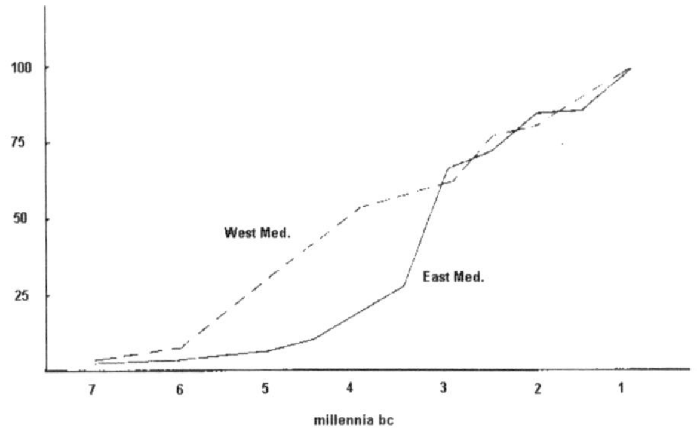

Figure 8.1 Cumulative Colonisation Graph (redrawn after Cherry 1981).

The high number of sites documented for this period easily justifies the conceptual link between island colonisation and the Neolithic. The presence of fewer Palaeolithic and Mesolithic than Neolithic sites on islands has been explained by a number of reasons, ranging from loss of evidence caused by rising sea levels, to the different nature of the evidence itself (seasonal camps as opposed to permanent structures), to lower population densities (fewer people leave fewer traces), to human ignorance or avoidance of the islands. Human presence on islands did increase during the Neolithic, however there is evidence of effective manipulation of resources also before then, in the form of animal species introduced to some islands for the purposes of hunting and replenishing extinct fauna. Deer (*Cervus dama*) were brought to Cyprus as early as the 8th millennium cal. BC (Davis 1984; Vigne 1996: 67; Peltenburg et al. 2001: 46), indicating that lack of resources on islands during the Mesolithic could be overcome – and that, though faintly visible, the foundations of island life in the Mediterranean were laid down earlier than the Neolithic (cf. Lewthwaite 1989). This evidence (combined with the increasing pre-Neolithic data that have become available in recent years) highlights that there can be different types of colonisation, not just in the form of permanent settlement, but also of repeated visitation, both of which require a high degree of knowledge of the islands and surrounding waters. These activities share similarities and differences, in terms of attachment or commitment to a specific place, and of responses to particular subsistence strategies. Nonetheless, colonisation before the Neolithic remains generally overlooked, with the result that causes and dynamics of island colonisation remain to be sought and explained for other periods.

Testing the models

Many models of island colonisation draw upon the work of Cherry (1981, 1990), which provides a focus for this paper. In 1981, Cherry created a plot of cumulative percentage of the islands in the eastern and western Mediterranean with evidence of occupation by a given millennium BC (*Figure 8.1*) (1981: 62). The graph depicted colonisation as a linear process, expressed as a cumulative percentage of islands occupied per millennium bc. In 1990, Cherry synthesised some significant developments that had taken place since 1981, but did not update the graph in the light of these new discoveries, with the result that his colonisation model is more than twenty years out of date. As a first step, I have amended the original graph using data that have become available since 1990 (*Figure 8.2*). The new graph displays some interesting similarities and differences compared to Cherry's original: overall, it confirms that colonisation in the western Mediterranean took place at a steadier and faster rate than in the east up until the 3rd millennium BC, when colonisation in the east took over. Importantly, however, it also shows the increasing evidence of human presence on islands before the Neolithic, particularly during the Mesolithic. Cyprus (Simmons 1989, 1991, 1998, 1999); Youra in the Cyclades (Sampson 1996, 1998); Lefkas, among the Ionian islands (Soyoudzoglou-Haywood 1999); and Brač (Bass 1998), in the Adriatic, can be added to the islands already listed by Cherry (1981: 43, 1990).

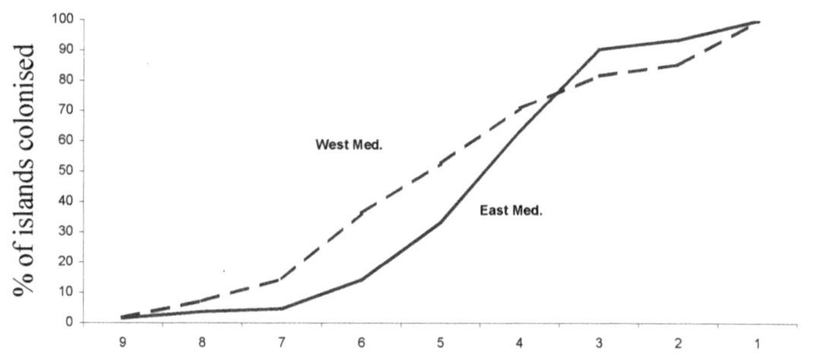

Figure 8.2 Reworked Colonisation Graph.

There is, however, an underlying problem with both graphs: they are 'cumulative' graphs, which means that the number of islands colonised during each millennium is *added* to those colonised during the following millennium, thus giving a false sense of long-term continuity. Their main defect is that they fail to incorporate abandonment. Our photographic memory of Cherry's original graph may be partly responsible for why we view Mediterranean island colonisation (which he conceived of in terms of permanent settlement) as a linear process. A non-cumulative plot (*Figure 8.3*) has the benefit that it does not add the number of islands colonised in the previous millennium to the following, but only accounts for how many *new* colonisation events take place during each millennium. There are problems with this type too, since it still does not take into account that these might be instances of re-colonisation of the same island following its abandonment. Thus, while it does not incorporate abandonment, a non-cumulative plot allows us to compare rates of colonisation between millennia. This by itself can produce interesting observations. For example, the graph (*Figure 8.3*) indicates that, in the eastern Mediterranean, a similar number of islands was colonised in the 7th and 2nd millennia cal BC, i.e. in the Early Neolithic and the Late bronze Age – and that this was not the case in the western Mediterranean.

The pan-Mediterranean record (by definition) can only show very general trends; thus it seems that explanation should lie not in overall modelling, but rather in patterns of regional development. A few archipelagos have been thoroughly investigated, and these offer the opportunity to investigate regional patterns of colonisation and abandonment (depending, of course, on whether gaps in the data correspond to abandonment or lack of research).

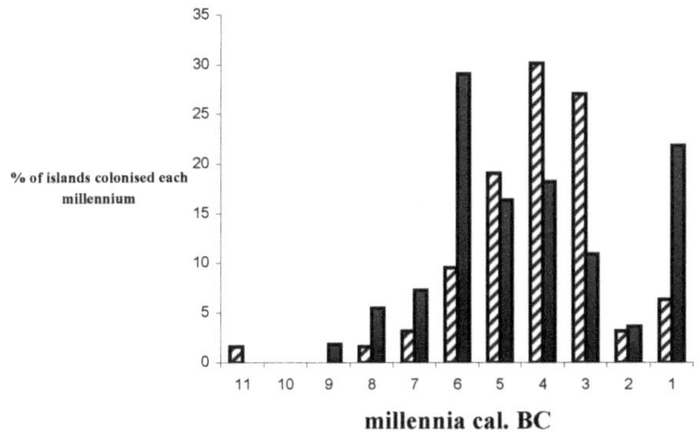

Figure 8.3 Non-cumulative Colonisation Graph.

Cycles of island colonisation

When this methodological framework is applied to island groups from different parts of the Mediterranean, the plots confirm an irregular rather than linear pattern of island colonisation. Depending on the data used, the graphs show either how many islands were colonised during each millennium (the Central Adriatic, the Ionian and Aegean, and the Tyrrhenian islands) (*Figures 8.4, 8.5 & 8.9*) (Bass 1998; Soyoudzoglou-Haywood 1999; Broodbank 1999, 2000), or how many were actually inhabited during each millennium, i.e. they incorporate abandonment (the Aeolian islands and the Spanish islands) (*Figures 8.6 & 8.7*) (Bernabò Brea 1957; Leighton 1999; Bellard 1995; Lull et al. 1999, 2002; Ramis and Alcover 2001; Ramis et al. 2002). The cumulative plot for the Tyrrhenian islands (*Figure 8.8*) is included as way of illustrating the potential danger of viewing the islands as gradually filled with no temporal gaps in their occupation. The non-cumulative plot for the same group (*Figure 8.9*) highlights that colonisation was a much more punctuated process and shows a temporal gap between the colonisation of the larger islands, Sardinia and Corsica, and of the smaller Tyrrhenian islands - such as Giglio, Pianosa, Palmarola, Ventotene, and Giannutri (Vaccarino 1935: 127; Buchner and Rittman 1948; Bronson and Uggeri 1970: 201; Brandaglia 1985; Rendini 1989; Ducci and Perazzi 1991; Aranguren et al. 1992; Della Ratta Rinaldi 1992; Canestrelli 1998: 9; Guidi and Piperno 1992: 310; Tykot 1994, 1996: 43).

While there is evidence of human presence in these islands from the early Neolithic into the Iron Age, the graph highlights that occupation was not continuous.

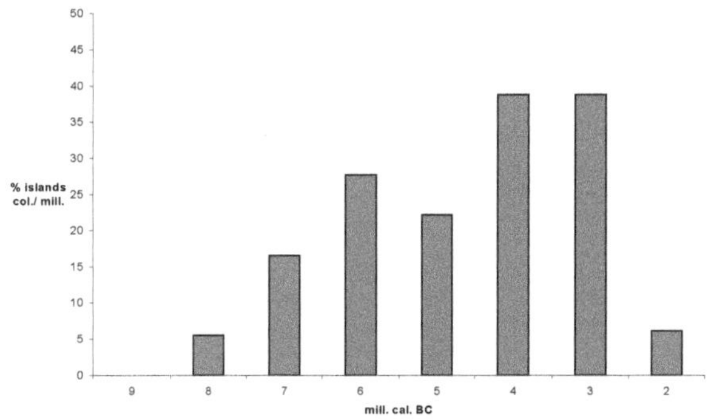

Figure 8.4 Central Adriatic Islands (rates of colonisation per millennium

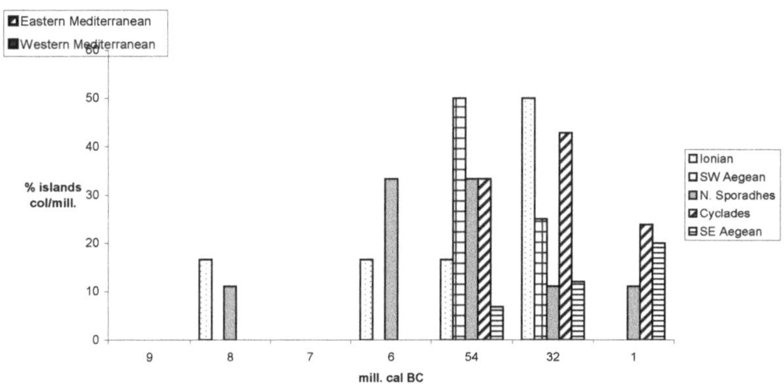

Figure 8.5 Ionian and Aegean Islands (rates of colonisation per millennium)

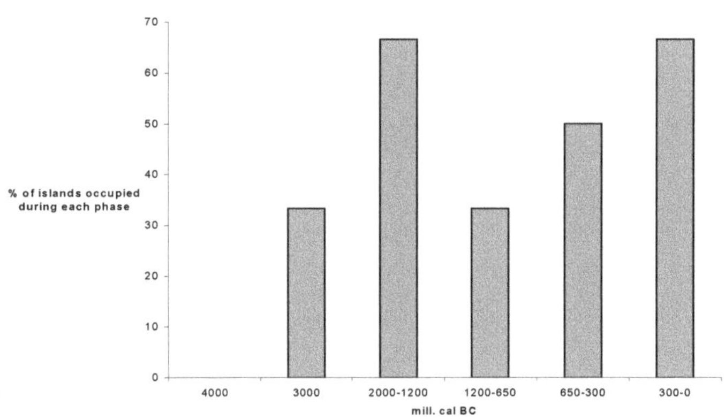

Figure 8.6 Aeolian Islands

The graphs display collectively the results of island projects carried out since the 1930s, and as such they are based on different types of data acquired under different research paradigms and strategies. Rather than representing colonisation (*sensu* Cherry 1981: 48), they give a representation of how human activity (including colonisation) varies spatially and temporally. At the same time, they suggest that viewing abandonment as settlement failure is over-reductive, since movement between islands may have been part of a successful strategy, aimed at maximising the use of resources through a process akin to rotation (however unsystematic). The uneven degree of archaeological exploration in different areas, and the loss of evidence from the islands caused by rising sea-levels render the definition of colonisation and abandonment processes problematic.

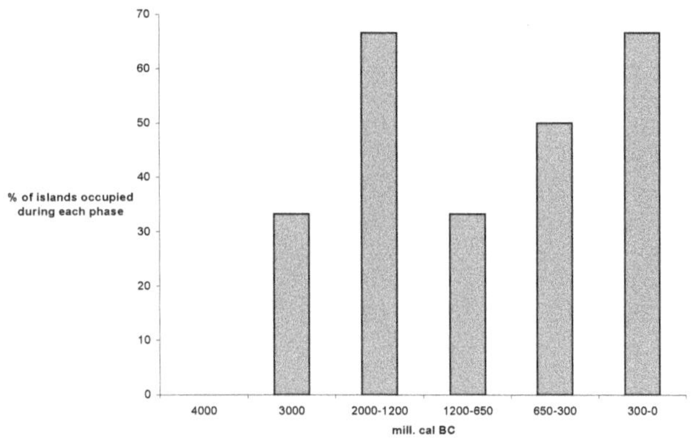

Figure 8.7 Balearic and Pitiussae Islands (includes abandonment)

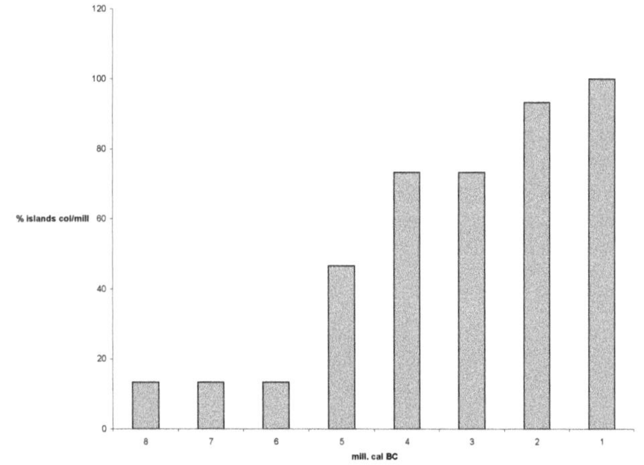

Figure 8.8 Tyrrhenian Islands (cumulative)

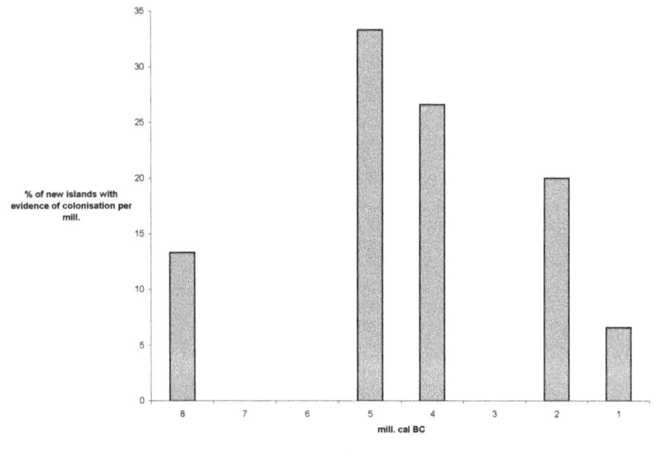

Figure 8.9 Tyrrhenian Islands (non-cumulative - rates of colonisation)

Nonetheless, it is encouraging to see that, while Cherry's 1981 sample was predominantly eastern Mediterranean (c. 70 islands investigated in the East as opposed to just 30 in the West), as a result of further research, my database now holds data from 55 western islands.

I am also investigating whether there is a significant relationship between depth of sea around the islands and age of remains found on them, which, if proven, would help devise a strategy to locate missing (i.e. submerged) evidence more effectively. Even with these provisos, the graphs demonstrate that human/island interaction was a *staccato* process, and that island life was hardly ever continuous. Indeed, in a number of cases, it can be proved that discontinuity in the archaeological record reflects abandonment rather than lack of research. With obvious caveats, the general trend towards depopulation of the small Mediterranean islands in the present (Baggioni and Hache 2000) offers a strong parallel to what may have happened in prehistory, when several islands, including large ones such as Cyprus, were abandoned and sometimes re-colonised (before, during and after the Neolithic). I am currently investigating whether cycles of occupation and abandonment in different island areas are in any way aligned or, in other words, whether similar patterns recur under specific geographical or historical circumstances.

Final thoughts

Island studies have been conducted at two levels, which are on parallel but separate tracks: while the models drawn from statistical analyses have tended to be excessively abstract, individual island histories have privileged the detail at the expense of the wider picture. The real depth of the similarities and anomalies highlighted by a statistical approach can only be understood by addressing them through detailed island histories and geographies, first individually and then comparatively, since locally contingent factors may be responsible for different/similar patterns. Archaeological survey has enormous potential in this respect, since it can provide both a synchronous and diachronic picture of occupation. More importantly, surveys can sometimes pick out subtle nuances which characterise different types of human activities on islands, highlighting that permanent settlement should not considered as the only useful category when studying colonisation or abandonment. The evidence that I have reviewed so far suggests that - even when we adopt traditional geographic regional divisions in the first instance - physical and cultural similarities and differences emerge clearly between island basins, suggesting at times the creation of different and rather subtle boundaries or groups altogether. Once we identify the dynamics of these patterns and establish their underlying causes, we will fully appreciate that human activity on islands can express itself through a variety of cultural manifestations, and - by questioning different colonisation paradigms – reinforce the idea that no single model can answer all our questions.

Bibliography

Ammerman A.J. & Cavalli Sforza L.L. 1973. "A population model for the diffusion of early farming in Europe", in C. Renfrew (ed.) *The Explanation of Culture Change.* London: 345-58.

Ammerman A.J. & Cavalli Sforza L.L. 1979. "The wave of advance model for the spread of agriculture in Europe", in C. Renfrew & K.L. Cooke (eds.) *Transformations: Mathematical Approaches to Culture Change.* London: 275-94.

Baggioni M. J. & Hache J-D. (eds.) 2000. *Quel Statut pour les Îles d'Europe? (What Status for Europe's Islands?) Islands Commission of the Conference of Peripheral Maritime Regions of Europe.* ISBN: 2-7384-9250-9. http://www.eurisles.com/statut_iles/EN/cadre.htm. www.eurisles.com. [Accessed 06/10/2003]

Aranguren B. M., Perazzi P. & Rendini P. 1991-1992. "Isola del Giglio: testimonianze dal Castellare del Campese" *Rassegna di Archeologia* 10: 674 ff.

Bass B. 1998. "Early Neolithic offshore accounts: remote islands, maritime exploitations, and the trans-Adriatic cultural network" *Journal of Mediterranean Archaeology* 11: 165-190.

Bellard G. C. 1995. "The first colonization of Ibiza and Formentera (Balearic Islands, Spain): some more islands out of the stream?" *World Archaeology* 26 (3): 442-455.

Bernabò Brea L. 1957. *Sicily before the Greeks.* London.

Brandaglia M. 1985. "Il Neolitico a ceramica impressa dell'Isola del Giglio: L'industria Litica" *Studi per l'ecologia del Quaternario* 7: 53-76.

Bronson R. C. & Uggeri C. 1970. "Isola del Giglio, Isola di Giannutri, Monte Argentario, Laguna di Orbetello" *Studi Etruschi* 37: 201-214.

Broodbank C. 1999. "Colonization and Configuration in the Insular Neolithic of the Aegean", in P. Halstead (ed.) *Neolithic Society in Greece.* Sheffield Studies in Aegean Archaeology. Sheffield: 15-41.

Broodbank C. 2000. *An Island Archaeology of the Early Cyclades.* Cambridge.

Buchner G. & Rittman A. 1948. *Origine e passato dell'Isola d'Ischia.* Napoli.

Canestrelli A. 1998. *Elba. Un'isola nella storia.* Pisa.

Cherry J.F. 1981. "Pattern and Process in the Earliest Colonization of the Mediterranean Islands" *Proceedings of the Prehistoric Society* 47: 41-68.

Cherry J.F. 1990. "The first colonization of the Mediterranean islands: a review of recent research" *Journal of Mediterranean Archaeology* 3(2): 145-221.

Davis S. 1984. "Khirokitia and its mammal remains: a Neolithic Noah's Ark", in A. Le Brun et al. (eds.) *Fouilles récentes à Khirokitia (Chypre) 1977-1981.* Éditions Recherche sur les Civilisations 41. Paris: 147-62.

Della Ratta Rinaldi F. 1992. "Un insediamento della media eta del bronzo a Ventotene" *Latium* 9: 5-34.

Ducci S. & Perazzi P. 1991. "Notiziario, Parte I: Isola di Pianosa – La Scola" *Studi e Materiali. Scienza dell'Antichità in Toscana* 6: 308-309.

Evans J.D. 1977. "Island archaeology in the Mediterranean: problems and opportunities" *World Archaeology* 9: 12-26.

Gaffney V., Kirigin B., Petrić M. & Vujnović N. 1997. "The Adriatic Islands Project. Contact, Commerce and Colonialism 6000BC – AD 600. Volume 1. The archaeological heritage of Hvar, Croatia". BAR International Reports, Oxford.

Gaffney V., Kirigin B., Hayes J., Kaiser T., Leach P. & Stančič Z. 2000. "The Adriatic Islands Project: contact, commerce and colonization 6000 BC- AD 600", in R. Francovich & H. Patterson (eds). *Extracting Meaning from Ploughsoil Assemblages. The Archaeology of Mediterranean Landscapes* 5: 185-198.

Guidi A. & Piperno M. 1992. *Italia Preistorica.* Roma, Bari.

Guerrero V. M. 2001. "The Balearic Islands: Prehistoric Colonization of the furthest Mediterranean islands from the mainland" *Journal of Mediterranean Archaeology* 14(2): 136-158.

Leighton R. 1999. *Sicily before History. An Archaeological Survey from the Palaeolithic to the Iron Age.* London.

Lewthwaite J. G. 1989. "Isolating the residuals: the Mesolithic basis of man-animal relationships on the Mediterranean islands", in C. Bonsall (ed.) *The Mesolithic in Europe.* Edinburgh: 541-55.

Lull V., Micó R., Rihuete C. & Risch R. 1999. *La Cova des Càrritx y la Cova des Mussol. Ideología y Sociedad en la Prehistoria de Menorca.* Barcelona.

Lull V., Micó R., Rihuete Herrada C. & Risch R. 2002. "Social and Ideological Changes in the Balearic Islands during the Later Prehistory", in W. H. Waldren and J. A. Ensenyat (eds.) *World Islands in Prehistory: International Insular Investigations. V Deia International Conference of Prehistory.* Oxford: BAR International Series 1095. 117-126.

MacArthur R.H. & Wilson E.O. 1967. *The Theory of Island Biogeography.* Princeton.

Peltenburg E., Colledge S., Croft P., Jackson A., McCartney C. & Murray M.A. 2000. "Agro-pastoralist colonization of Cyprus in the 10th millennium BP: initial assessments" *Antiquity* 74: 844-53.

Peltenburg E., Colledge S., Croft P., Jackson A., McCartney C. & Murray M.A. 2001. "Neolithic Dispersals from the Levantine Corridor: a Mediterranean Perspective" *Levant* 33: 35-64.

Peltenburg E., Croft P., Jackson A., McCartney C. & Murray M.A. 2002. "Well established colonists: Mylouthkia 1 and the Cypro-Pre-Pottery Neolithic B", in S. Swiny (ed.) *The Earliest Prehistory of Cyprus: from colonization to exploitation.* ASOR: 61-93.

Ramis D. & Alcover J. A. 2001. "Revisiting the Earliest Human Presence in Mallorca, Western Mediterranean" *Proceedings of the Prehistoric Society* 67: 261-69.

Ramis D., Alcover J. A., Coll J. & Trias M. 2002. "The Chronology of the First Settlement of the Balearic

Islands" *Journal of Mediterranean Archaeology* 15 (1): 3-24.

Rendini P. 1989. *Attività archeologica all'Isola del Giglio: primi risultati in terra e in mare*. Cortona.

Sampson A. 1996. "Excavations at the cave of the Cyclope on Youra, Alonessos", in E. Alram-Stern (ed.) *Die Ägäische Frühzeit, 2. Serie. Forschungsbericht 1975-1993, 1: Das Neolithikum in Griechenland mit Ausnahme von Kreta und Zypern*. Vienna: 507-20.

Sampson A. 1998. "The Neolithic and Mesolithic occupation of the cave of the Cyclope, Youra Alonessos, Greece" *Annual of the British School at Athens* 93: 1-22.

Simmons A.H. 1989. "Preliminary report on the 1988 test excavations at Akrotiri Aetokremnos, Cyprus" *Report of the Department of Antiquities of Cyprus* 1989: 1-5.

Simmons A.H. 1991. "Humans, island colonization and Pleistocene extinctions in the Mediterranean: the view from Akrotiri Aetokremnos, Cyprus" *Antiquity* 65: 857-869.

Simmons A.H. 1998. "Test excavations at two Aceramic Neolithic sites on the uplands of Western Cyprus" *Report of the Department of Antiquities of Cyprus*: 1-16.

Simmons A.H. et al. 1999. *Faunal Extinction in an Island Society: Pygmy Hippopotamus Hunters of Cyprus*. New York.

Soyoudzoglou-Haywood C. 1999. *The Ionian islands in the Bronze Age and Early Iron Age, 3000-800 BC*. Liverpool.

Tykot R.H. 1994. "Radiocarbon dating and absolute chronology in Sardinia and Corsica", in R. Skeates & R.D. Whitehouse (eds.) *Radiocarbon dating and Italian Prehistory. Accordia Specialist Studies on Italy*. London: 115-145

Tykot R.H. 1996. "Obsidian procurement and distribution in the Central and Western Mediterranean" *Journal of Mediterranean Archaeology* 9 (1): 39-82.

Vaccarino B. 1935. "Isola di Giannutri. Vigna Vecchia" *Notizie degli Scavi pubblicate dall'Accademia dei Lincei 1935*: 127.

Van Andel T.H. & Runnels C. N. 1995. "The earliest farmers in Europe" *Antiquity* 69: 481-500.

Vigne J-D. 1996. "Contribution des peuplements de vertebras insulaires à la connaissance de la navigation préhistorique en Mediterranee », in *Pour qui la Méditerranée au 21eme siecle? Navigation, echanges et environment en Mediterranee.* Actes du Colloque scientifique, Montepellier France 11-12 Avril 1996: 65-76.

Waldren W. H. 1982. *Early Settlement in the Balearic Islands*. Deya Archaeological Museum and Research Centre (DAMARC) Series 13. Deya de Mallorca.

Waldren W. H. 1992. *Radiocarbon and other Isotopic Age Determinations from the Balearic Islands. A Comprehensive Inventory*. Deya Archaeological Museum and Research Centre (DAMARC) Series 26. Deya de Mallorca and Oxford.

Adventures in Fields of Flowers: Research on contemporary saffron cultivation and its application to the Bronze Age Aegean.

Jo Day - Trinity College Dublin

Background

Anyone familiar with Minoan art will be aware of the large number of representations of crocuses – on ceramics, on wall paintings, in faience, perhaps on sealstones, and also in Linear B. The exact species of crocus shown has been a hotly debated topic for many years, with most experts favouring either *Crocus cartwrightianus* or *Crocus sativus* (see Amigues 1988 for a summary). They are both purple, autumn-flowering plants, the only difference being that the stigmas of *C. cartwrightianus* are about 1 cm. shorter than those of *C. sativus* (Mathew 1999: 21 - 22). The DNA of both plants is very similar, and it has been suggested that *C. cartwrightianus* is perhaps the wild ancestor of the cultivated *C. sativus* (Grilli Caiola 1999: 32; Mathew 1982: 56). Indeed, today on the island of Santorini, *C. cartwrightianus* is still gathered for its saffron (Tzachili 1994). However, the debate over which of these two species was depicted is redundant for the purposes of this paper, as both of them can provide saffron.

Saffron comes from the dried stigmas, or female reproductive organs, of the flower. Each flower has only three of these, which flop out over the top of the petals. This partly explains why saffron is so expensive, and it has been estimated that it takes at least 160,000 flowers to produce one kilo of the spice (Humphries 1996: 83). Nowadays we tend to use saffron in cooking, mostly for its yellow colouring properties. However, through the ages, it has been credited with a variety of different properties. These range from dye for clothes (famously the yellow robes of Buddhist monks), to dye for skin and nails, as apparently used by Cleopatra, to numerous medicinal properties. It is still important in Ayurvedic and Chinese medicine (Abdullaev & Frenkel 1999: 104). The Herbals of the Middle Ages reported a variety of effects, for example Culpeper reports that saffron

> *"refreshed the spirits, good against fainting fits and palpitations of the heart; strengthened the stomach, helped the digestion, cleansed the lungs, and was good in coughs."* (Stacey 1973: 5).

It was also reported in Tournefort's Herbal that too much saffron could cause death by immoderate laughter! (ibid.)

Today saffron is grown commercially in five main regions: Iran, Spain, Morocco, Kashmir and Greece. In October 2002 I spent some time in the appropriately named village of Krokos, near Kozani, in Western Macedonia. This is the saffron-producing heartland of Greece, with an annual output of between six and eight tons (Boutzina 1999:13). Saffron has been cultivated here since the 1700's, when it came back with merchants who had been trading with central Europe (Boutzina 1999: 25 - 26). At that time, saffron was much more widely cultivated in Europe than now, even in England and possibly Ireland.[1]

Saffron cultivation and processing at Kozani, Greece

The altitude, soil and climate of Kozani are suited to crocus growing, and it has now become a major source of income in the region. In 1971 a co-operative was set up with the aim of keeping the profits in the region, and encouraging younger people to stay in the area. Everybody who farms saffron now must market their crop through the co-op, thereby ensuring a certain standard of product. Now '*Krokos Kozani*s' is a recognised trademark, has an ISO certificate of quality, and has brought an increase in wealth to the area.

The aim of my trip to Krokos was to see if new insights could be gained into saffron cultivation in the Bronze Age by examining the methods used today. Although the use of analogy in archaeology has come under fire at times, Johnson has pointed out that those based on natural laws, which concern the physical world, and plants and animals, are safer than those based on cultural variables (Johnson 1999: 55 - 56). Essentially, whether growing crocuses today in Macedonia, or three hundred years ago in England, or several thousand years ago in the Aegean, certain conditions must be met for the crop to be a success. There are ties to a specific seasonal cycle. Also certain conditions must be fulfilled if good quality saffron is to be obtained from the flowers; again these apply equally to today or the past. Furthermore, saffron is one of the few commercial crops in the world today where its cultivation remains non-mechanised (although its processing has always made use of whatever labour-saving devices are current), and so offers information about traditional agricultural practice. It must be stressed however, that direct comparisons between Krokos and the Bronze Age Aegean are not being drawn; rather this ethnographic research was more a springboard for further ideas.

In July or August the corms[2] are planted, buried at least 10 cm deep into the soil to ensure that they mature properly.[3] A field will produce a good crop for seven

---

[1] For details of saffron cultivation in England and Ireland at this time, see Douglass 1732.

[2] Corm is the botanically correct term for the underground storage organ of the crocus. Whereas bulbs store food in their leafy scales, a corm is basically a thickened stem.

[3] All the following information on saffron cultivation comes from discussions with farmers and observations in Krokos, unless otherwise acknowledged.

years, but then the flowers start to fail and the plants are dug up. This field is not used for saffron again for at least ten years. The plants flower in October during a brief period of about three weeks. Each corm will then produce between two and five flowers.

There is no successful mechanisation of the harvesting process yet. Each corm is planted by hand, and each flower is picked by hand, snipped off between thumb and forefinger just below where the petals join the stem. The flowers are gathered into the pickers' aprons (who seem to be mostly women), and then transferred into baskets or sacks for transportation. This is to ensure that they do not get squashed, as then the separation of the stigmas becomes much more difficult. It is slow and backbreaking work, but the fastest picker can harvest perhaps thirty thousand flowers in a day (Boutsina 1999:13). The flowers are brought back to the farmers' homes that same evening for processing. This must be done as soon as possible after harvesting, as once the flowers start to wither, separation of stigmas is extremely difficult. This work can go on late into the night, depending on the amount harvested that day.

Figure 9.1 Jorgos separates the petals from the stigmas

In Krokos the petals are separated from the stigmas using an ingenious device. A handful of flowers are thrown onto a large rotating drum. At the base of this drum is a spiky mat, to which the stigmas and stamens adhere. Wind from an electric fan blows the petals off these interior parts, as the flowers are only loosely held together with a very small bit of stem (*Figure 9.1*). The petals are collected up by a brush inside the drum and swept into a waiting basket. In the earlier times a goatskin was used instead of the mat, and a set of bellows blew the petals away. The petals are discarded, and purple piles of them in various stages of decay can be seen lining the roads out of the villages, or in trails leading up to houses where processing is going on. In the past apparently they were used for dying wool and cotton. Now an investigation is being carried out as to whether the volatile oils within them are of any medicinal use (Fatehi *et al* 2003).

Once the petals have been removed, it is time to sort and dry the spice. A first sorting is done to remove large foreign bodies, like caterpillars and stray petals. Then the pile of stigmas and stamens is put on top of a mesh and shaken. Most of the stamens fall through, while the stigmas remain on top. These stigmas are then gone through again by hand, to remove all remaining yellow stamens (*Figure 9.2*). If the spice is to be dried, it is now placed on wire trays in ovens for seven or eight hours. In the 18th century in England they used small charcoal kilns for drying (Douglass 1732: 6 - 7), but it can be dried just by the sun, sometimes under glass.

Figure 9.2. Athena and Chrystoula sort the wet saffron.

The method of drying does seem to affect the properties of the saffron however – sun-drying does not have such strong colouring action (Basker 1999: 97). While I was in Krokos, they were also experimenting with freezing as a way of preserving the saffron, and so while still 'wet' it is put in small boxes, taken down to the co-op, and checked and weighed. The amount is recorded so each farmer gets correct payment (around €585 per kilo at time of visit). From here, the product is packaged, and distributed.

Saffron in the Aegean Bronze Age

At this point it is appropriate to review briefly what scholars have said on the role of saffron in the Aegean Bronze Age. Evans (1921: 265) believed that saffron was an attribute of the Great Goddess, who may have presided over a saffron industry on Crete. Marinatos (1976: 34) interpreted the Xeste 3 fresco from Akrotiri as

> *"crocus is being gathered in a Mediterranean meadow in summer. The plant, which is precious for its healing and colouring properties, is regarded and honoured as a divine gift. Young maidens in splendid festive attire gather the stamens of the blossoms in baskets... At all events, the representation shows a festive celebration, and more precisely a festival where women are the only participants".*

Doumas (1983: 76) later wondered whether this fresco did indeed show ritual activity, or the owner of the house was a merchant involved in saffron trade, who chose to depict the source of his wealth on his walls. Saffron has also been interpreted elsewhere as economically important, as a luxury item for trade (Morgan 1986: 30 - 31). Rackham (1978: 757) suggested that the gathering was happening in a garden, whereas Amigues (1988: 228) believes the flowers shown are in the wild.

Nanno Marinatos (1984: 65, 80 - 81) suggests that the scenes are to do with girls' rites of passage, citing saffron's effectiveness against menstrual cramps as part of her evidence. Recently a scenario has been envisaged where the women of Thera harvested saffron for its vitamin A content, which is essential for healthy eyesight (Rehak 2002: 50). The author sees this reflected in the red lines of the eyes of males on frescoes, whom he believes were denied saffron by the women (ibid.). A final suggestion is that poultices of saffron were used to soothe the eyes of Therans who would have suffered from wind-blown grit, perhaps explaining why the young woman in the Priestess fresco has yellow eyelids (Young Forsyth 2000: 159). Due to the brevity of this paper, unfortunately it is not possible to discuss these ideas further.

Summary

Although extremely brief, this outline demonstrates the amount of differing opinions and ideas emerging on saffron in the Bronze Age. It is important to query whether anything I learned in Krokos is actually useful or helps me to say anything new about saffron in the Aegean Bronze Age. Was my trip simply, as Bahn (1989: 52) says of ethnoarchaeology, *"an excellent means of getting an exotic adventure holiday in a remote location"?* Is it possible to use all this information on saffron farming in western Macedonia, or am I doomed, to

> *"return to your desk and use these brief studies to make sweeping generalisations about what people in the past and in totally different environments must have done"* (ibid.)?

Although recently more commentators have begun to realise that it is the stigmas rather than the stamens, which are made into saffron, this issue is often still unclear. I can categorically state that it is *only* stigmas, which are dried, that form true saffron. Similarly, when faced with statements, such as that the stigmas only become red once dried (Rehak 2002: 37), again I can say that this is not true, having sat at a table sorting wet, red stigmas into the small hours. Misconceptions about such basic facts should not be allowed to proliferate.

With regard to the Xeste 3 fresco, people have tended to read it as a continuous narrative of saffron picking and offering the crop to a goddess, but it should be pointed out that the separating of stigmas from petals, and their drying, is not shown. Does this indicate that the whole flower was not picked in the Bronze Age, only the stigmas? Presumably saffron was dried, as wet stigmas rot quickly, so why is this aspect of the process also overlooked in art? This issue may, of course, have more to do with our assumptions about reading the fresco as a continuous narrative than about saffron processing.

Arising from this however, I have become curious about drying methods used in the Bronze Age, and wonder about what kind of evidence we might hope to find to shed more light on this aspect of the spice's preparation. Recent research by the '*Flavours of Their Time*' project[4] has found traces of saffron (isophorone) in some vessels from Chrysokamino in Crete, in a context that is believed to be a herbal remedy workshop (Beeston 2002). Hopefully, wider use of chemical analyses will add to our knowledge of vessels which contained saffron.

Learning about the great effort and amount of flowers needed to produce one kilo of saffron has made me reconsider the saffron information in the Linear B tablets. Saffron is unique in being the only spice measured by weight, like gold and bronze, rather than by dry volume like the other spices (Ventris & Chadwick 1973: 51). The standard measures of saffron are 3.4 grams, 20.8 grams and 250 grams. It can be worked out that to make 250 grams of saffron you require at least 37,500 flowers, or a minimum of 7500 corms which could take up at least 75 square metres of land. Of course, this does not have to be one large plot. Indeed, Sarpaki (2001: 205) sees these large amounts as indicative of collecting wild saffron from a large catchment area.

I think that the very reason she uses for this argument, the large amounts of saffron, actually suggests a more

[4] This exhibition is published as Y. Tzedakis & H. Martlew (eds.) 1999. *Minoans and Mycenaeans: Flavours of Their Time*. Athens.

intensive form of cultivation, and that saffron cultivation in the Bronze Age was a specialised form of agriculture, rather than just the fruits of somebody's garden or walk in the hills. Certainly such a useful commodity, with potential as a dye, as medicine, as flavouring for food and drink, could bring riches to those who control its production (as indeed it has done in modern Krokos). Perhaps Evans (1935: 718) was not so far wrong when he suggested that one of the sources of wealth for the lords of Knossos was saffron.

Acknowledgements

I am grateful to the Irish Research Council for the Humanities and Social Sciences for funding this research; thanks also to all in Krokos who helped with my enquiries, especially Rita Gola in the saffron co-op.

Bibliography

Abdullaev F. I. and G. D. Frenkel. 1999. "Saffron in Biological and Medical Research", in M. Negbi (ed.) *Saffron - Crocus sativus L.* Amsterdam: 103-114.

Amigues S. 1988. "Le Crocus et le Safran sur une Fresque de Théra" *Revue Archéologique* 2: 227-242.

Bahn, P. 1989. *Bluff Your Way in Archaeology*. Horsham.

Basker D. 1999. "Saffron technology", in M. Negbi (ed.) *Saffron - Crocus sativus L.* Amsterdam: 95-102.

Beeston R. F. 2002. "New Finds from Chrysokamino". Paper delivered at the conference *Minoans and Mycenaeans: Flavours of Their Time - New Work and New Discoveries*, 16th November 2002, Birmingham Museum and Art Gallery.

Boutsina E. L. 1999. *Krokos – Safran*. Krokos.

Douglass J. 1732. "An Account of Saffron: The Manner of its Culture and Saving for Use, with the Advantages it will be to this Kingdom" *Published Proceedings of the Royal Dublin Society*, Dublin: 3-14.

Doumas C. 1983. *Thera: Pompeii of the Ancient Aegean.* London.

Evans A. J. 1921. *The Palace of Minos at Knossos*, Volume I. London.

Evans A. J. 1935. *The Palace of Minos at Knossos*, Volume IV (2). London.

Fatehi M., Rashidabady T., & Fatehi-Hassanabad Z. 2003. "Effects of Crocus sativus petals' extract on rat blood pressure and on responses induced by electrical field stimulation in the rat isolated vas deferens and guinea-pig ileum" *Journal of Ethnopharmacology* 84 (2-3): 199-203.

Grilli Caiola M. 1999. "Reproduction Biology of Saffron and its Allies", in M. Negbi (ed.) *Saffron - Crocus sativus L.* Amsterdam: 31-44.

Humphries J. 1996. *The Essential Saffron Companion.* London.

Johnson M. 1999. *Archaeological Theory: An Introduction.* Oxford.

Marinatos N. 1984. *Art and Religion in Thera.* Athens.

Marinatos S. 1976. *Excavations at Thera VII.* Athens.

Mathew B. 1982. *The Crocus*. London.

Mathew B. 1999. "Botany, Taxonomy and Cytology of C. sativus L. and its Allies", in M. Negbi (ed.) *Saffron - Crocus sativus L.* Amsterdam: 19-30.

Morgan L. 1986. *The Miniature Wall Paintings of Thera.* Cambridge.

Negbi M. (ed.) 1999. *Saffron - Crocus sativus L.* Amsterdam.

Rackham O. 1978. "The Flora and Vegetation of Thera and Crete Before and After the Great Eruption", in C. Doumas (ed.) *Thera and the Aegean World I; Papers presented at the Second International Scientific Congress, Santorini, Greece, August 1974.* London: 755-764.

Rehak P. 2002. "Imag(in)ing a Women's World in Bronze Age Greece", in N. Sorkin Rabinowitz and L. Auanger (eds.) *Among Women: From the Homosocial to the Homoerotic in the Ancient World.* Austin: 34-59.

Sarpaki A. 2001. "Condiments, Perfume and Dye Plants in Linear B: A Look at the Textual and Archaeobotanical Evidence", in A. Michailidou (ed.) *Manufacture and Measurement: Counting, Measuring and Recording Craft Items in Early Aegean Societies.* Meletemata 33, Athens: 197-265.

Stacey H. C. 1973. *Walden's Saffron - Crocus sativus.* Saffron Waldon.

Tzachili I. 1994. "Archaies kai sygchrones krokosyllektries apo to Akrotiri tis Santorinis" *Ariadne* 7: 7-33.

Ventris M. & J. Chadwick. 1973. *Documents in Mycenaean Greek.* Cambridge.

Young Forsyth P. 2000. "The Medicinal Uses of Saffron in the Aegean Bronze Age" *Echos du Monde Classique/Classical Views* XLIV, n.s. 19: 145-166.

Votive niches in funerary architecture in Cyrenaica (Lybia)

Eugenio Di Valerio, Igor Cherstich, Mauro Carinci, Francesca Siciliano, Giordano D'Addazio, Angela Cinalli
Università G. d'Annunzio Chieti, Italy

This paper treats the votive niches cut into the Graeco-Roman rock-cut tombs of Cyrene (Shahat-Libya). The work is part of a wider project undertaken by Chieti University in the Necropolis of Cyrene, and is just one of several interrelated activities which include the topographical survey, the mapping of tombs around the ancient city, the typological study of burials and the interpretation of funerary symbols.

This paper represents the efforts of six members of our team - each indiviudal author examines a different aspect of the votive niches, which combined together provides an overview of their development and significance in Cyrene. The analysis can be separated into two parts: firstly, the typology of the niches and their evolution are examined by DiValerio, Cherstich and Carinci. Secondly, the features once contained within the niches, such as funerary statues, marble portraits, small tombstones, are examined by Siciliano, D'Addazio and Cinalli. Our research is still in progress, but as shown by these preliminary findings we are moving towards an improved understanding of the burial customs in the Cyrenean Necropolis.

Developments in the Roman period (E. Di Valerio)

Cyrene was founded in around 631 BC by colonists coming from Thera (Bonacasa 2000: 19) and is located on a calcareous marl plateau surrounded by deep valleys (*widian*). Along the roads outside the city walls, there are numerous tombs dating from the Greek Archaic period to the Roman period. In total, the necropolis was in use for almost ten centuries (sixth Century BC - fourth Century AD).

The tombs are either built or rock-cut (Cassels 1955: 9-21). The built tombs are more frequent in the plains, whereas the rock-cut tombs are generally along the cliffs and rocky surfaces of the *widian*. Rock-cut tombs are certainly the best-preserved funerary monuments, but today their appearance is quite different from that of antiquity. During the Roman period in particular, the appearance of the necropolis changed completely.

In 96 BC, after the testament of *Ptolemy Apione*, Cyrenaica became Roman and from 75 BC was ruled by the Romans (Vismara 1989: 10, 83). The first burials influenced by Roman customs began in this period. The re-use of earlier tombs was typical of the newcomers, especially those of the middle social classes. The result was that many Greek sepulchres were re-opened, modified, enlarged and re-sealed (sometimes more than once), generation after generation (Stucchi 1975: 227-229).

In this context, the eclecticism of Roman art was characterised by a lack of symmetry and structural harmony. This general tendency is reminiscent of the so-called "plebeian art" that expresses itself in Roman historic relief, in which the intention of storytelling is more important than aesthetic taste (Bianchi Bandinelli 1969).

The Roman modifications on earlier tombs can be seen inside as well as outside these tombs. The first set of modifications include enlargements of the chambers, the building of *arcosolia* (a type of large niche with a circular vault used as a burial) and niches for various purposes, such as for cinerary urns, ossuaries, votive gifts, etc. The *arcosolia* are certainly a Roman innovation, and are sometimes decorated with frescoes or rock-cut shells. The second set of modifications relate to rearrangements of the façades, usually involving the cutting of niches for the insertion of portraits or inscriptions.

The wide distribution of Roman funerary portraits (e.g. *Figures 10.4-10.5*) in this Greek city is probably connected not only with Roman burial customs, but also with the diffusion of the rough schematic portraits. The insertion of these portraits in particular, transformed the appearance of some façades completely. In fact, the earlier mouldings, friezes and paintings were disregarded. The façades lost their organic unity and their symmetrical arrangement.

Sometimes the architectural details were not completely destroyed, and an attempt was made to preserve them. For instance the niches for the portraits in the metopes of doric friezes, or in the middle of the façade were cut in order to maintain a sense of axiality. In the earlier Greek period, the architectonic elements were harmoniously inserted in the façades, because the tomb was closed and the dead were venerated from the outside. In the Roman period however, there was a change: the sepulchres were accessible and the symmetry in the façade became less important than the necessities of cult.

Classification of the niches (M. Carinci)

There are few publications about the myriad of niches in Cyrene's Necropolis. This section is an attempt to explain the function of these niches and their classification. The façade niches are very important because they provide architectonic and archaeological information. J.C.Thorn's exhaustive study and forthcoming publication have been of great help in our research and classification (Thorn forthcoming: 103-117). Given below are the results from two different approaches to the study of the form and function of the niches. The classification includes six main types, and in some cases sub-types (*Figure 10.1*) (Thorn forthcoming: 104-105).

Type A was employed for the so-called "funerary goddesses"; well-dated female busts that are numerous in Cyrenaica from the end of the sixth century BC. Type A has different sub-types. Type A1 was made for the earliest aniconic busts, cut at the shoulders. It dates from between the end of the sixth and the beginning of the fifth Centuries BC. Niches of type A2 are more deeply excavated from the rock and probably contained the more naturalistic aniconic busts, which Beschi dates to c. 460 BC. Type A3 was probably still used for aniconic (but well proportioned) busts dating to the fifth Century BC. Type A4 has an asymmetric outline and it seems possible that it was made for a certain type of statue showing a single hand taking off the veil or mantle from the face, a style which was very popular in the fourth Century BC.

Type B includes quadrangular shallow niches for containing inscribed funerary slabs. Generally, this sort of niche was not closely linked to the original and the earliest projecting façades, so we can safely assume that they are Roman in date.

Type B niches were often associated with niches for small funerary Roman portraits: represented here by Type C. This type of niche is unique as it sometimes contain holes, which were probably used to affix the portraits with metal supports, only now occasionally represented by mortar found within the cavity.

Type D was employed for holding cinerary urns, and have two important sub-types: D1, which was cut to hold a separate urn, and D2 in which the urns were directly relief-cut into the rocky façade. Moreover, they can have very different shapes. Examples of D1 in the western necropolis for instance, show variability in the urn types they contain: they certainly contained a hydria and an amphora, and have been dated by Thorn to the Classical or Hellenistic periods.

Type E is represented by the ossuary-niche. It is possible to distinguish them from the cinerary urn niches due their dimensions, probably required due to the size and quantity of human bones stored within them – however this is hypothetical, as no actual bones have been found inside this type of niche.

Finally, type F contained generic votive offerings of various types. Therefore these niches were very different in shape and size. In particular we mention here examples of triangular niches in the western necropolis, which are cut with additional holes, which were in turn connected to urns or ossuaries. These holes were probably intended for liquid offerings, which could flow through the hole and directly into the urn.

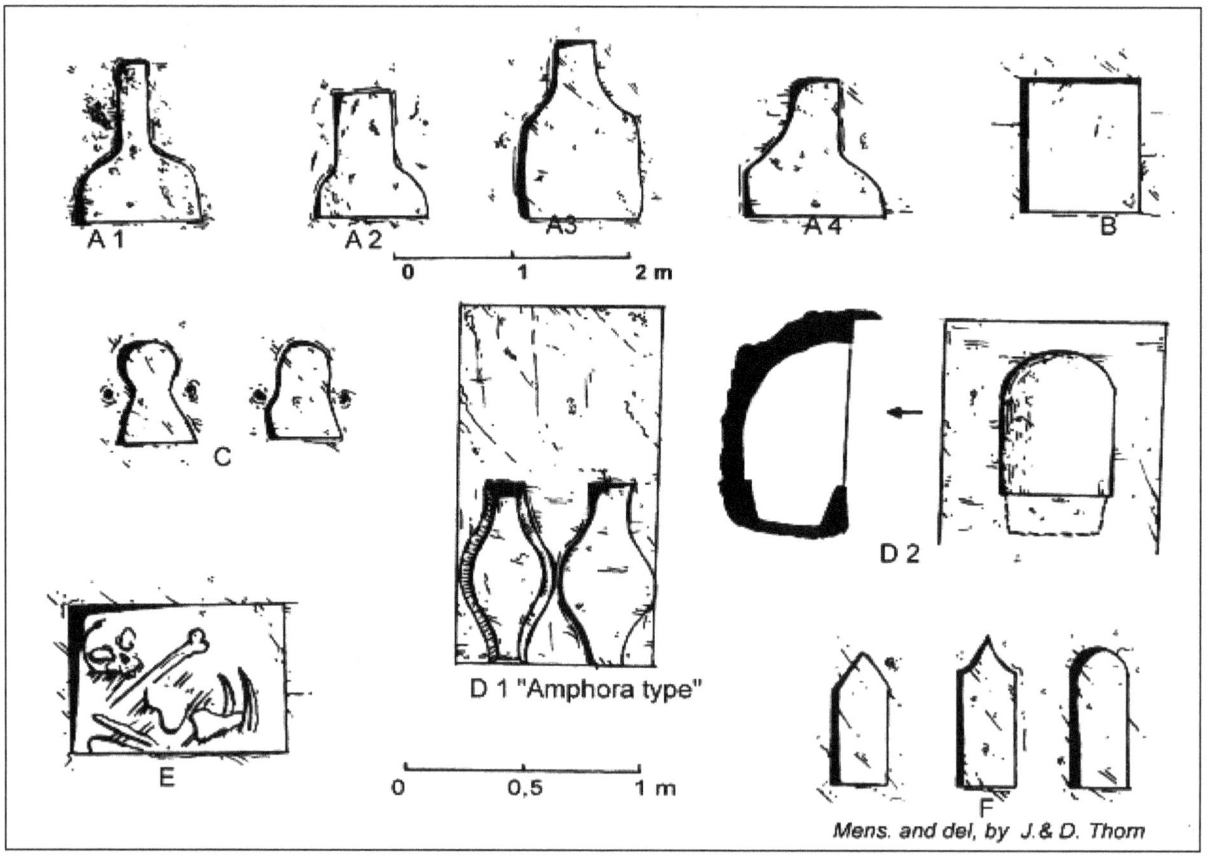

Figure 10.1. Classification of votive niches.

Burial customs: ossuaries and cinerary urn niches (I. Cherstich)

Only two types of niches provide evidence regarding the treatment of the dead: ossuaries and niches for cinerary urns (*Figure 10.1* Types D and E). In fact, while the function of all other types is to contain something that is ritually connected with the dead inside the tomb, these two types are, to all intents and purposes, two different forms of burial. The necropolis indicates that varied forms of death-treatment existed, but certainly in Cyrenaica inhumation was the most common burial custom.

This is quite reasonable if we consider that inhumation was, in general, the Greek burial preference (Kurtz & Boardman 1971: 26-27), whereas in the nearby region of Tripolitania for instance, cremation is the most common treatment, probably due to Punic influences. However, there are still a few Cyrenaican examples of cremation: firstly the cremations in the tumuli of Battus and Messa which are the earliest tombs in Cyrenaica, dating to the late seventh – early sixth Centuries BC (Rowe *et al* 1956: 4-6), and secondly the cinerary urns found within niches of Types D1 and D2, dated to the Classical or Hellenistic period (using Carinci's classification above).

Type D niches are usually inside the tombs, at the rear of some *loculi*. For instance, two examples of D1 type are visible in Rowe's tomb M 16 (Thorn forthcoming: 104-105, 116, fig. 369). These niches, in terms of shape and size, are ideal for containing an amphora, such as the Panathenaic ones found by Rowe in tomb M 15 (Thorn forthcoming: 104-105,116, fig.369). D1 niches can also have the shape of a hydria, as in tomb M 3, where the niche was later converted into a D2 type; in fact it has small, drilled holes probably used for a wooden panel which may have once been decorated with a painted portrait (Thorn forthcoming: 104-105,116, fig.369)

Therefore, we can argue that cremation, after being present in the earliest tombs (the two tumuli mentioned above), survived as a far less usual practice. The typical Cyrenaican tomb has numerous *loculi* used for inhumations (see L. Cherstich's paper in this volume), but very few cinerary niches (or perhaps none), showing that in each group, only a few people received this different burial rite. This could be due to social or ethnic motivations, i.e. perhaps they were people from different Greek countries with different burial practices.

We can draw some comparisons with Alexandria, which is geographically nearby, where cinerary urn niches are uncommon, and inhumation (as in Cyrene) is the most common form of treatment. In fact, the few cremations are in pots known as Hadra Hydriae, which were manufactured in central Crete. Consequently it is widely assumed that they hold the remains of citizens who died abroad (mercenaries, traders and ambassadors) rather than local residents (Morris 1992: 52-54).

Concerning Cyrenaican ossuaries (such as tomb N 4) there is a possibility that some of them relate to Jewish burials. (Thorn forthcoming: 177). Although Cyrene had a Jewish community powerful and large enough to cause the revolt in 115-117 AD, there is only a single Hebrew inscription known (an epitaph of uncertain provenience). No tombs have typical Jewish symbols such as the *menorah*, *ethrog* (cedar's fruit) or *shofar* (the horn for reuniting the people).

Rabbinic sources indicate that graves were normally marked to prevent ritual defilement, however the ossuaries have neither symbols nor epitaphs (Noy 1998: 82). However, in the quarry XV of Tocra (another Greek city of Cyrenaica) there are ossuaries very similar to those of Cyrene, which are marked with Jewish epitaphs in Greek (Rowe *et al* 1956: 43-56). There are also secondary burials in ossuaries in Judaea (Noy 1998: 75), so a link is quite possible. Furthermore we must consider the importance that preserving the bones of earlier burials had in the Jewish tradition. This is testified in rabbinic texts such as the *Mishnah* in which "tomb-niches" are mentioned (Davies 1999: 103). The fact that we find these ossuaries in burial areas used by gentiles (non-Jewish people) is not unusual, as not all the ancient communities of the Diaspora had their own burial places (Noy 1998: 78).

Cyrenaican Funerary Goddesses (F. Siciliano)

Among the main finds related to the façades of the tombs there are the so-called "funerary goddesses" (*Figure 10.2*) which are very typical and unique in Cyrenaican funerary sculpture (Beschi 1969-1970: 133-341). They consist of busts of normal size, although larger busts are not uncommon. A major feature here is that the face is not characterised and is represented by a plain surface (Chamoux, 1953: 294).

This type of funerary bust is generally considered to mark the start of a development in the iconography of funerary goddesses as protectors of the tombs, from an initial representation in the form of a small column to more developed female figures. This aniconic type can be dated, in a schematic classification, to between the end of the sixth and the beginning of the fifth Centuries BC. They were cut at the shoulder, wear a *polos*, and have a small veil covering just the back of the head and the shoulders. This type is connected with Type A1 niches (*Figure 10.1*). During the first half of the fifth Century BC a new type appears, still aniconic, but with a wider representation of the bust, and is connected with Type A2 niches (*Figure 10.1*).

For the second half of the fifth Century BC, we contribute evidence of a new typological development, which becomes more naturalistic in representation with a large and thick veil covering the figure almost completely, and a single hand holding one side of the veil, but with the face remaining aniconic. This type is connected with the Type A3 niches.

Figure 10.2. Cyrenaican 'funerary goddess' bust

For the first time in the fourth century BC, appear examples of a statue with one hand uncovering the face from the veil, or with the face directly uncovered, revealing a detailed representation of the facial features (Chamoux, 1953: 294-295). Initially this sort of sculpture was probably used as a substitutive representation of the body in the role of the Greek κολοσσὸs ("substitute"): a sort of substitute body of the dead person, employed as symbolic σεμα ("sign").

This was certainly the initial function of the aniconic columns that gave birth to their use as symbolic representations. The transformation into female funerary goddesses was probably a local syncretistic interpretation of Kore-Persephone (Chamoux 1953: 300), which was represented as a bust indicating the goddess in the moment of her ανοδοs (elevation from the earth).

The original location of most of the statues is unfortunately, not well known, because they are rarely found *in situ*. They were certainly always located outside tombs, becoming part of the façade itself (Collignon 1911: 204-205). In some cases, such as for the so-called "temple-tombs", they could be situated on the rooftop of the tomb. On sarcophagi, the location of statuettes clearly appears to have been on the top of the lids, where small squared bases have been found. However, it is often difficult to associate the busts chronologically with particular kinds of tomb architecture, because they were often reused in later periods and in different locations (Beschi 1969-1970: 140-141).

Funerary inscriptions (G. D'Addazio)

In general, the bases of "funerary goddesses" possess brief inscriptions indicating the name of the dead person, in the nominative or genitive cases, and always with an indication of the patronymic in the genitive. Among the numerous inscriptions of bases, a good example is the base with the inscription ΑΦΡΟΔΙΤΙΑ / ΙΗΕΝΟΣ, with the name Afroditia in the nominative case and the father's name Zenos with the Doric form of the Z expressed as I.

In addition to finds closely related to the façades and the niches of the Cyrenaican tombs, are also anthropomorphic tombstones. Both the tombstones and bases are useful for the study of funerary inscriptions, which are numerous in the region. These are a homogeneous group and attempt to roughly represent the dead person, either in an iconic or aniconic form. Often the tombstone bears an inscription, which is important for the study of personal names, and in particular for the study of the Doric inflections in Cyrenaican Greek.

Iconic tombstones were most common initially, even including an attempt to characterise facial characteristics in the form of rough portraits. The tombstones were accompanied with short inscriptions, generally limited to the name of the dead person and his father's name (Bacchielli & Reynolds 1987: 489-522). In the later examples, the faces no longer seem to be characterised, leaving more space to the epigraphic field. However, it must be said that the former type appears more commonly at Cyrene, Apollonia and Ptolemais, while the latter type is typical of Tocra. It is therefore possible that the two types were both used simultaneously without any chronological distinction.

A good example of this type, dating to the first Century AD, comes from Lamluda (*Figure 10.3*), bearing the inscription:'ΘΕΥΤΙ / ΜΑ ΦΙΛ / ΙΩ (ετων) μά':"*Theutim a,* dead when she was 40 years old", is mentioned in the nominative and as the daughter of *Filios*. Very typical of the local Doric dialect is the form in – ω̣ for the genitive, instead of the form – ου, and the diphthong Θευ, instead of Θεο Bacchielli & Reynolds 1987: 496, fig. 44)

The study of personal names for this region is fundamental to our understanding of the links and the relationships between the Greek and indigenous population. Literary sources (Pindar Pythica 9, 117-123; Herodotus 4, 159-160; Thucydides 7, 50,2; Pausanias 1,7,1; Diodorus Siculus 3, 49, 4,17,4-5) provide lists of the local tribes with ethnic descriptions, mentioning Macai, Nasamones, Marmarides and Ghiligami. However epigraphic sources are not so numerous. The so-called Diagram of Ptolemy I Soter (322-321 BC) is one of the first documents (Klaffenbach *et al* 1938: IX, 1) and often shows Greek names in association with local patronyms or indigenous names with Greek fathers (Gasperini 1987: 403 – 413).

Figure 10.3. 1st Century AD inscribed iconic tombstone from Lamluda

Funerary portraits (A. Cinalli)

When Cyrenaica became Romanized, between 96 and 74 BC, and in spite of the difficulties in the introduction of Roman art and funerary customs (Stucchi, 1975: 165, 227-229), the use of the funerary portraits was introduced at an early stage and originally re-elaborated according to the strong Hellenic character and local traditions.

In the Roman period, the earliest Greek tombs were reused with a different arrangement. The Romans modified the original architecture of the tombs by carving niches on the façades, and by adding architraves and mouldings, for the placement of portraits of the deceased (Cassels 1955: 1 ff.; Pls. I-XIII). These modifications often destroyed or damaged the earlier decorative motifs. In some cases the funerary portraits could also be directly sculptured in relief within the rocky façade of the tomb. They were usually made from marble, but owing to the lack of this material, the Romans employed architectonic decorations, "funerary goddesses", and other portraits and statues. Their dimensions varied from between 15 cm in height to life-size scale.

The famous bronze head from Cyrene (not from a funerary context), now in the British Museum, is the chief document of Lybian somatic features, visible on most funerary portraits: curly hair, prominent cheekbones and lips, almond shaped eyes with the upper lid more curved than the lower one forming a straight line, a low forehead and a short moustache (Huskinson 1975: 33 ff.).

The Cyrenaican funerary portraits (*Figures 10.4-10.5*) generally represent young persons. This could depend on the choice of the client to be portrayed whilst still alive in their most ideal condition (Rosenbaum 1960: 5 ff.), or it relates to a miscomprehension in the local interpretation of realistic Roman portraiture (Bacchielli 1987: 472-475).

Figure 10.4. Cyrenaican funerary portrait (left)
Figure 10.5. Cyrenaican funerary portrait (right)

Moreover, in spite of the individual features, we often have the impression of lifelikeness, due to ties of blood or to the same workshop production, as these examples were made as a series. In my opinion it is likely that many portraits were only blocked out and kept in the workshop ready to be rendered in more detail when ordered.

The funerary portraits have typical "background strips", which run between the neck and cheeks, with a surface which corresponds to the level and colour of the façade surface, obliterating the impression of depth. However, in some cases the portrait foreside appears to jut out from the façade. The bust consisted of a head, neck and a small part of the chest; the eyes were painted (and not drilled), which was usual from the Hadrianic period.

These features find parallels with the Egyptian mummy-portraits from Alexandria (Bacchielli 1977b: 100-110) and the Fayum, which were painted with similar iconographic characteristics and had the same cut of the bust. Moreover, affinities are possible with portraits from Palmyra and with a series of busts from Crete (Beschi 1976: 385-387).

The close similarity of the Cyrenaican funerary busts with the Egyptian portraits is not surprising: they are a witness to the links between the two countries and of the common historical tradition shared from the Ptolemaic age onwards. During the period of Romanization, Cyrene and Alexandria both received the seed of the Roman portrait style. This was with independent, but also concomitant consequences, due to Alexandria's dominance of the region. Alexandria had imposed a huge influence on Cyrene since the advent of Hellenism. Other parallels also attest to the liveliness of trade and the possibility of mutual artistic contacts.

Bibliography

Bacchielli L. 1977a. "Un ritratto cireneo nel Museo Nazionale Romano e alcune osservazioni sui busti funerari per nicchia" *Quaderni di Archeologia della Libia* 9: 77-92.

Bacchielli L. 1977b. "Un ritratto cireneo in gesso nel Museo greco-romano di Alessandria" *Quaderni di Archeologia della Libia* 9: 97-110.

Bacchielli L. 1987. "La scultura *libya* in Cirenaica e la variabilità delle risposte al contatto culturale greco – romano" *Quaderni di Archeologia della Libia* 12: 459 – 522.

Bacchielli L. & Reynolds J. 1987. "Catalogo delle stele funerarie antropomorfe'(Appendice)", *Quaderni di Archeologia della Libia* 12: 489 – 522.

Barker G., Lloyd J., Reynolds J.(eds.) 1985. *Cyrenaica in Antiquity,* Oxford.

Beschi L. 1969-1970. "Divinità funerarie cirenaiche" *Annuario della scuola archeologica d'Atene* e *delle Missioni Italiane in Oriente* XLVII-XLVIII: 133-341.

Beschi L. 1976. "Un supplemento "cretese" ai ritratti funerari romani della Cirenaica" *Quaderni di Archeologia della Libia* 8: 385-397.

Bianchi Bandinelli R. 1969. *Roma. L'Arte Romana nel centro del potere*. Milano.

Bonacasa N. 2000. *Cirene.* Milano.

Cassels J. 1955. "The Cemeteries of Cyrene" *Annual of the British School at Rome XXIII*: 1-43.

Chamoux F. 1953. *Cyrène sous la monarchie des Battiades*. Paris.

Collignon M. 1911. *Les statues funéraires dans l'art grec*. Paris.

Davies J. 1999. *Death, burial and rebirth in the religions of antiquity*. London and New York

Gasperini L. 1987, "Echi della componente autoctona nella produzione epigrafica cirenaica", *Quaderni di Archeologia della Libia* 12: 403 – 413.

Huskinson J. 1975. *Roman sculpture from Cyrenaica in the British Museum*. London.

Klaffenbach G., Robert L., Tod M.N., Hondius J.J.E. 1938. *Supplementum Epigraphicum Graecum* Vol. 9. Leiden.

Kurtz D. & Boardman J. 1971. *Greek Burial Customs*. Oxford.

Longo O. 1987. *Elementi di grammatica storica e di dialettologia greca*. Padova

Morris I. 1992. *Death-ritual and social structure in classical antiquity*. Cambridge.

Noy D.1998. "Where were the Jews of Diaspora buried?", in M. Goodman (ed.), *Jews in a Graeco-Roman World.* Oxford: 75-89.

Rosenbaum E. 1960. *A Catalogue of Cyrenaican portrait sculture*. London

Rowe A., Buttle D., Gray J. 1956. *Cyrenaican Expedition of the University of Manchester 1952.* Manchester.

Stucchi S. 1975. *Architettura Cirenaica*. Roma.

Thorn J.C. forthcoming, *Rowe's Cyrenaican expedition.*

Vismara C. 1989. *L'Organizazzione dell'Impero.* Roma.

Ars Fullonia. Interpreting and contextualising Roman fulling

Miko Flohr - University of Nijmegen

What would it be like to live in a town in Roman Italy? What would we see when walking through the streets? What would the town smell and sound like? At the beginning of the new millennium, archaeologists are becoming increasingly interested in the thoughts and feelings of the people who made the objects they investigate: they want to interpret what past people saw, heard and felt and how these determined their perception of the world they lived in.

Life in the ancient cities of the Italian peninsula has been the subject of archaeological discourse ever since the beginnings of archaeology in the 18th century, but only recently, our focus has begun to shift from categorizing and describing material remains towards interpreting the social processes that created them. Several publications from the last decade, such as the one for Pompeii (cf. Wallace-Hadrill 1994; Dickmann 1999), shed an entirely new light on objects and sites that have been intensively studied for the last two centuries.

This interpretive approach, however, has not yet succeeded to overcome one of the most tantalising problems concerning our perception of urban life in Roman Italy, which is mainly determined by an emphasis on the domestic situation of the urban elites. Archaeologists working on ancient urban communities still mainly concentrate on the material remains related to the private environment of the upper classes; there is only a marginal role for the vast majority of the urban population, who spent their lives working day and night to earn their living. We view antiquity through the eyes of the happy few, without asking ourselves whether this perspective would actually provide the best insight into the life in ancient cities.

An alternative way of approaching urban society may be to investigate the socio-economic context of urban workshops. Economic activities are fundamental to every society: everyone is, in some way or another, involved in the economy. Commercial or industrial production had a central role in the social landscape of ancient cities and it has left many visible remains in the archaeological record. An investigation inquiring how production was organized and embedded in its urban context, thus may add a lot to our perception of Roman urban societies.

This paper presents the first results of research on one particular kind of workshop: the *fullonica. Fullonicae* were involved in the finishing of woollen cloth and clothing. They had a well-organised layout with a complicated system for the supply and disposal of water, often relatively well preserved. In certain excavations (e.g. Pompeii and Ostia) some highly informative specimens have been brought to light. A considerable amount of literary and epigraphical data facilitates the interpretation of the material evidence. This research aims to interpret the organisation of the production process and to establish the position of *fullonicae* in their urban environment. The present article concentrates on the identification of *fullonicae* and the organisation of the production process in the larger establishments of Ostia and Pompeii.

How can we identify a *fullonica*?

The material remains of *fullonicae* have never been thoroughly compared with each other and there has been little discussion on how to define *fullonica.*[1] Thus, it is hard to identify material remains as a *fullonica*. We may define *fullonica* as a place where *fullones* (fullers) performed their daily professional activities. However, though this definition may be justified by the linguistic origin of the word, it does not make the identification of establishments any easier. Apart from a few Pompeian *fullonicae* identifiable by *graffiti* on their façades referring to *fullones*, there are no establishments preserving any epigraphical evidence.[2] Most workshops can only be identified by analysing their material remains.

Typical for large *fullonicae* is a complex of several large basins. These basins, in most cases three or four, were built on a lower level and thus accessible from the surrounding working space. They were connected by a system of tubes and overflows (*Figure 11.1*, label 3). The complex was supplied with water by one single outlet in one of the utmost basins. Some basins were equipped with low benches. These have often been interpreted as steps, but probably they also had a role in the fulling process. An indication of this may be found in the *fullonica* of Stephanus at Pompeii. Here, the owner decided that the small bench he already had in one of his basins did not suffice, and built a new bench next to the other on a slightly lower level (*Figure 11.2*). There would have been no ground for this extension if the benches were simply used as steps. Most *fullonicae* have basins with, as well as without benches.

The presence alone of one or more basins is not enough for the identification of a *fullonica*, even though these are a typical feature of the latter. It is not hard to imagine a basin designed for some other goal as well. To identify a *fullonica*, we need a complex of interconnected basins, which were easily accessible, with one single supply of

1 Discourse on the subject includes Moeller 1978, Pietrogrande 1978, Uscatescu 1994 and, more recently, Bradley 2002. Unfortunately, none involves a satisfying interpretation of the material data.

2 E.g. the *fullonica* of Stephanus (I 6, 7), the *fullonica* of Vesonius Primus (VI 14, 21.22) and the *fullonica* of Mustius (VI 15, 3).

water and working benches in some of them. However, every building with a complex of basins satisfying these criteria also shows the remains of so-called 'treading stalls', little niches surrounded by low walls (*Figure 11.3, 11.4*). At the bottom of each stall, a small vat was placed or built in. In Ostia, these vats were made of terracotta, whereas at Pompeii, fullers used portable vats made of lead or wood.[3] Typical for Pompeian *fullonicae* are little jars built into the walls surrounding the niches. On excavation, some of these still contained a white, clayish substance usually interpreted as 'fullers' earth' (e.g. Moeller 1978: 46) (see below).

Treading stalls seem to have been typical of *fullonicae* and they constitute the most useful criterion for fullery identification. A painting from the *fullonica* of Veranius Hypsaeus (VI 8, 20.21.2), now in the National Museum of Naples (De Caro 1996: 258-259), shows that fullers used treading stalls to tread and scrub the cloth and to wring it out. Every *fullonica* had treading stalls, though it is not always possible to trace them, since some *fullonicae* also made use of wooden stalls.[4] Concrete treading stalls can be recognized by several characteristic features, such as remains of water-resistant plaster on the walls and on the floor, remains of a prefabricated vat or an imprint of such a vat in the bed of mortar. The presence of treading stalls makes it possible to identify five *fullonicae* at Ostia and at least seven at Pompeii.[5]

Working in a *fullonica*

Roman fullers treated woollen cloth with several substances. The elder Pliny (Plinius Maior *NH* 28.91; 35.195-198) mentions that they used urine and 'fullers' earth'. Similar substances were used by fullers in the Middle Ages and later to refine newly woven cloth by manipulating the individual hairs until they made up a thin layer on the surface of the cloth (cf. Jenkins & Ponting 1982: 23). The cloth becomes smoother and insulates better. We may assume that in antiquity, fullers pursued the same goal.

The process started with a treatment of the cloth in the treading stalls with the chemicals mentioned above. The way in which the cloth was treated was as important as the use of urine and fullers' earth: the combination of treading, scrubbing and wringing made the wool fluffy and felty. After this first phase, the cloth was rinsed. During the process of rinsing, the cloth was combed with a brush made of thistles, as we know from remarks of the elder Pliny and from analysis of sediments in the drain of a recently excavated *fullonica* in Barcelona, which contained remains of thistles (Juan-Tresseras 2000). The same sediment also contained traces of lavender, suggesting that the cloth could have also been treated with substances to create a more pleasant smell.

The washing and rinsing were carefully organised and spatially separated. The treading stalls had a separate system of discharge to prevent the mixture of water and chemicals from ending up in the rinsing vats. In the *fullonica* of Stephanus at Pompeii (I 6, 7), a narrow gutter was built on top of the wall between two basins, carrying the water from the treading stalls on the east side of the rinsing complex to the drain at the northwest side of the working space (*Figure 11.1*). In the rinsing complexes of large *fullonicae*, the water flowed from the first basin through the second and the third basin to the drainpipe. This implies that the fresh water in the first basin was less polluted than that in the second basin, which in its turn quality of the water facilitated a spatial differentiation was cleaner than the water in the third. The decreasing within the rinsing process: the cloth was cleaned in steps, gradually proceeding from the most polluted towards the cleanest basin.

This differentiation is also implied by the layout and positioning of most rinsing complexes. Working benches can only be found in basins near the drainpipe (*Figure 11.1*, label 3). In some Pompeian *fullonicae*, the basin where the fresh water entered the complex was higher and not accessible.[6] In the three major *fullonicae* in Ostia, there are treading stalls on all sides of the rinsing complex, but not on the side of the outlet of the water supply. Presumably, the clothes were rinsed and combed in the basins near the drainpipe. The clean water in the first basin was reserved for a final soaking.

After the rinsing, the cloths were finished. Important for our interpretation of this phase are two Pompeian frescos. A painting from the *fullonica* of Veranius Hypsaeus (VI 8, 20.21.2) shows the cloths being hung out, combed, inspected for irregularities, folded up and pressed together (De Caro 1996: 258-259). A frieze in the largest *oecus* in the House of the Vettii (VI 15, 1) depicts similar operations. It is not certain in what order these operations took place, but we may imagine that the process ended with folding up and pressing. In the *fullonica* of Stephanus at Pompeii, remains of a press have been found next to the entrance of the shop, far from the treading stalls and the rinsing complex, which indicates that the finishing, too, had its own separate location in the

---

3 These portable vats have not been found, but their existence is suggested by the fact that most stalls lack a bottom.

4 E.g. wooden stalls have been depicted on a frieze in the House of the Vettii at Pompeii (VI 15, 1).

5 For Ostia is worth mentioning the two *fullonicae* in the Via della Fullonica (II, 11, 1), and the workshops in the Via degli Augustali (IX 13), along the Cardo (VIII 5, 3) and behind the temple of the Fabri Navales (III 2). At Pompeii, apart from the four establishments mentioned above (supra n.3), we may identify *fullonicae* at I 4, 7, V I, 2 and VI 16, 3.4.

6 This was the case in the *fullonica* of Veranius Hypsaeus (VI 8, 21.22) and the *fullonica* of Stephanus (I 6, 7).

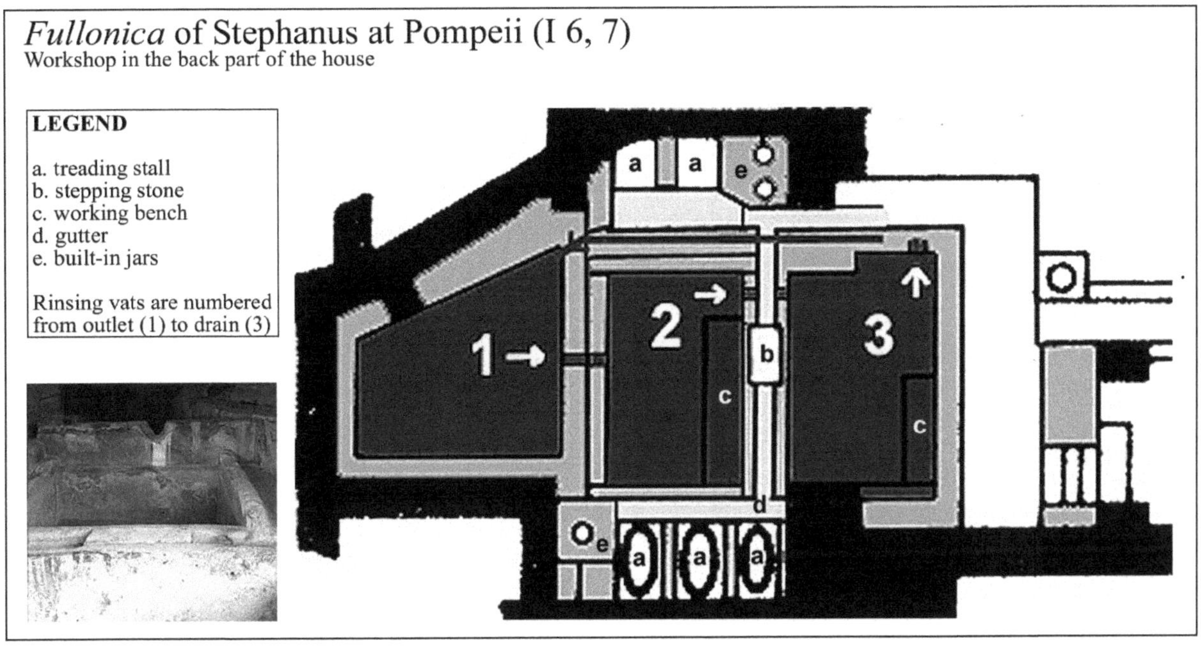

Figure 11.1. Fullonica of Stephanus at Pompeii.

Figure 11.2. Secondary bench in the fullonica of Stephanus at Pompeii.

Figure 11.3. Treading stalls in the fullonica of Stephanus

workshop (cf. Moeller 1978: 42). This is also suggested by the layout of some fullonicae at Ostia, which reserved one of the sides of the rinsing complex for the finishing.

Fullonicae and the city

From the beginnings of Latin literature, Roman authors occasionally refer to fullers.[7] These references, though often somewhat belittling or deriding, emphasize that fullers were an integrated part of Roman society from the second century B.C. onwards. It has often been asked why fullonicae, which probably caused an appalling odour and must have been "proverbially unpleasant places to be around" (Bradley 2002: 35) could be located in almost every place in a Roman town, and why the Romans did not rather ban them to the outskirts of the urban area.

---

[7] E.g. Plautus, *As.* 907, *Aul.* 508, *Pseud.* 779; Seneca, *Epist.* 15.4, Martialis 3.59, 6.93, 12.59.

However, if we analyse the establishments at Ostia and Pompeii, it appears that these *fullonicae* cannot have caused much nuisance to their urban environment. The large Ostian fulleries were surrounded by thick walls and at least partially covered by a roof preventing smells from reaching the nearby street. In the small *fullonicae* of Pompeii and Ostia, the treading stalls, which probably caused most of the nasty smells, were mostly located at the back of the shop or even in a separate room. The larger Pompeian *fullonicae* have been built in the garden areas of private houses and in distance from the street to prevent public nuisance.

Apparently the smells and sounds these workshops produced did not even disturb the rest of the house: all houses with a *fullonica* in their back yard also maintained domestic functions. They had residential rooms with a decoration that probably date from after the insertion of the workshop and, on excavation, contained objects of a domestic nature (Flohr 2003: 448). In the *fullonica* of Vesonius Primus, a side branch of the water pipe feeding the *fullonica* in the back yard ends in a beautiful marble *impluvium* that is either contemporary, or predates the workshop. Contrary to what is often assumed (Bradley 2002: 36; Wallace-Hadrill 1994: 138), the presence of a *fullonica* did not make a house uninhabitable.

Conclusion

The *fullonicae* of Ostia and Pompeii show the way in which fullers tried to improve the efficiency of the fulling process by spatially separating the various phases of the fulling process, and by designing a logical route of the product through the various parts of the workshop. It may be concluded that *fullonicae* were an integrated part of Roman towns and could be even situated in a domestic context. Future research will have to make clear to what degree the many differences between identified establishments were caused by historical developments or socio-economic factors and which roles *fullonicae* and *fullones* could play in their urban communities.

Figure 11.4. Remains of treading stalls in fullonica II xi 1 at Ostia.

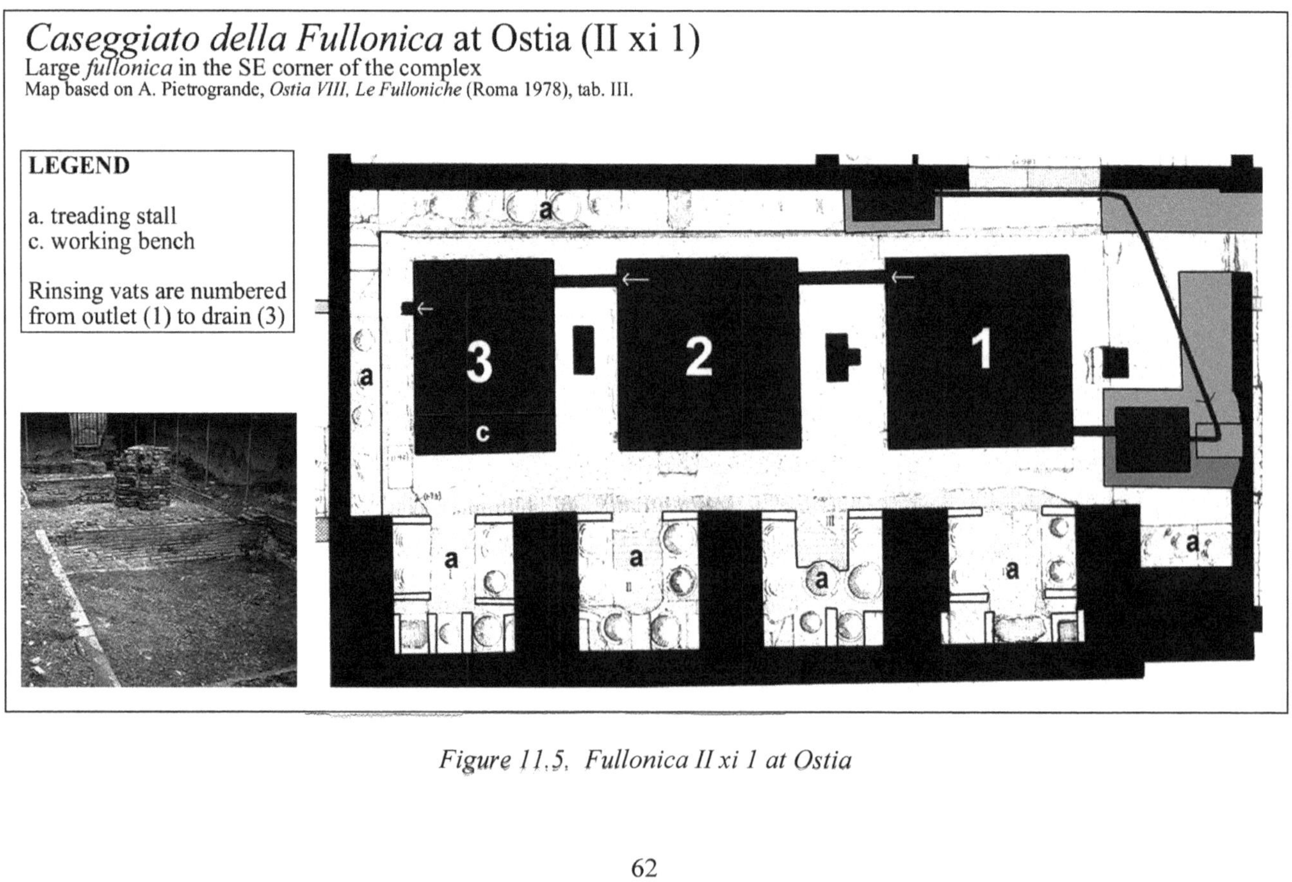

Figure 11.5. Fullonica II xi 1 at Ostia

Bibliography

Bradley M. 2002. "'It all comes out in the wash': looking harder at the Roman *fullonica*" *Journal of Roman Archaeology* 15: 21-44.

De Caro S. 1996. *The National Archaeological Museum of Naples*. Napoli.

Dickmann J-A. 1999. *Domus Frequentata. Anspruchsvolles wohnen im pompejanischen Stadthaus*. München.

Flohr M. 2003. "*Fullones* and Roman Society. A Reconsideration" *Journal of Roman Archaeology* 16: 447-450.

Jenkins D. & Ponting K. 1982. *The British Wool Textile Industry 1770-1850*. London.

Juan-Tresseras J. 2000. "El uso de plantas para el lavado y teñido de tejidos en época romana. Análisis de residuos de la *fullonica* y la *tinctoria* de *Barcino*" *Complutum* 11: 245-252.

Moeller W. 1978. *The Wool Trade of Ancient Pompeii*. Leiden.

Pietrogrande A. 1978. *Scavi di Ostia VII. Le Fulloniche*. Roma.

Uscatescu A. 1994. *Fullonicae y Tinctoriae en el mundo Romano*. Barcelona.

Wallace-Hadrill A. 1994. *Houses and Society in Pompeii and Herculaneum*. Princeton.

GIS Study of the Rural Sanctuaries in Abruzzo: Preliminary Report.

Domenico Fossataro, Debora Lagatta & Vanessa D'Orazio
Università G. d'Annunzio Chieti - Italy

Introduction

This paper is intended as a preliminary report of a wider research and cataloguing project that investigates the rural sanctuaries of Abruzzo (Central Adriatic Italy). In particular, this project looks at the topographical setting of sanctuaries, their specific function and role within the landscape, relationships with other classes of sites, and their positioning in relation to the local road network.

The aim of the project is to build a comprehensive database suitable for cataloguing these sites. The typological criteria recorded include details on architectonic and archaeological remains, as well as topographical features of the surrounding area. The extensive participation of students and scholars of Chieti University with the project offers the possibility of a number of different research approaches. This is already providing interesting results at a preliminary stage, although it is certainly too early to draw any definitive conclusions without continued surveys and excavations in the future.

This paper firstly provides a brief background to the project, including a synthesis of the distribution of the sanctuaries in the region, and interpretations of their role and meaning. Secondly, two case studies from northern Abruzzo and another two from southern Abruzzo demonstrate how typological and topographical analysis can be applied to sanctuaries of different scale, function and regional location.

Organization of the research and role, location and function of the rural sanctuaries (D. Fossataro)

This project intends to create a schematic database to synthesise the substantial amount of excavation and survey data currently available. Even at this preliminary stage, it is possible to examine the topographical similarities and differences of these sanctuaries in some detail. Moreover, the use of Geographical Information Systems (henceforth GIS) allows us to construct a detailed catalogue of the location of the main sanctuaries and their architectonic or monumental remains. The Access database is compatible with GIS, allowing for each site to be analyzed in terms of its monumental resources and topographical setting. This can be easily modifed and enlarged over the course of future surveys, excavations and research. The integrated recording system also allows us to run queries and plot specific information directly onto thematic maps (e.g. *Figure 12.1*), allowing a diachronic and synchronic study of the sanctuaries in the region.

The numerous rural sanctuaries in Abruzzo always played an important role, both in pre-Roman and Roman times, and especially in the Hellenistic period. This was certainly because in mountainous areas they served a religious function, as well as being political and economic centres within particular territories. They were generally located in hilltop positions, easily accessible but at the same time defensive, always very close to springs or river valleys and along important tracks or at nodal points of the local road network. The examples cited in this paper served as central sanctuaries within the *pagus* (the tribal territorial unit), or were closely related to smaller villages in the region.

The topography of the region is characterised by river valleys and upland plateaus, which have been used since prehistory as natural tracks and were later exploited as Roman roads and *tratturi*[1]. The role of transhumance was extremely important for this region. The *tratturi* served not only for the movement of flocks, but also goods, people and culture. Transhumance is also clearly attested in the area both by literary sources (for instance Varro, De Re Rustica II) and by archaeological sources such as the two reliefs from Pretoro and Sulmona, with representations of shepherds and their flocks (Mancini 1997: 20-28).

It is important to note that due to their position and aggregative function of the sanctuaries, they frequently became essential meeting points for a scattered population organised into the tribal territorial units or *pagi*. The sanctuaries were often used as political centres or markets within the pagi, and along routes. This is true for Iuvanum (*Figure 12.1*, no.4), an important market-sanctuary in southern Abruzzo, and situated along a *tratturo* originating in Apulia and continuing towards the north of the region, to Umbria, as well as to Latium and northern Campania.

In the Hellenistic period this political role is well attested by many finds, especially by the presence in some sites, such as Iuvanum (*Figure 12.1*, no.4), Quadri (*Figure 12.1*, no.1) and Pietrabbondante, of theatres directly linked with the sanctuaries and temples. In these cases the theatres were used for political assemblies. For this reason the Romans destroyed some of them, as in the case of Pietrabbondante which was badly damaged soon after the Social War.

The GIS study of the sanctuaries is now providing increased levels of information, especially about their hierarchical distribution. It seems possible that every

[1] The 'tratturi' were natural roads and valleys connecting different areas and regions, which were used seasonally during periods of transhumance, i.e. for the movement of flocks.

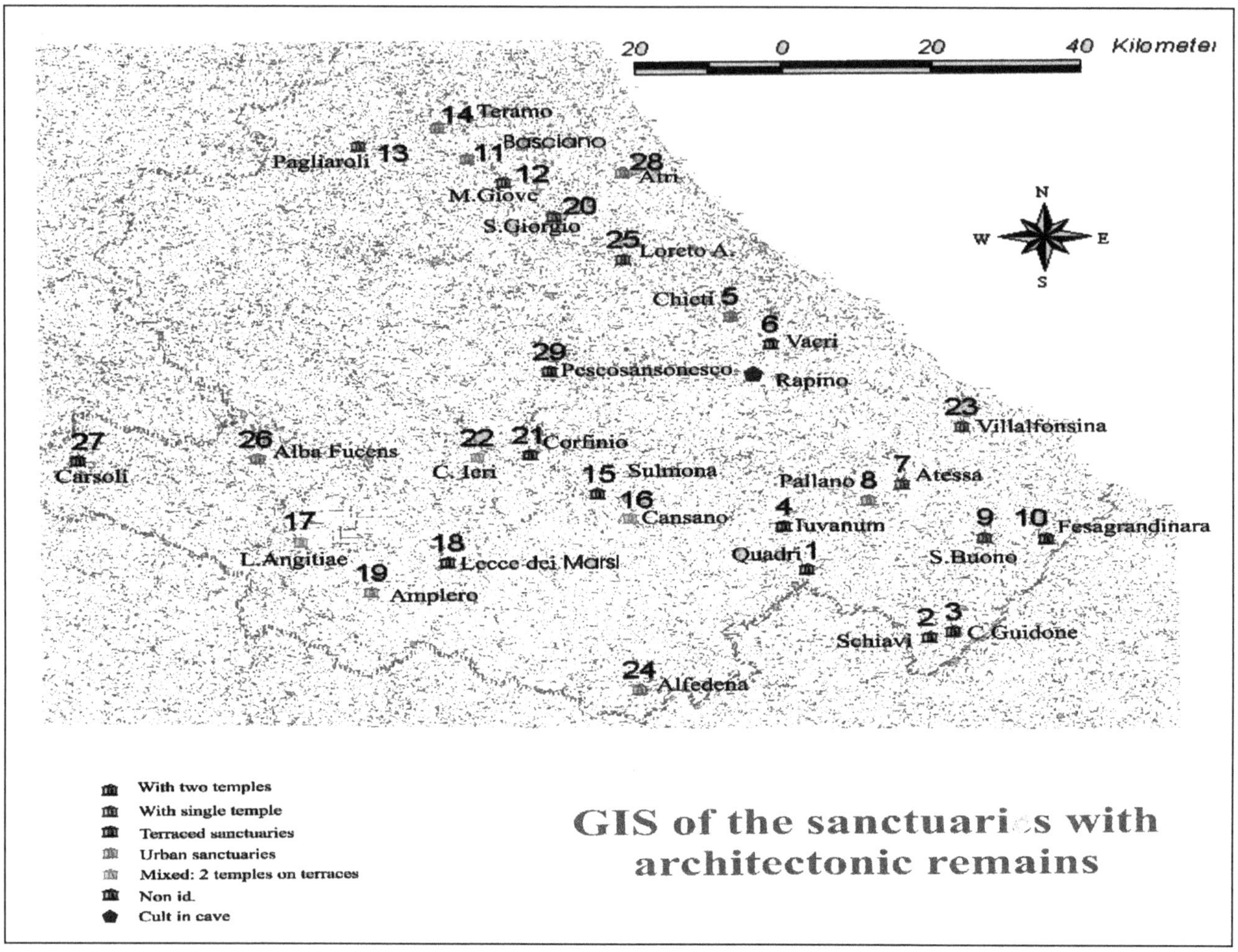

Figure 12.1. Map of Abruzzo (Italy) showing location of sites mentioned in the text

Sabellian or *Safin*[2] tribe had a main sanctuary, which can be considered the religious and political centre *par excellance* within the *pagus*. Smaller sanctuaries were also located in boundary areas similar to the better-known 'religious belts' of sanctuaries attested for Greek colonies, which acted as religious frontiers.

Looking at the GIS of the main sanctuaries, it is clear that the Hellenistic sanctuaries with a double temple enclosed by a *temenos* (*Figure 12.1*), as in the examples of Vacri and Schiavi (*Figure 12.1*, nn.2 and 6) examined below (as well as Molise), are typical of southern Abruzzo, taking direct architectonic influence from the Campanian area. The terraced sanctuaries (*Figure 12.1*, nos.15, 16, 22) are distributed in inner Abruzzo, in a small area along the Roman road known as *Via Valeria*, connecting Abruzzo directly with Latium and Rome. For this reason, they were probably strongly influenced by the more famous terraced sanctuaries of Latium. The hilltop sanctuaries with single temples dating to the late Hellenistic period, as in the case of Monte Giove and Colle San Giorgio (*Figure 12.1*, nos.20, 12) seem to be more typical of coastal areas, especially in northern Abruzzo, where they are regularly located along a secondary track of the local road network.

Using GIS, the distribution of these sites within the region seems to be quite regular and capillary for southern Abruzzo; whereas the distribution is not as regular for northeastern Abruzzo, which shows a high density of sanctuaries. This is probably due to a different form of settlement organization in this area in republican times. The northwestern area of the region is completely lacking this type of site. There are two possible explanations for this. The first explanation could be based on an *argumentum ex silentio*, and therefore the absence could be due to the poor quality of data and lack of surveys in the area. On the other hand, the second explanation relates to the remote and isolated nature of this mountainous area, which could have generally had a smaller number of sanctuaries and settlements. However, it is still impossible at present to establish the real reason without further research.

[2] The Sabellians or Safin were the local Italic population, which was tribally organised. The Safin were Osco-Sabellian speakers, economically devoted to stock raising, traded in dairy products, practised agriculture especially within the hilly coastal area and plateaus, and exploited the forests in the mountainous countryside.

The Abruzzo region contains 90 sanctuaries, and the monumentalization of these sanctuaries dates mainly to the Hellenistic period (4th – 1st Centuries BC). However, archaeological finds dating from the 6th – 4th Centuries BC suggest the existence of earlier cults at many sites, and probably an earlier political function for the sanctuaries in the late Archaic period. For example, the Sanctuary of Grotta del Colle, near Rapino (*Figure 12.1*, pentagon symbol) is a famous example of an early cult place attested in a cave. Only 20% of the 90 sanctuaries survived the process of municipalisation in the Roman period[33].

Two examples from southern Abruzzo (D.Lagatta)

In southern Abruzzo, there are many rural sanctuaries dating to the 3rd-2nd Centuries BC. They are always located along the main tracks of the local road network, indicating the importance of the sanctuaries in the territory, and the importance of easy accessibility.

The sanctuary of Vacri (*Figure 12.1*, no.6 and *Figure 12.2*) is situated in a boundary area and it was probably the main sanctuary of the local *pagus* system, like the ones at Rapino, Chieti and Lanciano situated at the same distance from Vacri (see *Figure 12.1*, no.6). In addition, there were probably smaller sanctuaries more closely related to villages.

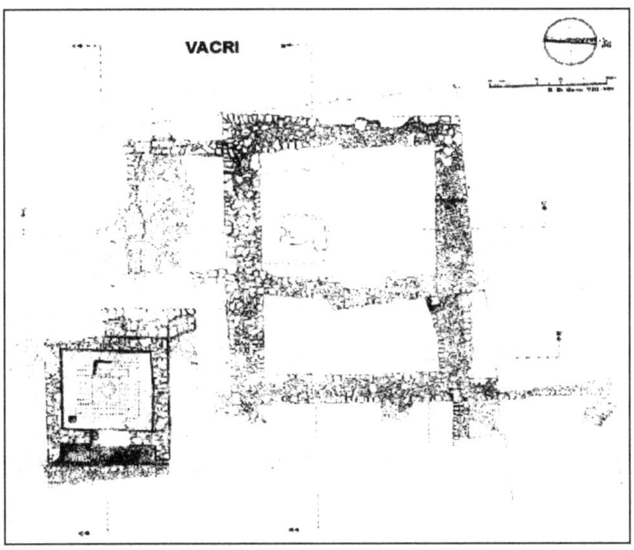

Figure 12.2. Plan of the sanctuary at Vacri

The site has a hilly position between two river valleys at the crossing point of four important roads and tracks. According to the excavators, between the 5th-4th Centuries BC, there was evidence for open-air cult activity at the site, attested by remains of a pavement and a fire installation, associated impasto pottery and three small bronzes representing Heracles (Papi 1997; Campanelli, Faustoferri 1997).

The monumentalisation of the area with a double temple (*Figure 12.2*) is attributed to the late Hellenistic period. The first temple, on a high podium, dates to the 3rd Century BC, and the second and smaller temple decorated with plaster walls and a pavement in *opus signinum* dates to the beginning of the 1st Century BC.

Among finds at Vacri is the architectonic frieze in terracotta representing Heros riding the panther (popular in Abruzzo), and female figures alternating with palmettes and lotus flowers. The votive offerings found during the excavation of this sanctuary were both varied and numerous, suggesting both local production and the importation of products.

The end of the 1st Century BC saw the decline of the Vacri sanctuary, which was renovated with the construction of a new building, probably dating to the Flavian period as attested by diagnostic finds including coins.

The sanctuary of Schiavi (*Figure 12.1*, no.2 & *Figure 12.3*) has a mountainous position dominating the river valleys of Sente and Trigno. A settlement found nearby dates to the end of the Bronze Age and the beginning of the Iron Age, and is related to a necropolis that was in use from the end of the Bronze Age up until the Imperial Roman period. A secondary track of the main *tratturi Ateleta-Biferno* and *Celano-Foggia* crosses the area, connecting the sanctuary with the mountain and the coast and other important Samnite sanctuaries, such as Pietrabbondante and Vastogirardi in Molise (not shown on map).

The Sanctuary of Schiavi was the main one in the *pagus* and certainly had political and economic functions as well as religious ones. The first phase dates to the 3rd Century BC when the slope of the mountain was terraced and paved. A first temple was built between the 3rd – 2nd Centuries BC on a high podium with a central staircase, a doric frieze and four columns with ionic four-faced capitals. This latter feature appears to be uncommon in the region. The second phase dates to the 2nd Century BC, when the sanctuary was enlarged and a second temple was built without a podium. This appears to be similar to the small temple of Vacri. It has a reddish pavement, and bears an inscription with the name of the magistrate who paid for its construction.

Also, architectural decoration was present, including features such as *antepagmenta* with four frontal faces and

3 Soon after the Social War, at the beginning of the 1st Century BC, a new territorial and administrative organization was introduced in this region, as elsewhere, according to the Roman system. Therefore some of the earlier settlements, together with new centres became municipia, which acted for the territory as administrative, as well as political and economic centres. This new establishment introduced an urban reorganization, which changed the earlier settlements and territorial assets completely.

vegetal motifs, and *antefixes* with Potnia Theron, which are very popular in the region[4].

A large votive deposit has also been found, mostly dating to the 3rd –2nd Centuries BC, with unusual and original terracotta offerings, a beautiful diadem with golden leaves, as well as small bronzes representing Heracles.

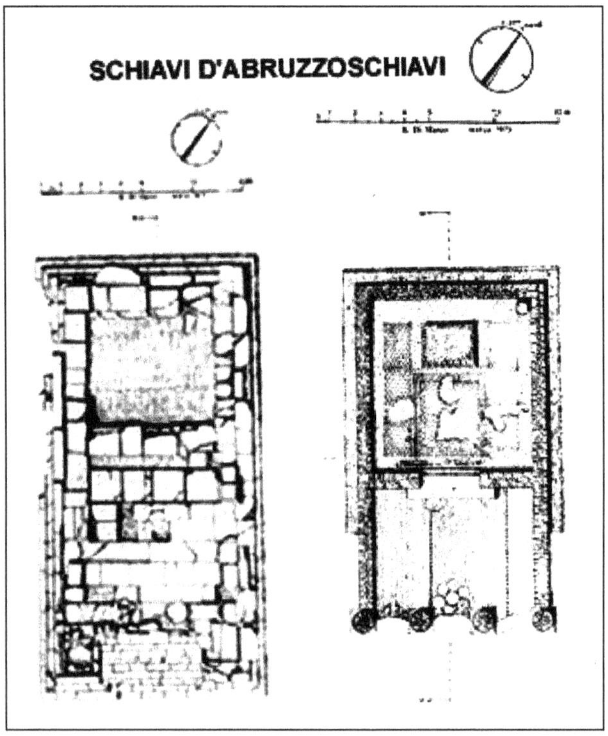

Figure 12.3. Plan of the sanctuary at Schiavi.

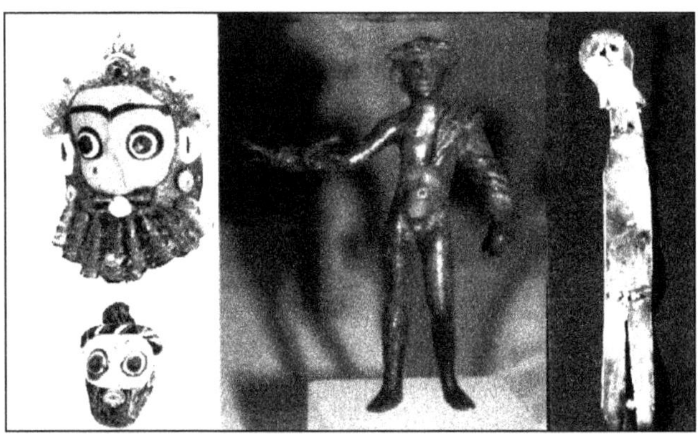

Figure 12.4. Finds from the necropolis and sanctuary at Monte Giove

Two cases in northern Abruzzo: Monte Giove and Colle S. Giorgio (Vanessa D'Orazio)

The two rural sanctuaries in Northern Abruzzo examined here are Monte Giove and Colle S. Giorgio (*Figure 12.1*, no.12 and 20). Both are within the territory of the Latin colony of Hatria, which was founded by the Romans in 289 BC.

The *tratturi* helped both cultural and commercial exchanges in the area, so much so that we find imported objects of a high artistic standard, including some that had travelled great distances. For instance, the Greek pottery and the Phoenician pendants found at Penna S. Andrea and Campovalano are witness to infrequent Greek and Punic imports, which could be either directly exchanged via the Adriatic Sea, or mediated by Etruscan trading networks. In particular the *tratturo Frisa-Rocca di Roseto* has many branches here and the main archaeological sites of the area are situated along them.

The first site to be examined has the evocative name of Monte Giove[5] (*Figure 12.1*, no.12 and *Figure 12.4*): remains of a sanctuary were found about thirty years ago on top of a hill. The necropolis at the base of the hill dates to between the 7th – 3rd Centuries BC.

The plan of the sanctuary is not yet clear. The temple, which dates to the Hellenistic period, contained many rooms, although the function of these rooms remains uncertain. The sanctuary is situated on a river terrace along a natural track, which in Roman times was monumentalised and became a proper road known as the *via Caecilia.*

In the necropolis nearby, nine female burials are attested, their funerary equipment consisting of ornamental objects such as fibulae, rings, but no spindles. Tomb no.8 appears to be richer than the others, and amongst the finds were two Phoenician pendants made from faience (*Figure 12.4*, left). So far, a nearby settlement has not been found, so it has been suggested that the necropolis close to the sanctuary probably belonged to a group of people connected with the cult centre.

However, earlier objects were also found in the sanctuary, which were closely related to cult activities. For example, a rare silver leaf figurine dating to the fifth Century BC (*Figure 12.4*, right) represents a man in schematic form. The figurine has parallels in Umbria, but mainly in bronze rather than silver. Another interesting find is a bronze statuette of Veiove throwing a thunderbolt (*Figure 12.4*, centre), which certainly dates to the Hellenistic period.

The sanctuary at Monte Giove was probably one of the main cult centres of the Sabellic people, those who called themselves the *Safin* in the first part of the 5th Century BC[6]. This is well attested in the famous inscribed stele from the area of Penna S. Andrea, a site in close proximity to Monte Giove. It is located within a boundary

4 Antepagmenta and antefixes are well known architectonic decorations of Greek, Italic and Roman temple roofs.

5 That is Hill of Iovis, mentioning directly in the name a possible religious function.

6 See note 2 of Fossataro's text.

area between the territories of Hatria and Interamnia along a Roman road, suggesting its use as one of the boundary sanctuaries of the religious belts mentioned above.

Turning finally to the second sanctuary of northern Abruzzo, that is Colle S. Giorgio (*Figure 12.1*, no.20, and *Figure 12.5*). It was built at the top of a hill, very close to a natural spring, along a *tratturo* and a Roman road, which is also attested in the famous map known as *Tabula Peutingeriana.*

Figure 12.5. Architectural and decorative fragments from Colle S. Giorgio

Little survives of the temple, because a Medieval Church was built over it. However, highly elaborate architectural decorations in terracotta were found in a votive deposit nearby, and can be attributed to the temple. Among the terracotta finds were many fragments of the pediment suggesting that representations of the main gods of the Roman Pantheon once adorned it. The frieze fragments (*Figure 12.5*) are particularly rich in decorative form. Representations include the *Potnia Theron* or the *Despotes Theron, Heros* riding panthers (Azzena G.,1987; Iaculli G.,1981; Messineo G.- Pellegrino A.,1986), and the so called Flower-woman. In addition, more simple slabs were decorated with vegetal motifs. Most of the representations are paralleled with examples found in Abruzzo, such as at Chieti, Vacri, Alba Fucens, Schiavi d'Abruzzo, Atri and Pagliaroli (Azzena 1987: Campanelli, Faustoferri 1997).

Acknowledgements (Vanessa D'Orazio)

I would like to thank here Prof.E. Fabbricotti and Dr.O. Menozzi for their advice and corrections, and my colleagues D. Fossataro and D. Lagatta, for their collaboration and help. Thanks also to the photographer G. Lattanzi for allowing me to use his professional photographs, and to C. Agostini for his help in the preparation of a CD for the conference. Moreover, I would like to thank also a great number of persons who encouraged and advised me in this research, such as Dr. S. Lapenna, Dr. V. D'Ercole, Prof. G. Iaculli, Dr. V. Tordone, Dr. L. Cherstich, I. Cherstich and M. Carinci.

Bibliography

Azzera G. 1987. *Atri.* Roma.

Barbetta S. 2000. "La Via Caecilia da Roma ad Amiternum", in E. Catani & G.Paci (eds), *La Salaria in età antica.* Macerata: 47-64.

Bispham E., Smith C. 2000. *Religion in Archaic and Republican Rome and Italy.* Edinburgh.

Bradley G. 2001. *Ancient Umbria: states, culture, and identity in central Italy from the Iron Age to the Augustan era.* Oxford.

Campanelli A. & Faustoferri A. 1997. *I Luoghi degli dei. Sacro e natura nell'Abruzzo Italico.* Pescara.

Coarelli F. & La Regina A. 1984. *Abruzzo e Molise.* Bari.

D'Ercole V. 1986. "Penna Sant'Andrea. Necropoli e santuario", in *La valle del medio e basso Vomano, II, 1* Report "Cassa di Risparmio della Provincia di Teramo", De Luca Editore Srl, Roma 1983: 131-135.

D'Ercole V. & Grassi B. 2000. "Necropoli protostoriche abruzzesi a sud della Salaria" in E.Catani & G.Paci. (eds.) *La Salaria in età antica.* Macerata: 193-266.

D'Ercole V., Papi R. & Grossi G. 1990. *Antica terra d'Abruzzo.* L'Aquila.

Dench E. 1995. *From Barbarian to New Men. Greek, Roman and Modern Perceptions of peoples from the Central Apennines.* Oxford.

Fabbricotti E. (ed.) 1990. *Iuvanum. Atti del I Convegno di Studi. Chieti Maggio 1983.* Chieti.

Fabbricotti E. (ed.) 1996. *Iuvanum. Atti del II Convegno di Studi. Chieti Marzo-Aprile 1992.* Chieti.

Franchi Dell'Orto L. & Messineo G. 1993. "Viabilità antica e il toponimo Valle Siciliana. La via Cecilia", in *La valle Siciliana o del Mavone, I, 1* Report "Cassa di Risparmio della Provincia di Teramo", De Luca Editore Srl, Roma 1983: 113-122.

Guidobaldi M.P. 1995. *La romanizzazione dell'Ager Praetutianus.* Napoli.

Guidobaldi M.P. 2000. "La via Caecilia: cronologia e percorso d'una via publica Romana", in E.Catani & G.Paci (eds.) *La Salaria in età antica.* Macerata: 277-292.

Iaculli G. 1984. "Architettura templare preromana: la decorazione", in *Storia come presenza,* Report "Rotary Club – Pescara" and "Cassa di Risparmio di Pescara e Loreto Aprutino", Pescara 1984: 19-22.

Iaculli G. 1975. "Terracotte architettoniche da Colle S. Giorgio", *Archeologia Classica* vol. XXVII, Roma 1977, "L'Erma" di Bretschnider: 253-266

Iaculli G. 1983. "Note sulle terrecotte architettoniche d'Abruzzo" *Quaderni dell'Istituto di Archeologia e Storia Antica 3.* Viella (Roma) 1982/1983: 57-84.

Iaculli G. 1993. *Il tempio italico di Colle S.Giorgio.* Penne.

Iaculli G. 1994. "Chieti-Civitella. La decorazione a stecca". *Rivista di Antichità.* Anno III, 1 Napoli 1994, Loffredo Editore: 157-174.

Lapenna S. & Ruggieri M. 2001. *Terra di Confine tra Marrucini e Carnicini.* Chieti.

La Regina A. 1986. "Penna Sant'Andrea. Le stele paleosabelliche". In *La valle del medio e basso Vomano, II, 1*. Report "Cassa di Risparmio della Provincia di Teramo", Roma 1986, De Luca Editore Srl: 125-130.

Staffa A.R. 1998. *Loreto Aprutino*. Pescara 1998.

How monkeys evolved in Egyptian and Minoan art and culture

Cybelle Greenlaw – Trinity College Dublin

Numerous faience and limestone figurines attest to the existence of baboon cults even in early dynastic times, and excavations at the Naqada III period royal cemetery at Hierakonpolis have revealed the remains of a number of animals, including baboons (Adams 2000: 171-2). During the Old Kingdom, the Egyptians' fascination with baboons and other monkeys grew and evolved into new forms. By the First Intermediate Period, monkeys also began to appear in Minoan glyptic art, and by Late Minoan IA, wall paintings from Crete and Thera also began to depict monkeys engaged in a variety of activities. The first part of this paper, then, examines a few of the monkey's changing roles in Egyptian art and culture up to the early New Kingdom, when monkeys reached their zenith in Minoan iconography. The second part consists of a brief exploration of the ways in which the image was adopted and changed by this Aegean culture.

Before undertaking an examination of the monkey's roles, however, it may be helpful to consider the three species most commonly depicted in Egyptian art: the hamadryas baboon (*Papio hamadryas*), the anubis baboon (*Papio Anubis*), and the vervet (*Cercopithecus aethiops)*. The first is one of the smaller species of baboon, striking for its pink face and red hindquarters. Males have a massive cape, and their fur becomes a silvery grey as they mature. The anubis, on the other hand, is a large Savannah baboon. Both males and females have olive-brown fur and dark faces and hindquarters. The vervet is a small species of guenon, known for its long, flexible tail. Males are a light grey, while females are greenish-brown. Both have dark faces and hands. It is worth noting that none of these monkeys currently exists in Egypt, and it is unlikely that they ever did. A fourth species that does inhabit the Upper Nile Valley, and may also be found in Egyptian art, is the patas monkey (*Erythrocebus patas*). This is a hardy, long-tailed simian with red fur and a white stomach.[1]

The early figurines mentioned above are believed to represent the baboon deity, Hedj-Wer, or the Great White One (Friedman 1998: 16). Early in the Old Kingdom, this god merged with the lunar god, Thoth, whose cult centre was Hermopolis (Lurker 1980: 121). The species depicted in the figurines is difficult to determine. While Thoth's baboon was unquestionably the hamadryas, skulls found at Hierakonpolis have been identified as anubis baboons (Linseele 2003). By the end of the Old Kingdom, Thoth was most often depicted in his anthropomorphic, ibis-headed form, but baboons were used as visual metaphors to represent qualities associated with him, such as wisdom, knowledge, judgment, writing and excellence (Wilkinson 1999: 157).

Other species did not represent deities, but all were commonly depicted in tomb chapel scenes and kept as pets. Some important themes visible in these scenes are monkeys with their keepers, monkeys under their master's seat, and monkeys engaged in a variety of human activities (Vandier d'Abbadie 1964: 147-77). The first category involves servants holding monkeys by a leash or simply accompanying the animals. The guardians may be ordinary men or dwarves. Although scenes such as the painted relief from the 5th Dynasty tomb of Ti portray the monkey as twice the size of its dwarf handler, it is clear from the morphology of the animal that one of the smaller species is depicted. In fact, dwarves are not portrayed with baboons at all (Dasen 1993: 116). The vervet's usual spot, though, seems to have been under the master's chair. Numerous scenes show the creatures sporting collars, bracelets and anklets and munching on fruit placed before them in baskets, while their master sits above. The last category comprises monkeys engaged in diverse human activities for the sake of entertainment. For the most part, these are scenes of music and dance.

After the collapse of the Old Kingdom, monkeys did not figure in tomb chapel scenes again until the 12th Dynasty and Second Intermediate Period, where scenes are most prevalent in the tombs of nomarchs at Beni-Hasan and public officials at el-Kab. This decline was probably the result of troubles at the end of the Old Kingdom and social changes brought about by the 12th Dynasty kings. The rise of a middle class may have caused monkeys to lose some of their appeal as symbols of luxury used to ornament wealthy households (Vandier d'Abbadie 1965: 12). In funerary practices, however, the myth of Osiris was growing in importance, and Thoth's role in it may have paved the way for the baboon's appearance in many different media during the New Kingdom, which will now be considered.

In his lunar aspect, Thoth is often represented in the form of a blue faience baboon, wearing the lunar disc and carrying the wedjat eye. While the animals strongly maintained their connection with the moon, they were also closely connected to the sun. This association can be found even in the Pyramid Texts, but the image of the baboons raising their arms to greet the sun became more prevalent in the New Kingdom. Sun hymns, such as this, describe their functions:

> *"The baboons that announce Re when this great god is to be born again about the sixth hour in the netherworld. They appear for him after they have come into existence. They are at both sides of this god until he rises in the eastern horizon of*

[1] For an excellent study of the hamadryas baboon, see Kummer 1995. Osborn and Osbornova (1998) provide a good general survey of other species depicted in Egyptian art.

the sky. They dance for him, they jump gaily for him, they sing form him, they sing praises for him, they shout for him. When this great god appears before the eyes of [all humankind] then these hear the speech of jubilation of the Wetenet-country [=baboons?]. They are those who announce Re on heaven and earth" (Houlihan 1996: 96).

Te Velde remarks that the jumping and singing of the baboons may denote ecstatic behaviour, as does portrayal of the baboons as ithyphallic (1988: 130-1). Baboons also gained importance in funerary iconography during the New Kingdom. The human head of Hapi, one of the four sons of Horus represented on canopic jars, was replaced with that of a hamadryas from the 19th Dynasty. Vignettes of spell 125 from the Book of the Dead depict Thoth in his anthropomorphic, ibis-headed form as an advocate of the dead, recording the results as the heart of the deceased is weighed in the balance, but often his presence is reinforced in the form of a baboon sitting on top of the balance.

After the weighing of the heart, Thoth functioned as an interpreter for the deceased. Te Velde cites this tradition in evidence:

> *"I am Thoth and I speak to you the language of Re as herald. They spoke to you before my words were understood...I give god's offerings to the gods and invocation offerings to the blessed dead. I am Thoth who ascribes truth to the Ennead and all that comes from my mouth comes into being like (all that comes from the mouth) of Re..."* (Velde 1988: 136).

He was, therefore, an intermediary between the human and divine realms, who aided Egyptians in their most important transitional phase, the rite of passage from this world to the next.

Monkeys also make a great come-back in tomb chapel scenes in the New Kingdom, especially in scenes of tribute bearers bringing baboons and vervets from Nubia and Punt. Additionally, many women's objects, such as hairpins, bowls and kohl pots involve monkey figures (Vandier d'Abbadie 1966: 143-201).

By the first quarter of the 18th Dynasty, the period roughly corresponding to Late Minoan IA on Crete, the monkey had become more prevalent in the art of Aegean civilisations as well. It should be noted, though, that there is no direct physical evidence that monkeys were ever imported to Crete or Thera, so iconographical evidence is our only means of understanding the Minoan attitude toward these exotic creatures. Perhaps the earliest known Minoan representation of a monkey is an ivory knob seal from Trapeza, dating from Early Minoan II-III. It depicts a long-tailed monkey, sitting in a human pose. Later, monkeys appear on a small number of sealstones and six wall paintings from Crete and Thera, most of which date to the Late Minoan period.

Most scholars accept that the image was adopted from Egypt and that the monkeys in wall paintings are a type of *cercopithecus*, but the identification is not so straightforward. Some of the monkeys do have these characteristics, and Egypt was certainly the most powerful influence on the Minoan adoption of simian imagery. However, in many cases, artists produced hybrids, and, in at least one painting, an Asian species, the rhesus macaque (*Macaca mulatta*), may be depicted. Its westernmost habitat corresponds to modern-day Afghanistan (MacDonald 1984: 89), and the species could have been introduced to the Aegean via Mesopotamia.

Before analyzing the wall paintings individually, it is worth remarking that Aegean monkeys are always blue, and all seem to have white cheeks, stomachs and brow ridges. These conventions represent a departure from their more realistic Egyptian counterparts, but it seems possible that the use of blue may have been adapted from the Egyptian lunar figurines. Additionally, Cretan monkeys have realistic bodies, while those of Theran monkeys are highly anthropomorphic. The gender is never identifiable.

The earliest painting, dating from Middle Minoan IIIA, is the Saffron Gatherer from the Early Keep at Knossos. Evans originally restored it as a blue boy, probably because of the arm bands and belt. As previously mentioned, though, pet monkeys were often portrayed wearing these items in Egyptian tomb chapel scenes. The species is difficult to identify because of the fragmentary nature of the painting, but it does have the slender limbs of a small monkey, such as a vervet. Because the simian appears to be reaching for a crocus, it can be compared with a scene from Xeste 3, room 3a from Thera.

This painting, like all those remaining to be discussed, dates to Late Minoan IA and depicts a highly anthropomorphic monkey with an elongated pink muzzle, perhaps a hamadryas, handing saffron to a seated female figure. The fact that a griffin stands behind her suggests that she is a goddess. In the nearby rocky landscape, girls of various ages are picking crocuses. The scene is set in a room that was probably designed for religious functions, and, as Marinatos notes, each girl is depicted in a different stage of the crocus gathering process (Marinatos 1987: 123). One the north wall, a girl is seated on a rock, holding her bleeding right foot, which may symbolize blood lost during menstruation, childbirth, or more likely, the loss of virginity. This notion is reinforced by the fact that saffron has been used to ease pain associated with these events (ibid.: 132). The girl's hairpin has also fallen off, releasing her back lock, which Davis suggests may have been a symbol of virginity (Marinatos 1986: 402). The scene appears to be a rite of passage, and Marinatos is probably correct in asserting that the monkey is making the offering of saffron on behalf of the girls (Marinatos 1987: 124). A similar function, as we have seen, was also associated with Thoth in his role of advocate of the deceased. In each case, humans are shown in a liminal state, and both cultures, therefore, seem to have required an intermediary figure to intercede for them at uncertain periods of their lives.

In the House of Frescoes at Knossos, monkeys are shown in a naturalistic, but not realistic, landscape, hunting for and eating eggs. These monkeys are usually identified as a type of *cercopithecus*, but I would suggest that their morphology and tail length are more reminiscent of rhesus macaques. In any case, the presence of papyri in the landscape does reflect an Egyptian influence. This scene finds a counterpart in the Beta 6 painting from Thera, in which long-tailed monkeys jump and swing through a rocky landscape. Despite their anthropomorphic bodies, their activities are entirely simian, and no human is present in either painting. Although landscape scenes exist in Egyptian painting, monkeys are not present in them. This combination seems to be an entirely Aegean creation.

The last two Theran paintings are very fragmentary, but certain aspects are worthy of comment. In the fragments from Xeste 3, room 4, a monkey is playing a harp or lyre and at least one other is engaged in a sword fight or dance. Marinatos suggests that because this room is connected to Xeste 3, room 3a by pier and door partition, the monkeys may be holding a contest in honour of the goddess (Marinatos 1984: 113). It seems more likely, however, that the scene is borrowed from the Egyptian theme of monkeys as entertainers.

The fragment from the North Magazines depicts a monkey with a vervet-like face, raising its arms in the manner of the hamadryas in its role of Sun Greeter, as it faces an architectural structure, probably an altar. In this variation of the Egyptian theme, the monkey retains its function as greeter, but loses its association with the sun. Again, these monkeys are genderless, so clearly the Minoans had no interest in adopting the ithyphallic aspect of the animals.

Finally, one can say that the iconography of monkeys and their significance to these cultures should not be considered static and unchanging. Just as new themes arose in Egyptian art over time and in accordance with social changes, new variations emerged when the image was transferred to the Aegean. Although the monkey's behavior has no doubt remained a constant from the Predynastic Period to the present, changing human reactions have created a visible artistic and cultural evolution.

Bibliography

Adams B. 2000. *Excavations in the Locality 6 Cemetery at Hierakonpolis 1979-1985.*

British Archaeological Reports 903: Oxford.

Dasen V. 1993. *Dwarfs in Ancient Egypt and Greece*: Oxford.

Davis, E. 1986. Youth and Age in the Theran Frescoes. *American Journal of*

Archaeology, 90, 399-406.

Friedman F. (Ed.) 1998. *The Gifts of the Nile: Ancient Egyptian Faience*: London.

Houlihan P. 1996. *The Animal World of the Pharaohs*: Cairo.

Kummer H. 1995. *In Quest of the Sacred Baboon*: Princeton.

Linseele V. 2003. Weird Animals from the Elite Cemetery. *Archaeology's Interactive Dig* (accessed on6/10/03)<http://www.archaeology.org/interactive/hierakonpolis/animals.html>.

Lurker M. 1980. *The Gods and Symbols of Ancient Egypt*: London.

MacDonald D. (ed.) 1984. *All the World's Primates*: New York.

Marinatos, N. 1984. *Art and Religion in Thera*: Athens.

Marinatos N. 1987. An Offering of Saffron to the Minoan Goddess of Nature. *Boreas* 15: 123-132.

Osborn D. and J. Osbornova 1998. *The Natural History of Egypt*: Warminster.

te Velde G. 1988. Some Remarks on the Mysterious Language of Baboons. In J.H. Kamstra et al. (eds.) *Funerary Symbols and Religion*: Kampen.

Vandier d'Abbadie J. 1964. Les Singes Familiers dans l'Anicenne Egypte I:

l'Ancienne Empire. *Revue d'Egyptologie* 16: 147-177.

Vandier d'Abbadie J. 1965. Les Singes Familers II: le Moyen Empire 17: 177-88.

Vandier d'Abbadie J. 1966. Les Singes Familiers III: le Nouvel Empire 18: 143-201.

Wilkinson R. 1999. *Symbol and Magic in Egyptian Art*: London.

The central place of religion in Chalcolithic society of the southern Levant

Eva Kaptijn – Leiden University, Netherlands

Relativism

In the beginning of the last century social scientists realised that the people of the cultures they were studying did not perceive the world in the way they did themselves. Each culture orders the world around it in a unique fashion; it makes its own categorisation. Their society is based on this order, but their way of ordering the world is also based on society. This is possible because both society and categorisation change continuously and mutually influence each other. Categorisation, ideology, religion, etc. will therefore differ between cultures. The relativism that emerged at the beginning of the 19th century fully realised and emphasised this problem. In the course of several decades different types of relativism, ranging from moderate to absolute relativism, have developed. While moderate relativists hold that cultural variability will produce different social and psychic understandings among people, absolute relativism states that everything is culturally determined and that cultural variation is limitless.

Although these seem to be statements on a continuous scale differing in intensity, there is an important difference. Moderate relativists consider it possible to come increasingly closer to the cognition of another culture. They argue that it is possible to discriminate between hypotheses about other cultures and to select the best one. Absolute relativism regards this as impossible and argues that each statement is as good as the next one. According to this view social sciences will always remain completely subjective. Such a view would mean the end of social research. Most social scientists therefore hold a moderate view; they recognise the problems and restrictions highlighted by relativism but look for ways to get some understanding of other societies.

One of the ways to achieve this is to accept the assumption of the psychic unity of human beings. This assumption means that next to physical similarity, the workings of the human mind are also identical among people of different race and culture. It is also assumed that the human mind has not changed in the last 10.000 years. The human mind poses limits to the abilities of human beings. The possibilities of human behaviour are therefore not endless. Society consists of people. All people are subject to the same limitations of their mind. Henceforth society is also limited in its possibilities. The properties of an aggregate are indeed not the same as those of its individual constituents, but still variation has its boundaries. The variation of society is therefore limited by human cognition.

Cognitive archaeology

In archaeology cognitive research has been undertaken in two ways. The first approach searches for cognitive theories in other disciplines to use them in archaeology. This is a processualist approach and studies cognition through an empiricist methodology. Therefore this type of cognitive archaeology must be refuted. The second approach is a reaction from post-processual archaeologists. They argued that processual archaeology had dehumanised history through the exclusion of values, ideas and beliefs from archaeological research. By explicitly focusing on these cognitive aspects they try to make archaeological reconstructions of society more holistic. The study of cognitive aspects of society should be executed in the same rigorous way as the investigation of other topics and has to be equally based on evidence. It is attempted here to study Chalcolithic society in this manner.

Ideology and religion are inextricably linked and mutually influence each other. Ideology expresses the basic values held by a society. These values are generated by and reflected in the economic and political organisation of that society. The social movement, institution, class or group that is in power in that society will use ideology to avoid change and maintain its social position. Ideology is expressed through rituals, behaviour, traditions and objects. Ideological meaning is therefore embedded in the material objects archaeology is concerned with. Through study of the material culture and the socio-economic and socio-political structures visible within it some insight is gained into the ideology of a society. There is however no direct relationship between ideology and the socio-economic and socio-political structure of a society. Ideological values are not necessarily put into practice. It is also possible that one artefact expresses several meanings depending on the situation and the observer. This is inherent to the heterogeneous nature of society.

Just like ideology religion is reflected in the material remains of a society. Religious beliefs are expressed in rituals. In the performance of rituals material objects such artefacts or monuments can play a role. Certain artefacts are specifically made for use in rituals. These artefacts may express some of the meaning attached to them. Archaeology is able to touch upon religion through these artefacts. The first thing the archaeologist has to do is determine which materials can be regarded as having functioned in religion. Unfortunately the only way to do this is still to look for combinations of elements that cannot be explained in a utilitarian way. Three religious rooms were identified in this way; i.e. in Gilat, Tuleilat Ghassul and the compound of Ein Gedi. These rooms stand out from domestic contexts by a combination of architecture, inventory, layout, setting and activities. Most of the artefacts discovered in these rooms were identified as having religious meaning. Additionally graves were also considered as religious because the same religiously interpreted artefacts were found and because a definition was used in which religion attempts to ascertain survival. Death is therefore usually incorporated into religion, as it is an important part of each society.

Similarities within this assemblage of religious artefacts can be detected. If several artefacts share certain characteristics it may be concluded that these reflect some demarcated meaning. These characteristics can therefore be defined as religious symbols. For Chalcolithic religion four symbols could be identified. In this article the identification and meaning of these symbols will not be elaborated on.

Through the adoption of a definition of religion that focuses on religion as a means of enforcing survival and well-being it was attempted to get some insight into the meaning of symbols and beliefs held in religion. It was assumed that through religion, people attempted to influence the aspects of life that they were dependent on for their survival and well-being. If Chalcolithic society is sufficiently understood it is possible to identify some of these essential aspects and perhaps recognize them in archaeological remains which have been identified as religious.

Religion in Chalcolithic society

In the remaining part of this article the central position religion took within Chalcolithic ideology will be elaborated. The definitions and methodology described in the above will be used.

Egalitarian Society

Chalcolithic society was principally egalitarian. This is visible in burials, houses and villages. Burials, together with houses, are one of the few means archaeology has that can insights into the individuals that constitute society.

All architectural units in the Chalcolithic are simple broad rooms, e.g. rectangular houses with the door in one of the long sides. Except for the three above-mentioned rooms none have revealed an exceptionally rich inventory or elaborate building techniques. The three rooms are markedly different in this respect. Furthermore villages have a local orientation and most do not seem to have had any regional significance. Virtually all artefacts were locally produced. Again the three rooms form an exception.

In the Chalcolithic of the southern Levant different burial methods were used. Most burials are secondary internments in ceramic ossuaries or stone containers, both usually containing one individual. The ossuaries take the form of jars or boxes. Both can be elaborately decorated with paint or ceramic applications. The majority of these decorations can be grouped in symbols. These ossuaries and containers occur in groups in caves and stone circles. These differences are however related to regions and not to individuals. Burials in caves are restricted to the coastal plain and western hills of Samaria. The stone circles are only found in the area surrounding Shiqmim. Differences in type of burial can therefore not be considered as an indicator of social differentiation within a community.

Physical anthropological research has pointed out that both men and women, adults and children were buried. No distinctions were made according to sex or age. It can therefore be concluded that no horizontal differentiation is visible in Chalcolithic society. Neither are there indications for vertical differentiation. Although burials are usually accompanied by grave gifts none of the graves has a marked abundance or wealth of gifts. All graves are connected to only one or two items. So neither graves, nor houses, nor villages show indications of social stratification. The type of artefacts that were buried together with the dead are similar to those identified as religious through their occurrence in the religiously-interpreted rooms.

Religious status objects

These artefacts are important to this discussion because they show that Chalcolithic society did contain the parameters to make it a differentiated society. Each of these items have one or more characteristics that make it a possible object of status, e.g. an elaborate specialised production technique was used, it was made of a rare material, or it had travelled a long distance to reach the place where it was found. Moreover all religiously interpreted artefacts have these characteristics and none have been found in domestic contexts. It seems that status objects were restricted to the religious domain.

One of the artefacts that are usually found in graves and in the religiously-interpreted rooms is a small open ceramic vessel: the V-shaped bowl. Pottery production analysis has shown that these bowls were made using a slow-turning pottery wheel (Roux & Courty 1997). This is an innovation of the Chalcolithic period and was primarily used for this pottery type. The use of this technique would not have meant a reduction in manufacturing time. It was an elaborate technique that could only have been performed by specialized craftsmen. The term specialized denotes here that the technique was not open to everybody. Additionally, petrographic analysis of the clay of the V-shaped bowls of Abu Hamid has shown that 90% of the clays derived from the Beersheba area, although suitable clays were locally available (Roux & Courty 1997: 35). The remaining 10% of the bowls were produced locally, so the technique was known. It seems that bowls were imported because some special significance was attached to them. These bowls have travelled over a considerable distance and their production was not open to everyone. Neither characteristic was found in any other pottery type.

Copper artefacts are found in both graves and in the religious rooms. Copper appears for the first time in this period. A distinction is usually made between utilitarian artefacts, like adzes and chisels, made of pure copper in an open mould, and ceremonial artefacts with strange forms and made from a copper-arsenic alloy using the 'lost wax'-technique. Recent finds have demonstrated that the distinction between these categories is not as rigid as was once argued (Kerner 2001: 146-148). Moreover the utilitarian artefacts cannot have been used as such because of the softness of pure copper. None of the artefacts has revealed traces of use. Furthermore, the so-called easy method of open mould casting would still have been complicated and only open to a group of specialists. Therefore the entire group of copper artefacts can be regarded as status objects connected to religion. They have namely never been found in domestic contexts. Specialists made them. They have travelled great distances because places of production of pure copper are only found in the south and the artefacts appear over the entire country. Also they were made from a rare material; copper is only found in the Wadi 'Araba and copper-arsenic alloy only as far away as Transcaucasia.

Flint discs are also regarded as religious artefacts as they only occur in religiously interpreted contexts. These so-called tabular scrapers are flat, intensively worked, round or toothed discs. A specific tabulated flint that only occurs in the southern Negev was used for their production. Only one site has been found where these discs were made. They do however occur over the entire country. The technique needed for their production is complicated and not everyone will have been able to produce them. Although they are called scrapers they cannot be used as such. Flint artefacts found in domestic contexts were all locally produced from local flint.

Other artefacts that occur regularly in graves and religiously interpreted contexts are the so-called pillar figure and the basalt vessel. The first is a basalt pillar ending in a bowl and has a nose protruding from one of the sides. Many have eyes, horns, ears and a beard. The basalt vessel has the same form as the ceramic V-shaped bowl. The type of basalt used only occurs in the Golan and in northeastern Jordan. Basalt artefacts are nevertheless also found in the coastal plain and Beersheba Valley, irrespective of the difficulty of transporting these heavy objects. Basalt is hard and difficult to work. Recently it was discovered that a small number of the basalt vessels were not made of basalt but of phosphorite (Gilead & Goren 1989: 10). Phosphorite cannot be visually distinguished from basalt, but is less hard, lighter, considerably easier to work and at many places locally available. Nonetheless the majority of these artefacts is made from basalt. There was a more important non-economic reason that resulted in the preference of basalt over phosphorite. Basalt artefacts used in domestic contexts were made of small basalt boulders locally available in wadi beds.

Conclusions

It can be concluded that the objects identified as religious based on their occurrence in the three religiously interpreted rooms and in graves all bear the characteristics of status objects; they were made from rare materials with elaborate specialist techniques and have travelled over considerable distances. No other artefacts of Chalcolithic society have these characteristics. Chalcolithic society was egalitarian. Villages had a local orientation and artefacts were made from local materials. Nevertheless contact existed as is shown by the religious artefacts. The only architectural units that had a regional significance were the three religious rooms. Provenance studies of the artefacts found within show that objects from several regions were brought to these rooms (Alon & Levy 1989). So although no indications of social differentiation have been found the prerequisites were present. From the material now available it must be concluded that the only domain of Chalcolithic society that had a regional significance and can be connected with status objects is religion.

Bibliography

Alon D. & T.E. Levy 1989. "The Archaeology of Cult and the Chalcolithic Sanctuary at Gilat". *Journal of Mediterranean Archaeology* 2(2): 163-221.

Gilead I. & Y. Goren 1989. "Petrographic analyses of 4th millenium BC pottery and stone vessels from the Northern Negev, Israel". *Bulletin of the American School of Oriental Research* 275: 5-14.

Kerner S. 2001. *Das Chalkolithikum in der südlichen Levante. Die Entwicklung handwerklicher Spezialisierung und ihre Beziehung zu gesellschaftlicher Komplexität.* Orient-Archäologie Band 8, Berlin (diss. 1998).

Roux V. & M.-A. Courty 1997. "Les bols élaborés au tour d'Abu Hamid: rupture technique au 4e millénaire avant J.-C. dans le Levant-Sud". *Paleorient* 23(1): 25-43.

Archaeology's Well Kept Secret: The Managed Antiquities Market

Morag Kersel – University of Cambridge

In October of 2002, it was announced that a nondescript limestone box, looted from a cave in Jerusalem and held in the private collection of long-time collector Oded Golan, might have the earliest known archaeological references to Jesus (Gugliotta 2002; Legon 2002; Wilford 2002). As news reports of the Aramaic inscription[1] were released, the details of the ossuary's archaeological context were sketchy at best, causing many scholars to question the authenticity of the find. At the initial press conference, Hershel Shanks, editor of *Biblical Archaeological Review,* announced that the owner purchased the ossuary 15 years ago from an antiquities dealer in Jerusalem's Old City for $200-$700. In later news reports this was revised so that the date of purchase was sometime in the 1970s.

The date of purchase is fundamental to who actually owns the ossuary. If in fact the ossuary was acquired in 1987, it belongs to the State of Israel rather than Golan. In Israel it is legal to buy and sell artefacts from collections established before the 1978 Antiquities law; after that date all antiquities are property of the state. *"This could be something genuinely important, but we can never know for certain"* said Dr. P. Kyle McCarter, Jr., professor of Biblical and Near Eastern studies at Johns Hopkins University. *"Not knowing the context of where the ossuary was found compromises anything we might say and so doubts are going to persist"* (Wilford 2002). This recent discovery clearly demonstrates the destructive influence that looting of archaeological sites and subsequent trade in antiquities (whether legal or illegal) can have on our collective knowledge of the past.

The movement of artefacts from looter to smuggler, dealer, collector, and finally, to a public institution is the classic story of how international legal inconsistencies and the global market act in the process of laundering artefacts for eventual sale in the legitimate market. Principally there are two types of trade in antiquities: legal and illegal. Managed antiquities markets, which officially sanction the trade in artefacts, exist in only a few nations (e.g. Israel, Britain). In those countries, antiquities dealers function respectably, as in the art market, by purportedly recycling old collections and buying up legally available material for sale in the market place. The market is fuelled by consumer demand and the demand for antiquities is on the rise (Lowenthal 1998). Under existing conditions, this can only lead to theft from collections and, even more significantly, to an increase in the already widespread destruction of ancient sites and irreplaceable monuments. The reason is simple: there is a finite supply of antiquities legitimately available to trade. Moreover, as objects move into permanent non-circulating collections, the sanctioned supply diminishes.

A traditional market for illegally excavated antiquities is illustrated in *Figure 15.1*. From clandestine excavation to middlemen to dealers, the smuggling proceeds from the illegal exportation of the item and continues with shipment through one or more transit countries, where the artefact is "laundered". The now legal pieces are then transported to their final destination and potential purchasers. Today's antiquities trade functions within the context of legal and illegal markets, high levels of profitability, a limited supply and an ever-increasing demand (Adler & Polk 2002).

Figure 15.2 depicts the market model for Israel. Considered a managed antiquities market, which officially sanctions the trade in artefacts, the market in Israel does not conform to the traditional model for antiquities sales (*Figure 15.1*). There is no need for a transit market to launder the goods, because ostensibly they can leave the country legally. Goods pass directly from the source market to the destination market. This simplifies the entire process and eliminates an expensive step.

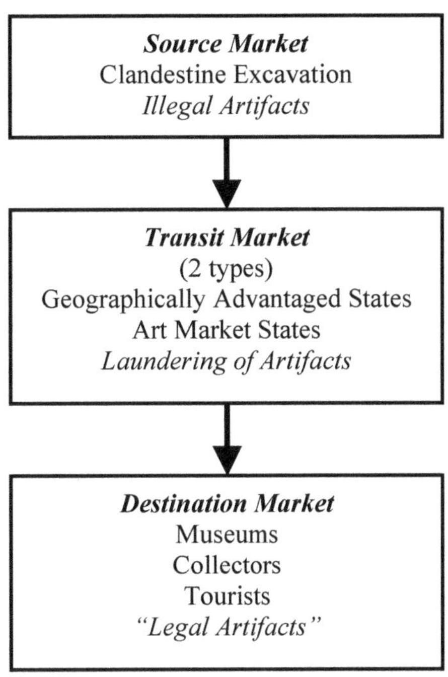

Figure 15.1: Flow of markets in the illicit trade in antiquities

There is support for legal international trade in cultural property, based on the belief that excessive restrictions on export encourage the growth of a black market, which is damaging to both artefacts and information about the human past. Some contend that by not legalizing the sale

[1] It has since come to light that the experts' panel convened by the Israel Antiquities Authority declared the ossuary inscription a fake (see Silberman & Goren 2003).

of antiquities, the market is driven underground and subsequently encourages looting of archaeological sites (Bisheh 2001; Borodkin 1995; Eisenberg 1994; Pearlstein 1996; Marks 1995; Merryman 1995; Shanks 2001). In a letter to the editor of the *Wall Street Journal*, private dealer Peter Marks (1995) stated, *"The establishment of an open and legal market is the only way to begin a new era of partnership and cooperation between countries and the international art market and preserve what is still in the ground".*

It has been a longstanding tradition to take home a piece of the Holy Land. For many there is something awesome and inspiring in holding a piece of history in one's hand, something akin to making physical contact with one's forefathers (Ilan *et al* 1989). Acquisition of artefacts of archaeological significance by tourists involves a complex set of dynamics and a well-developed system of markets where artefacts are commodified. Commodification is the process through which artefacts are transformed through market activities into goods with monetary value. It is this commodification of the past that has and continues to contribute to the indiscriminate destruction of the world's archaeological heritage.

Belk (1988) suggests that at any given moment in the contemporary western world around a quarter to a third of all adults are willing to identify themselves as collectors. The number is probably closer to half of all adults if we include all those who have had or will have a collecting experience (Pearce 1998: 1). As Clifford (1986: 238) puts it, *"in the West, collecting has long been a strategy for the deployment of a possessive self, culture and authenticity".*

The definition of collecting can be as complex as, *"the selective, active and longitudinal acquisition, possession and disposition of an interrelated set of differentiated objects (materials, things, ideas, beings or experiences) that contribute to and derive extraordinary meaning from the entity (the collection)"* (Belk *et al* 1990: 8). Or the definition may be as simple as, *"a collection exists if the owner thinks it does"* (Pearce 1998: 3).

Collecting is not solely about owning a piece of the past and collecting archaeological artefacts is not a new phenomenon. In its infancy, serious collecting was an integral part of scholarship. Passion for collecting motivated early excavations, and by the late 18th Century great national collections were being formed throughout Europe, which now constitute the holdings of many major museums. In that age, collecting was believed to be both a mode of science and a way to increase knowledge (Chase *et al* 1997: 31). In his study of archaeology as a discipline, Stiebing (1994) traces the topic of archaeology from a former pastime of dilettantes into a rigorous science. Today, archaeologists collect data, and more importantly, collect contextual data rather than objects, regarding the former as the intrinsically valuable object.

According to Renfrew (2000: 15), *"a significant cause of destruction of the archaeological record is looting: the illicit, unrecorded and unpublished excavation of artefacts from ancient sites".*

Elia (1993: 65) asserts that, *"collectors cause looting by creating a market demand for antiquities"* thereby inferring that collectors are the real looters, a sentiment echoed by Renfrew (1993). Once an artefact is removed from the ground, it can only be appreciated for its aesthetic value. Although it is not difficult to arrive at the conclusion that there is a causal relationship between collecting and the destruction of the world's archaeological heritage, realistically, collecting will never be completely eradicated. There will always be people willing to pay for objects that they desire to add to their collections.

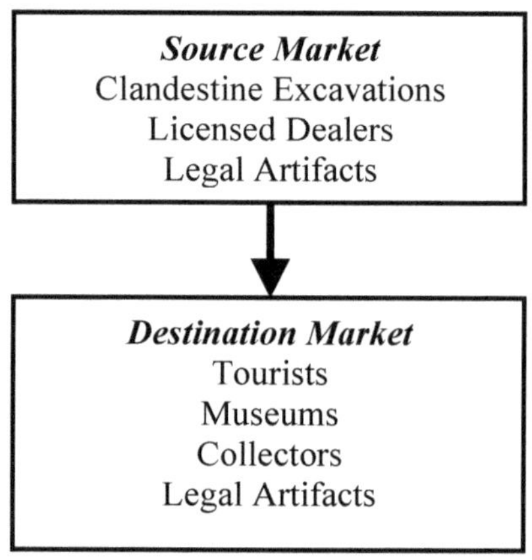

Figure 15.2: Market model for Israel

Antiquities markets can be broken down into three basic components: production, distribution, and consumption (Coe 1993: 273-277; Elia 1997: 86-88; Lyons 1998). The "production" portion of the system involves excavators and small dealers on the local level. In his paper on looting in Ghor es-Safi, Jordan, archaeologist Konstantinos Politis (2002) explains that the local unemployed population has learned the skill of tomb robbing and can distinguish between Byzantine and Bronze Age period materials (Politis 2002: 259). A great deal of time and effort is spent in recognising which artefacts will reap the greatest rewards. Sums paid for objects at this level are very low, but nevertheless meaningful to the subsistence diggers[2] involved in the illicit excavation (Matsuda 1998; McManamon and Morton 2000: 255). The larger dealers, galleries, international auction houses, and import firms make distribution possible. The most desired consumers in the antiquities market are the high-end collector and museums that lack specific policies

[2] Subsistence diggers are those individuals conducting illegal excavations in order to recover items that they then sell in the antiquities market using the proceeds to support them and their families (see Matsuda 1998).

prohibiting the purchase of unprovenienced[3] artefacts (Meyer 1973: 33-43; Renfrew 1993, 1995; Rose 1997; Tubb 1995; Watson 1997).

Illicit excavation has long been a problem in Israel and the territories of the Palestinian Authority, areas rich in archaeological resources (Twain 1869). Since the onset of the *al-Aqsa Intifada* (October 2000) and the subsequent closures of the occupied territories by Israel that prevents Palestinians from reaching jobs in Israel, looting has surged dramatically (Ephron 2001). The region's unrest has also ended the cooperation between Palestinian and Israeli antiquities policing efforts (Keyser 2002). In 2001 antiquities authorities on both sides of the border reported a rise in incidents of tomb robbing by 300 percent (Ephron 2001). Palestinian officials say that as much as half of the labour force is out of work and have turned to pillage in their own backyards as a way of surviving poverty. However, this does not lead to huge financial pay-offs for the looters. Diggers sell to middlemen, who take the goods to resell to the antiquities dealers in Jerusalem at a healthy mark up (Brodie 1998). Typically it is the middlemen who retain the lion's share of the profits, while the finders of the artefacts often receive less than one percent of the retail value of their discoveries (Borodkin 1995: 377). By and large the criminal penalties do not act as deterrents against pillage. In Israel, most clandestine excavators receive suspended sentences (Keyser 2002).

Currently in Israel there is a system of legally sanctioned dealers in antiquities, which constitute the distribution aspect of this process. It is legal to buy and sell artefacts in Israel that come from pre-1978 collections (Antiquities Law 5738). Two national laws established in 1978 (Antiquities Law 5738) and 1989 (Antiquities Authority Law 5749) were created to manage antiquities in Israel, oversee all excavations, establish a national body to administer antiquities and regulate the trade in antiquities. Dealers must apply for a license from the Israel Antiquities Authority (IAA) to deal in antiquities and export antiquities under this law. An official statement of authorization from the IAA should accompany all antiquities leaving the country. All antiquities uncovered in excavations are, by law, the property of the State of Israel; and all excavations require a permit. Digging without a permit is illegal. Nevertheless, the overwhelming majority of artefacts in the antiquity dealers' inventory are from illegal and unrecorded excavations (Ilan *et al* 1989: 41; Silberman 1989).

By law, antiquities dealers must register all transactions and all inventories (Antiquities Authority Law 5738-1978 Chapter 4 Section 17). In theory this should give the IAA an estimate of the volume of antiquities on the market. In reality many dealers dodge issues of questionable provenience by transferring registry numbers from item to item (Ilan *et al* 1989: 41). When an antiquities dealer in Israel is asked where his stock comes from, the answer is inevitably that they date to before 1978, when it was legal to excavate on private property and collect artefacts from the surface (Kersel 2000).

The IAA is also mandated to safeguard sites and prevent theft, through either the courts or the IAA robbery prevention unit (Antiquities Authority Law 5738-1978 Chapter 4). Yet the IAA robbery prevention unit is essentially powerless to stop the destruction. Not only is its responsibility vast—50,000 archaeological sites—the laws are not supportive of protection, because only the excavators and middlemen are acting illegally. It is perfectly legitimate to buy and sell the artefacts in a legally sanctioned shop. In 2000, estimates put the antiquities trade of some 80 licensed dealers at close to $5 million per year (Sontag 1999: A4).

In an article on looting in Israel, archaeologists David Ilan, Uzi Dahari and Gideon Avni (1989) asserted that the current antiquities laws in Israel were not working to protect against pillage. Quoting a survey conducted by the robbery prevention unit, they noted that the majority of purchases from antiquities dealers were made by individual tourists buying a single pottery lamp, jug, coin or glass vessel (Ilan *et al* 1989: 41). The authors suggested that outlawing the trade in antiquities could remove the main incentive to plunder—the market for relatively commonplace objects bought primarily by tourists (Ilan *et al* 1989: 41).

It is an unusual period in the antiquities business in Israel and the Palestinian Authority. On the one hand, more and more goods are surfacing in the shops in Jerusalem, which have been unearthed since the current wave of hostilities began. On the other hand, the violence has scared away tourists, a key clientele for the dealers in Jerusalem. With no tourists and a weak market, many dealers have sent away the middlemen without purchasing the artefacts.

In an article on managed markets, lawyer William Pearlstein declared, *"it should theoretically be possible to reconcile the competing demands of source nations and market participants through their voluntary participation in a managed, self-policing antiquities market"* (Pearlstein 1996: 142). He suggested that source nations would have to allow the controlled export of all antiquities except those which have a significant link to the culture, or history of the exporting nation (Pearlstein 1996: 142)[4]. This proposition was echoed by Shanks (2001: 135) when he suggested that the "low end" items like coins, lamps and pots from recent excavations and in

[3] The *provenance* (a term most often used by art historians) of an object includes the original location and context of an object as well as the history or ownership, a key part in the evaluation of an object's value. *Provenience* (most often used by anthropologists and archaeologists) is a term used to mean the original location and context of an object.

[4] The Israel Antiquities Authority is currently considering an initiative to trade sherds on an open market (Barkat 2003).

storehouses could be placed in the market. According to Shanks (2001: 135) this flood of artefacts may put the looters out of business.

According to O'Keefe (1997: 67), an influx of objects on the markets is just as likely to stimulate additional requests as it is to satisfy the current demand. Gerstenblith (2001) states that arguments in favour of encouraging free trade in ordinary commodities centre on the basic premise that the world is becoming an integrated market and that by allowing goods to circulate freely there will be more economic productivity, more jobs will be created, and the economies of different nations will be stabilised and enhanced (Gerstenblith 2001: 226). However, none of these reasons apply to the free trade in cultural goods because archaeological materials are not manufactured and cannot be produced to meet market demand without excavation (legal or illegal).

In Israel and the territories of the Palestinian Authority, where tourists are encouraged at every opportunity to feel connected with their ancestors by taking home a piece of the Holy Land, buying a simple oil lamp or piece of Roman glass is an affordable way to accomplish this goal. If the demand exists for a given commodity, and it is coupled with substantial profit motivation for the supplier, trade will occur. For trade to occur in the antiquities market, looting of archaeological sites is a key component in supplying goods for sale. If there is to be a solution to the looting problem, changing public attitudes towards the collecting of antiquities is the initial step.

Bibliography

Primary Sources
Israel
Antiquities Law 5738-1978
Antiquities Authority Law 5749-1989

Secondary Sources

Adler C. & Polk K. 2002. "Stopping this awful business: the illicit traffic in antiquities examined as a criminal market" *Art, Antiquity and the Law* 7 (1): 1-19.

Barkat A. 2003. "Antiquities authority considers trading finds on open market".
http://www.haaretz.com/hasen/pages/ShArt.jhtml?itemNo=346879&sw=antiquities (accessed on 05/10/2003).

Belk R. 1988. "Possessions and the extended self" *Journal of Consumer Research* 15: 139-168.

Belk R., Wallendorf M., Sherry J. & Holbrook M. 1990. "Collecting in a consumer culture" *Highways and Buyways*. Utah.

Bisheh G. 2001. "One damn illicit excavation after another: the destruction of the archaeological heritage of Jordan", in N. Brodie, J. Doole & C. Renfrew (eds.) *Trade in illicit antiquities: The destruction of the world's archaeological heritage*. Cambridge: 115-118.

Borodkin L. 1995. "The economics of antiquities looting and a proposed legal alternative" *Columbia Law Review* 95: 377-417.

Brodie N. 1998. "Pity the poor middlemen" *Culture without context: The Newsletter of the Illicit Antiquities Research Centre* Issue 3, Autumn.

Brodie N. & Doole J. 2001. "Illicit antiquities", in N. Brodie, J. Doole & C. Renfrew (eds.) *Trade in illicit antiquities: The destruction of the world's archaeological heritage*. Cambridge: 1-6.

Chase A., Chase D. & Topsey H. 1997. "Archaeology and Ethics of Collecting", in K. Vitelli (ed.) *Archaeological ethics*. Walnut Creek: 30-38.

Clifford J. 1986. "Objects and selves – an afterword", in G. Stocking (ed.) *Objects and others: Essays on museums and material culture. History of Anthropology*, vol. III. Madison WI: 236-246.

Coe M. 1993. "From *huaquero* to connoisseur: the early market in pre-Columbian art", in E. Boone (ed.) *Collecting the Pre-Columbian past: A symposium at Dumbarton Oaks, October 1990*. Washington DC: 271-290.

Elia R. 1993. "A seductive troubling work" *Archaeology* 46 (1): 64-69.

Elia R. 1997. "Looting, collecting and the destruction of archaeological resources" *Non-renewable Resources* 6 (2): 85-98.

Ephron D. 2001. "The tomb raiders".
http://www.msnbc.com/news/585402.asp?cp1=1 (accessed on 15/06/2001).

Eisenberg J. 1994. "Editorial" *Minerva* 5 (2): 2

Gerstenblith P. 2001. "The public interest in the restitution of cultural objects" *Connecticut Journal of International Law* 16 (2): 197-246.

Gugliotta G. 2002. "Stone box may be oldest link to Jesus" Washington Post.
http://www.washingtonpost...p-dyn/A61782-2002Oct21.html (accessed on 22/10/2002).

Ilan D., Dahari U. & Avni G. 1989. "Plundered! The rampant rape of Israel's archaeological sites" *Biblical Archaeological Review* (March/April): 38-41.

Kersel M. 2000. *We Sell History: Issues in the Illicit Trade of Antiquities and Cultural Repatriation.* Master's thesis, University of Georgia.

Keyser J. 2002. "Holy land's ancient sites hit by looting: archaeologists say region's troubles lead to desperate thefts".
http://www.msnbc.com/news/782447.asp?0si&cp1=1 (accessed on 19/07/2002).

Legon J. 2002. "Scholars: oldest evidence of Jesus?".
http://www.cnn.com/2002/TECH/science/10/21/jesus.box/index.html (accessed on 22/10/2002).

Lowenthal D. 1998. *The heritage crusade and the spoils of history*. Cambridge.

Lyons C. 1998. "Book notes: Provenienza! Tombaroli, mercanti e collezionisti: L'Italia archaeological allo sbaraglio" *American Journal of Archaeology* 102: 215.

Marks P. 1995. "Antiquities markets should be open, honest" letter to the editor, *Wall Street Journal.*

January 9 at A15.

Matsuda D. 1998. "The ethics of archaeology, subsistence digging, and artefact looting in Latin America: point, muted counterpoint" *International Journal of Cultural Property* 7 (1): 87-97.

McManamon F. & Morton S. 2000. "Reducing the illegal trafficking in antiquities", in F. McManamon & A. Hatton (eds.) *Cultural resource management in contemporary society: Perspectives on managing and presenting the past.* New York: 247-275.

Merryman J.H. 1995. "A licit international trade in cultural objects" *International Journal of Cultural Property* 4 (1): 13-60.

Meyer K. 1973. *The plundered past: The story of illegal international traffic in works of art.* New York.

O'Keefe P. 1997. *Trade in antiquities, reducing destruction and theft.* London.

Pearce S. 1998. *Collecting in contemporary practice.* Walnut Creek CA.

Pearlstein W. 1996. "Claims for the repatriation of cultural property: prospects for a managed antiquities market" *Law and Policy in International Business* 28 (1): 123-150.

Politis K. 2002. "Dealing with dealers and tomb robbers: the realities of the archaeology of the Ghor es-Safi, Jordan", in N. Brodie & K. Walker Tubb (eds.) *Illicit antiquities: The theft of culture and the extinction of archaeology.* New York: 256-267.

Renfrew C. 1993. "Viewpoint: collectors are the real looters" *Archaeology* 46 (3): 16-17.

Renfrew C. 1995. "Introduction", in K. Tubb (ed.) *Antiquities: Trade or betrayed: Legal, ethical and conservation issues.* London: xvii-xxi.

Renfrew C. 2000. *Loot, legitimacy and ownership.* London.

Rose M. 1997. "Rotten apples: a revealing look at Sotheby's and the antiquities market" *Archaeology* 50 (3): 64-70.

Shanks H. 2001. "How to stop looting", in N. Silberman & E. Frerichs (eds.) *Archaeology and Society in the 21st Century: The Dead Sea scrolls and other case studies.* Jerusalem: 132-137.

Silberman N. 1989. *Between past and present: Archaeology, ideology, and nationalism in the modern Middle East.* New York.

Silberman N. & Goren Y. 2003. "Faking biblical history" *Archaeology* September/October 56 (2): 20-29.

Sontag D. 1999. "Stealing millennial loot, from 2 millenniums ago" *NY Times.* November 12 at A4.

Stiebing W. 1994. *Uncovering the past: A history of archaeology.* Oxford.

Tubb K. (ed.) 1995. *Antiquities: Trade or betrayed: Legal, ethical and conservation issues.* London.

Twain M. 1869. *Innocents abroad or the new pilgrims progress: Being some account of the Steamship Quaker City's Pleasure excursion to Europe and the Holy Land: with descriptions of countries, nations, incidents and adventures, as they appeared to the author.* San Francisco.

Watson P. 1997. *Sotheby's: Inside story.* London.

Wilford J. 2002. "'Jesus' Inscription on stone may be earliest ever found".

http://www.nytimes.com/2002/10/22/science/22JESU.html (accessed on 22/10/2002).

New images of the Erechtheion by European travellers

Alexandra Lesk – University of Cincinatti

The building known as the Erechtheion was built in the second half of the fifth century BC on the Athenian Akropolis and once housed the oldest and most important cults and relics in Athens, namely those of Athena Polias, Erechtheus, Poseidon and Hephaistos, to name but a few. There are essentially four façades to this complex and multi-level temple, and it is these façades that are illustrated by the many travellers to Greece. The East Façade consists of a hexastyle porch with distinctive Ionic columns with the lotus and palmette necking band. The South Façade is primarily a plain wall with the famous Maiden Porch at its west end. The West Façade includes a side view of the Maiden Porch, four engaged columns perched upon a wall pierced by a small door, and the side view of the huge tetrastyle North Porch with its elaborate Ionic columns towering over the ancient city centre. The primary access to the Erechtheion was through the impressive North Door. This temple once had Parian marble figures attached to the grey Eleusinian limestone frieze blocks by clamps. They have all since fallen off the building, many of them having been carted off as manageable souvenirs by the early travelers to Athens.

Having obtained a general acquaintance with this most beautiful of buildings surviving from antiquity, we now turn to the evidence for the state of the building in the 18th and 19th centuries when western travellers began to visit Greece regularly. These travellers often kept detailed journals and, since there were no cameras, spent days – even months and years – carefully sketching the monuments they encountered. Many of these journals were published with engraved versions of their drawings, as there was a real appetite for such literature, especially in Britain. This paper summarises what happened to the Erechtheion from the mid 18th century onwards based on these travellers' descriptions and illustrations, and highlights the new evidence I have found and how it refines our understanding of the changes to the temple.

Paton *et al* conducted the first comprehensive study of the travellers' descriptions and illustrations during the early 20th century by collecting all the known images and descriptions of the Erechtheion up to the First World War (1927: 536-567). In the course of my research for the block-by-block analysis of what happened to the Erechtheion over time and modelled in CAD (Lesk 2004a), I consulted these images in person and searched for previously unexamined images and accounts in the libraries, museums and archives of Western Europe. This article presents and analyses a selection of these discoveries for the first time. Overall, the travellers' descriptions offer a different kind of information. They remark upon the general state of the building - for example, how many maidens were standing at the time in the South Porch - but more interesting is their aesthetic reaction to the building, their opinions on Lord Elgin, and their relationships with other travellers visiting Athens at the same time.

Some time after paganism was finally snuffed out in Athens with the closure of the philosophical schools in 529 AD, the Parthenon and the Erechtheion were converted into Christian basilica churches. At the end of the Fourth Crusade in 1204, the Franks took over Athens and installed themselves on the Akropolis. The Frankish noblemen used the Propylaea as a palace, the Parthenon was converted into a Catholic Cathedral and the Erechtheion was turned into a grand residence, probably for the Latin bishop. In 1453, the Turks brought an end to the Byzantine Empire by sacking Constantinople. Three years later, Athens surrendered. It is unclear how the Erechtheion was used immediately after the takeover of the Akropolis, but in the later seventeenth century, travellers were told it served as a harem (Spon 1678: 159-160; Wheler 1682: 364-365).

The first scholarly travellers to visit Athens and document her monuments were James Spon and George Wheler in 1676. They identified the Erechtheion based on the second century A.D. traveller Pausanias' description of the temple being "double" (I.26.6). After scrambling around the ancient hovels on the south side of the temple, they beheld the maidens of the South Porch "enclavées dans un mur," (Spon 1678: 160) or "embedded into a wall." (Wheler 1682: 365). In 1749, future librarian to George III, Richard Dalton, was the first draughtsman to illustrate the actual state of the temple relatively accurately, minus its known accretions (Dalton 1752). These famous drawings show that one of the maidens from the South Porch was already missing, probably having been destroyed during the Venetian siege of 1687, which more famously blew up the Parthenon. The west view also shows that the spaces between the columns in the North Porch were filled in. These walls hid a gunpowder magazine within.[1]

Not long after Dalton, the intrepid architect-travellers, James Stuart (1762-1789) and Nicholas Revett, raced against the Frenchman Julien-David Le Roy (1758) to publish elevations and architectural details of the Erechtheion. Their books represent the first modern

[1] There is a long history of storing gunpowder in the masterpieces of the Akropolis with tragic results. A gunpowder magazine blew up the Propylaea in 1640 killing the family of the Disdar, the Venetians bombed the stockpiles in the Parthenon in 1687. And in 1827, the Turks fired on the North Porch of the Erechtheion.

transmission of authentic Greek architecture to the West and became the basis of the Greek Revival movement. William Pars offers one of the earliest glimpses of some of the post-Classical accretions to the building in his watercolour of 1765, an engraving of which was included in Stuart and Revett's when the temple was transformed into a residence during the Frankish Period. Antiquities of Athens.[2] The most important feature is the crumbling ruin of the "North Addition" attached to the North Wall, which was built when the temple was transformed into a residence during the Frankish period.

The earliest of the previously unexamined depictions is by the Frenchman, Louis-François Cassas from around 1785 (*Figure 16.1*). Cassas was the artist in the entourage of the Comte de Choiseul-Gouffier, the French Ambassador to the Ottoman Porte. This painting is not overly helpful because Monsieur Cassas has taken the liberty to restore parts of the building that logic told him must be there. On the other hand, it does illustrate the remnants of the walls that formerly enclosed the Maiden Porch mentioned by Spon and Wheler in 1676, otherwise omitted in the earlier depictions such as Dalton's mentioned above. The excavations taking place in the foreground disclose the raison-d'être of his patron's interest in the Akropolis, namely to "Enlevez tout ce que vous pourrez...tout ce qu'il y a de pillable" (Legrand 1897: 57).[3]

Figure 16.1. L.-F. Cassas, The Erechtheion from the Southwest. 1786. By courtesy of the Benaki Museum.

William Gell, a British artist and traveller, gives us the best picture of the Erechtheion in 1800-1801, just prior to Lord Elgin's activities on the Akropolis. In his most famous view of the temple from the north, Gell uniquely includes traces of Gothic arches in the walls between the columns of the North Porch. These represent the remnants of windows belonging to the Frankish house (Gell 1800a). Two of Gell's drawings of the Erechtheion have never been published. The first shows a new perspective of the Erechtheion: the interior of the East Porch (Gell 1800b, *Figure 16.2*). All six columns are visible and the north anta is intact for the last time. The second shows the interior of the Maiden Porch (Gell 1800c, *Figure 16.3*). The Maiden Porch had been excavated in 1789 by the French consul Fauvel and re-excavated most recently by Elgin's agent, Lusieri. Previously travellers remarked on how the podium was filled with rubbish. Gell's documentary style in his artwork is paralleled in prose in an unpublished diary manuscript at the University of Bristol (Gell 1801). Gell's description of the Erechtheion is very detailed and mentions for the first time the loft in the North Portico where he saw the marble coffers of its ceiling. It was in this loft that hundreds of travellers would carve their marks later in the nineteenth century. Gell is also one of the few travellers who was less than impressed by the maidens, characterising them as not being "of the very first rate workmanship" but grants them as having "a singular effect from without" (Gell 1801: 67).

Figure 16.2. W. Gell, Inside the East Portico. 1800. Photo by author.

Figure 16.3. W. Gell, Interior of the Porch of the Maidens. 1800. Photo by author.

2 Pars. 1765. The Erechtheum from the Northeast. Reproduced as an engraving in Stuart and Revett 1789.

3 "Remove everything you can...everything there is that is pillageable."

In 1803, Elgin's agents removed Maiden #3 and replaced it with a pillar of masonry seen in many of the later images and dubbed "*Opus Elgin.*" Elgin also took the northernmost column of the East Porch and various other architectural fragments. A painting by Haygarth, unknown to the authors of *The Erechtheum*, is the first depiction of the East Porch since Elgin took the column (*Figure 16.4*). Its removal destabilised the rest of the North Wall, which soon crumbled to the ground.[4]

In 1805, following the trend set by Elgin, the Turkish garrison commander removed an architrave block from the West Façade, inscribed it in Arabic and placed in the new fortifications of the Akropolis. A new, though unfinished, drawing by George Basevi (1818b) shows the West Façade lacking its southernmost architrave block (*Figure 16.5*).[5] In total, there are four unpublished depictions of the Erechtheion by Basevi, a student of the great architect Sir John Soane, the most important showing the Erechtheion from the north (Basevi 1818a, *Figure 16.6*). The ground level to the east of the North Porch has risen significantly, and the remnants of the North Addition are still standing fifty-five years after Pars first drew them. This perspective offers new information about the cross-wall of the North Addition, namely an otherwise unknown pointed-arched niche in its west side, flanked by two other niches.

Figure 16.4. W. Haygarth, Erechtheum from the North. 1810-1811. By courtesy of the Gennadius Library.

An unpublished collection of Basevi's letters resides in the Library of the Soane Museum in London. These wonderful documents illuminate the competitive atmosphere amongst the young men on the Grand Tour for the best spots from which to draw the monuments on the Akropolis. Basevi wrote the following to his mother in 1818:

"I have been very successful in my choice of points of sight ... and they have been copying me, this among artists is not considered a correct thing to do, but I have never noticed it in any other way than by keeping my sketches when finished more to myself, and never asking to see theirs. They have understood me, and have set about point hunting everywhere, which affords Wilson and self great amusement.

Eastlake has carried this copying plan to a great extent, six or eight things he has made tale quale. He ought to be ashamed of it, as he is a regular established history painter, and the best of it is instead of being obliged to me for these hints for pictures they are Jealous of me. What a pack of fools all artists are with these mean ideas, they may copy all mine, if they would do it openly. I am not a painter nor wish for a reputation as such. I can never interfere with them nor they with me. So much for this nonsense...."[6]

Figure 16.5. G. Basevi, West Façade of the Erechtheion. 1818. By courtesy of the Trustees of Sir John Soane's Museum.

Not long after this rivalry, the past-time of painting the Erechtheion ceased completely for a period of about nine years owing to the Greek War of Independence. In 1821-1822, the Greeks besieged the Akropolis. Only one shot is supposed to have hit the Erechtheion, on the south window of the West Façade (Waddington 1825: 57). The Turks appear to have dismantled more of the walls in search of lead for shot.[7] The Greeks allegedly offered to give the lead to the Turks in order to prevent this dismantling.

4 Hobhouse (1813) reports that some of this destruction, particularly to the South Wall, can also be attributed to the search for lead during the short war between the Ottoman Empire and Great Britain in 1807-1809. The dome in the background is the mosque inside the Parthenon.

5 Hobhouse. 1809. *The Akropolis viewed from the Propylaea,* and not Pomardi, 1804-1805. *West End of the Pandrosion,* is the earliest to show the missing architrave block, contrary to Paton *et al* 1927: 553, note 6. The other depictions of the West Façade in Dodwell show the architrave as still in place.

6 Excerpted from G. Basevi to his Mother, 10 August 1818. Published by courtesy of the Trustees of Sir John Soane's Museum. Eastlake's painting is from the exact same view as Basevi's unfinished southwest perspective. C. L. Eastlake, 1818. *The Erechtheum* can be found in F.-M. Tsigakou 1981, pl. XIX.

7 A letter from Gropius to Blaquière dating to 25 April 1824 refers to this dismantlement, E. Blaquière 1825: 157.

Figure 16.6 G. Basevi, Erechtheion from the North. 1818. By courtesy of the Trustees of Sir John Soane's Museum.

The Greeks held the Akropolis until 1826 when the Ottoman army returned to besiege Athens. During this siege, Maiden #4 was hit and she toppled out of place. The west pilaster capital of the Maiden Porch was also completely shattered in this attack. As a result, much of the Porch's roof collapsed. The first detailed illustration of the Maiden Porch after this episode is another new depiction, this time by Danish artist and architect, Christian Hansen in 1835 (*Figure 16.7*). New masonry pillars have been added as a "quick fix" to support the remainder of the roof and a headless Maiden #4 has been propped up against "*Opus Elgin*."

Figure 16.7. C. Hansen, The Maiden Porch from the Southeast. 1835. Photo by author.

The temple suffered further damage during the conflict: The north and south columns of the West Façade fell down and the majority of the South Wall was dismantled, almost certainly raided for its metal clamps - this time by the Greeks (Paton *et al* 1927: 557). Most tragically, the west half of the North Porch collapsed when the northwest column capital was fired upon by the Turks early in 1827, killing eleven family members of the Greek commander, Gouras, who had been hiding in the loft above the vault of the powder magazine. Luckily, architecturally at least, the magazine itself did not explode.[8] The earliest depiction of the collapsed North Porch (and the destroyed North and South Walls) is a new drawing by Francis Arundale, made in 1834, which I found during a visit to the Department of Greek and Roman Antiquities in the British Museum (*Figure 16.8*). It shows how the three western columns, western entablature and northern entablature have been toppled to the ground. The three western monolithic ceiling beams and the four corresponding rows of coffers have collapsed, some of which came to rest on the roof of the vault.

Figure 16.8. F. Arundale, View of the Temple of Erectheus at Athens. 1834. Photo by author.

After the War of Independence, a visit to the remainder of the loft where Gouras' family had perished became a regular part of a visit to the Erechtheion. Travellers signed their names on the inside of the remaining portions of the North Porch, but very few travellers brought their art supplies (Lesk 2004b). In 1811, Charles Cockerell painted the loft and depicted it as a dark, airless garret, accessible only by a ladder in the northwest corner. In Copenhagen I found a much livelier and detailed depiction of the loft by Hansen made in 1835-6 (*Figure 16.9*). A man in a fez, most likely one of Hansen's Danish friends dressed up as a native, leans on a fallen coffer block resting on the roof of the vault and gazes at the view between the columns, now freed of the masonry between them.

The Akropolis was officially handed over by the Turks to Christopher Neezer, a Bavarian officer, in April 1833. Upon surveying his new domain, he said: "I entered the Acropolis and saw heaps of tumbled marbles. In the midst of the chaotic mass of column capitals, fragments of columns, marbles large and small, were bullets, cannon balls, human skulls and bones, many of which were near the slender Caryatids of the Erechtheion." (Papageorgiou-Venetas 1994: 225). There still is a cannonball among the long grass just west of the Maiden Porch.

The Greeks soon had dominion over their ruins and teams of eminent scholars, archaeologists and

[8] The Greeks had been planning to move the magazine to a new location for several years. See Paton *et al* 1927: 558-559, for a thorough consideration of whether the magazine had indeed been moved prior to the destruction of the North Porch.

architects from Greece, Denmark, Switzerland, Italy and Prussia set about the clearing of the post-Antique accretions on the Akropolis. A previously unknown oil painting by Martinus Rørbye from 1836 shows the Maiden Porch just prior to the first official exploration of the Erechtheion (*Figure 16.10*).

Figure 16.9. C. Hansen, Inside the North Portico of the Erechtheion. 1835. Photo by author.

Between 1837 and 1840, the new man in charge of the Akropolis, Kyriakos Pittakis, began putting the Erechtheion back together again. Pittakis received a great deal of criticism for being crude and unmethodological (Papageorgiou-Venetas 1994: 230). A new depiction by Johann Frey (1843) from the Museum of the City of Athens shows the random nature of the re-erection of the South Wall (*Figure 16.11*). Another new depiction, this time from 1847 by the English architect and founder of the British School at Athens, Francis Penrose, shows the re-erected western columns of the newly cleared North Porch carried out by the nascent Greek Archaeological Society (*Figure 16.12*). They could proceed no further with its restoration due to lack of funds. Pittakis re-erected Maiden #4. Her capital was missing, so the roof above her head was not replaced at this time. The southern column of the West Façade was also re-erected.

The Maiden Porch soon received further attention from the French architect Alexis Paccard who repaired the porch between 1846 and 1847 (Guillaume 1870). Two new depictions, by Cook and Martineau, show the Maiden Porch immediately after its restoration in 1850 (*Figures 16.13* and *16.14*). Maidens #1 through #5 are in place, but the roof is still missing. "Opus Elgin" has been replaced by a terracotta cast. In other renditions of the South Porch after this time, artists actually represent the difference in colour of the Elgin replacement (Werner 1877). Maiden #4 received a replacement capital, and the architraves, podium and crepidoma were repaired with new blocks and replaced. Italian sculptor Andreoli restored the missing parts of Maiden #6.

A series of previously unexamined Danish paintings shows the state of the Erechtheion just before and after the great storm of 1852 which toppled the recently restored West Façade. The first is by Laurits Winstrup (1851), painted just prior to the storm (Figure 16.15). The second is by Harald Stilling (1853), and is the earliest to show the effects of the storm (Figure 16.16). All the columns of the West Façade have fallen into the temple except for the lonely stub of the northernmost one.

Figure 16.10 M. Rørbye, Parti af Erechteion. 1836. Photo by author.

The 1850s were a time of the first scientific investigations into the building. An elevation by Winstrup (1851) is an early example of this kind of study pioneered by the great French architect Jacques Tétaz (Figure 16.17). The interior of the temple was fully excavated during the 1850s and 1860s and the crepidoma cleared all the way around as you can see in the last of the new depictions ofthe Erechtheion to be presented, one of several by the English architect, Spiers (*Figure 16.18*).

Figure 16.11. J. J. Frey, View of the Erechtheion and the Parthenon. 1843.
By courtesy of the Museum of the City of Athens.

Figure 16.12. F. C. Penrose, Portico of Minerva Polias. 1847. By courtesy of the British School at Athens.

Figures 16.13. H. Cook (1819-1980), The Erechtheion. 1846. By courtesy of the Museum of the City of Athens.

By the middle of the 19th century, photography was developing quickly and the Akropolis served as one of the earliest subjects for this nascent art form. As a result, the need for travellers to capture their visions of monuments on paper by prolonged gaze slowly began to diminish. The paintings and scientific elevations continued to be important, however, because the quality of the early photographs was low and did not show the details of the architecture. Balanos led the next large-scale reconstruction effort on the Erechtheion at the turn of the twentieth century. The North Wall, West Façade and entablature of the East Porch were repaired at this time with questionable accuracy. Seventy years passed before an attempt was made by the Committee for the Preservation of the Akropolis Monuments to rectify these problems and protect the monument from further damage. The Erechtheion continues to change today: The Akropolis Restoration Service is currently rebuilding the sanctuary of Pandrosus adjacent to the temple's West Façade.

Figure 16.14. E. H. Martineau, Athens, Acropolis, The Erectheum. 1850. By Courtesy of RIBA Library Drawings Collection.

Figure 16.15. L. A. Winstrup, The Erechtheion from the West. 1851. Photo by author.

Figure 16.16. H. C. Stilling, Erechtheion from the West. 1853. Photo by author.

Figure 16.17. L. A. Winstrup, The Interior of the North Wall of the Erechtheion. 1851. Photo by author.

Figure 16.18. R. P. Spiers, The Erechtheion from the Northeast. 1866. Photo by author.

Acknowledgements

I am grateful to the Kress Foundation for funding my travels in search of new images of the Erechtheion in institutions across Western Europe, and the Semple Classics Fund and University Research Council of the University of Cincinnati for supporting my research in Greece and the UK. I would also like to thank the Soane Museum, the British Museum, the Museum of the City of Athens, the British School at Athens, and the Royal Danish Academy of Fine Arts, for their generous help and permission to use the images in this article.

Bibliography

Basevi, G. 1818a. *Erechtheion from the North.* Pencil and Watercolour. London: Sir John Soane Museum, No. 44-3/15.

Basevi, G. 1818b. *West Façade of the Erechtheion.* Pencil on paper. London: Sir John Soane Museum, No. 44-3.

Blaquière, E. 1825. *Narrative of a Second Visit to Greece: including facts connected with the last days of Lord Byron ; extracts from correspondence, official documents,* &c. Vol. I. London.

Cassas, L.-F. 1785. *The Erechtheion from the Southwest.* Coloured engraving. Athens: Benaki Museum, No. 23981.

Cook, H. 1846. *The Erechtheion from the Southwest.* Lithograph. Athens: City Museum of Athens, No. 88.

Dalton, R. 1752. *A Series of Engravings, representing views of places, buildings, antiquities, etc., in Sicily, Greece.* Asia Minor, and Egypt. London.

Eastlake, C. L. 1818. *The Erechtheum.* Oil on canvas. Guildford: Sir Ellis Waterhouse Collection.

Frey, J. J. 1843. *View of the Erechtheion and the Parthenon.* Watercolour. Athens: Museum of the City of Athens, No. 532.

Gell, W. 1800a. *The Erechtheion from the North.* Watercolour. London: Department of Greek and Roman Antiquities of the British Museum, Gell Volume XIII: L. B. 55.

Gell, W. 1800b. *Inside the East Portico.* Watercolour. London: Department of Greek and Roman Antiquities of the British Museum, Gell Volume XIII : 121 (92).

Gell, W. 1800c. *Interior of the Porch of the Maidens.* Pen and ink. London: Department of Greek and Roman Antiquities of the British Museum, 1800, Gell Volume XIII : 106 (75).

Gell, W. 1801. "Diary of a Tour in Greece." Manuscript at the University of Bristol.

Guillaume, E. 1870. *Discours pour l'inauguration du tombeau d'Alexis Paccard.* Paris.

Hansen, C. 1835. *Inside the North Portico of the Erechtheion.* Sketch. Copenhagen: Collection of architectural drawings, Royal Danish Academy of Fine Arts, Kunstakademiets Bibliotek, Kbh.KAark. 14962.

Hansen, C. 1835. *The Maiden Porch from the Southeast.* Pencil sketch. Copenhagen: Collection of Architectural Drawings, Royal Danish Academy of Fine Arts, Kunstakademiets Bibliotek, Kbh.KAark. S 18542,31.

Haygarth, W. 1810-1811. *Erechtheum from the North.* View of the Interior of the East Porch and South Wall. Watercolour. Athens: Gennadeion, No 89.

Hobhouse, J. C. 1809. *The Akropolis viewed from the Propylaea.* Coloured Engraving. Athens: National Library.

Hobhouse, J. C. 1813. *Journey through Albania, and other provinces of Turkey in Europe and Asia, to Constantinople.* during the years 1809-1810. London.

Legrand, P.-E. 1897. *Revue Archéologique* 30 : 41-66, 185-201, 385-404.

Le Roy, J. D. 1758. *Les ruines des plus beaux monuments de la Grèce.* Paris.

Lesk, A. 2004a. "The Reception of the Erechtheion: A Diachronic Examination" *Ph.D. Thesis.* University of Cincinnati.

Lesk, A. 2004b. "Graffiti in the Erechtheion" *Abstracts: Archaeological Institute of America 104th Annual Meeting* 27: 23-24.

Martineau, E. H. 1850. *Athens, Acropolis, The Erectheum from the Southeast.* Pencil drawing. London: RIBA Library Drawings Collection, Front Study Room

Shelf D2.

Papageorgiou-Venetas, A. 1994. *Athens: the Ancient Heritage and the Historic Cityscape in a Modern Metropolis*. Athens.

Pars, W. *The Erechtheum from the Northeast.* 1765. Brushed drawing in grey and brown wash, with watercolour and bodycolour, some gum arabic, over graphite; on several sheets conjoined. London: Department of Prints and Drawings in the British Museum, No. MM-11-4.

Paton, J.M, L. D. Caskey, H. N. Fowler, and G. P. Stevens. 1927. *The Erechtheum*. Cambridge, Mass.

Penrose, F. C. 1847. *Portico of Minerva Polias*. North Portico from the East. Watercolour. Athens: House of Director of the British School at Athens, No. BSA-F-7.

Pomardi, S. 1804-1805. *West End of the Pandrosion.* From the Northwest. Engraving. Athens: E. Dodwell, *Tour Through Greece*, vol. I, between pages 356 and 357.

Rørbye, M. 1836. *Parti af Erechteion.* Maiden Porch from East. Oil. Copenhagen: Ny Carlsberg Glyptotek, on loan from a private collection, Room 238.

Spiers, R.Ph. 1866. *The Erechtheion from the Northeast.* Watercolour. London: Victoria and Albert Museum, No. E.2738-1918.

Spon, J. 1678. *Voyage d'Italie, de Dalmatie, de Grèce, et du Levant, fait aux années 1675 et 1676 par Jacob Spon et George Wheler.* Lyon.

Stilling, H. C. 1853. *Erechtheion from the West.* Den vestlige Side af Erechteion paa Athens Acropolis. Watercolour. Copenhagen: Collection of architectural drawings, Royal Danish Academy of Fine Arts, Kunstakademiets Bibliotek, Kbh.KAark. 17736a; Neg. no. 912.1993.

Stuart, J. and N. Revett. 1789. *The Antiquities of Athens.* Vol. II. London.

Tsigakou, F. M. 1981. *The Rediscovery of Greece: Travellers and Painters of the Romantic Era.* New Rochelle, NY.

Waddington, G. 1825. *A Visit to Greece in 1823 and 1824.* London.

Werner, C. F. 1877. *View of the Porch of the Caryatids on the Erechtheion.* Gouache on paper. Athens: Benaki Museum, No. 23958.

Wheler, G. 1682. *A Journey into Greece by George Wheeler, Esq. In Company of Dr. Spon of Lyons.* London.

Winstrup, L. A. 1851. *The Erechtheion from the West.* Pencil and watercolour. Copenhagen: Collection of Architectural Drawings, Royal Danish Academy of Fine Arts, Kunstakademiets Bibliotek, Kbh.KAark. 5853.

Winstrup, L. A. 1851. *The Interior of the North Wall of the Erechtheion.* Pencil and watercolour. Copenhagen: Collection of architectural drawings, Royal Danish Academy of Fine Arts, Kunstakademiets Bibliotek, Kbh.KAark. 6120b.

Mani: A Unique Historic Landscape in the Periphery of Europe

Konstantina Liwieratos – University College London

Introduction to Mani

Mani is almost unknown beyond the borders of Greece. Its dry mountainous lands stretch over the middle peninsula of the Southern Peloponnese in Greece, between the prefectures of Messenia and Laconia. The area is virtually inaccessible from the coast. It was very densely populated between the 6th and the 19th Centuries AD. Today, few permanent residents populate Mani, which is largely the reason why its precious historic landscape has remained almost unspoiled by changes in modern times.

Geography and geology

Mani has an altitude of 22° 28" and a latitude of 36° 23", which is the same as southern Sicily and Gibraltar. The peninsula is 995.1 sq.km in area (Komis 1995: 13), about 25 miles or 40 km in length and nowhere more than 10 miles or 16 km in width (Woodward 1905-6: 239). The area is dominated by the continuation of the Taygetos mountain range, with several peaks above 500m. Deep passes and small fertile plains between these peaks and a beautiful coastline with steep cliffs, natural harbours and white beaches complete the astonishing scenery (Moschou & Moschou 1978-9: 77).

Description of Mani's cultural heritage

Introduction

Mani has been continuously inhabited since the Middle Palaeolithic (Darlas 1999: 51-55). Scientific and archaeological research in the area is still in its infancy, but first results from surveys, excavations and publication can give some indication of this. Apart from the main problem that little scientific effort has been made to explore Mani, another consideration is that even archaeological remains published in the past might no longer exist due to the lack of maintenance. As a result personal knowledge and on the spot investigation are needed in order to describe Mani archaeologically. Many archaeological descriptions are a result of fieldwork combined with previous research on existing written sources.

Prehistoric Mani

Palaeolithic and Neolithic Periods (600000-3100 BC)

The study of the Palaeolithic period is a very recent chapter in Greek archaeology, where the scientific community has been focused on Classical Greece for decades. Mani is one of the areas of main interest in this new field (Darlas 1999: 12). Traces of Palaeolithic habitation are mostly to be found in caves. The oldest traces date back 100,000 years BC (Darlas 2002: 187). Open-air settlements existed as well, but these survive only in cases where they have been petrified, as in Lagadaki bay (Darlas 1999: 21). The rocky coastline on the west side of Mani is particularly rich in caves and prehistoric remains. Mani mainly consists of limestone and its erosion is the main reason for the formation of caves. This phenomenon is called 'karst', from the area in Slovenia where it was studied for the first time (Ibid. 1999: 19). Geological drifts and the shifting of the sea have contributed to the creation of a complicated network of caves. Some researchers estimate hundreds of caves and Darlas started a register for the west coastline, where by 2002 he had identified 80 caves (Sykka 2002: 14). The Ephorate of Paleoanthropology and Speleology has already undertaken a preliminary underwater survey of approximately 10 km of coastline, south of Stoupa along the western coast, and has identified many caves at a depth of -36m to -3m, which seem to be of scientific interest. In southern Laconian Mani, the same Ephorate has already identified 57 caves and explored four of them. Additionally, 50 rock-shelters have been located there, many of which are of palaeontological and historical interest (Giannopoulos 1993-4: 73). This kind of concentration of caves with geological and archaeological material is unique.

The oldest evidence for human presence in caves dates back to approximately 100,000 years (Darlas 1999: 44). In the cave 'Kalamakia', apart from evidence of settlement such as fire remains, human teeth, scalps and bones have also been found belonging to the Palaeolithic period. Bones of animals confirm the existence of elephants, leopards, foxes, wild boars, deer and other animals which have now disappeared from the entire area (Darlas 1999: 47). Collections of food and stone tools complete the picture of Palaeolithic Mani to this date. In the cave 'Apidima', apart from similar findings as the above, Middle Palaeolithic burials were also found, as well as an entire female skeleton (Pitsios 1985: 26-33).

Remains of Mani's Neolithic period are also abundant. Traces of habitation, burials, human remains, tools, jewellery, statuettes, vases and sherds have been found in a number of caves. The area most researched until now is that of Diros. A number of caves, many sites in the immediate or wider region and the Neolithic acropolis of Diros create the impression of a densely inhabited area (Papathanasopoulos 1971: 289-304). Based on the large quantity of stored food, pottery and obsidian tools, Papathanasopoulos suggests that the cave 'Alepotripa' of Diros was not used as a residence for the agricultural community, but as a trade and storage centre for the settlements spread around the entire gulf

(Papathanasopoulos 1983: 221-224). Some Neolithic findings from the caves of the Diros area are exhibited in a small museum in Diros bay itself.

Bronze Age Period (3100-1100 BC)

From the Helladic period (i.e. Bronze Age) onwards there are fewer remains due to a number of reasons. According to the evidence, people in this period concentrated more in the fertile plain of Laconia. Those who inhabited mountainous Mani had to deal with a scarcity in materials. Mani's inhabitants used and reused the same materials every time they built something new (Hope Simpson 1957: 239). People did not abandon ruins after a catastrophe; they 'recycled' stone, tiles, wood and metals.

Another reason why the Helladic period remains scarcely documented is that there have not been any systematic surveys or excavations in Mani. Looting may have been a problem also (Hope Simpson 1957: 234, 238-9). Mycenaean tombs have been found, as well as other places where traces of tombs could be observed. Consequently, it is possible that graves have been systematically emptied of their treasures and partly destroyed.

There is more information about the Late Helladic or Mycenaean phase, as it is the first era that is described in written sources. The legendary poet Homer, who is presumed to have lived in the 8th Century BC but narrated stories dating back to the Mycenaean period, refers to some towns in Mani in his epic poem, the Illiad. He mentions Kardamyli, Iri, Enopi, Messi, Itilon, Las and Kranai (*Il.* ii. 582ff and ix. 149ff., 291ff.). These towns are described either as participating in the Trojan War, or as being offered by Agamemnon to placate Achilles. Kranai is mentioned as the location where Paris and Helen first stopped in their flight (*Il.* iii. 443-6). There have been attempts to identify these sites as the predecessors of modern towns or villages in Mani, with the help of written sources of later periods, like those by Pausanias from the 2nd Century BC and Strabo, who lived in the transition between the 1st Century BC and the 1st Century AD. Pausanias, for example, identifies Homeric cities with towns he visited in the 2nd Century AD; he believes that the current Mezapos is Mycenaean Messa (Woodward 1905-6: 243). It is believed that traces of that period are still hidden in Mani's landscape (Hope Simpson 1966: 116).

Historical Mani

Geometric to Roman Periods (1100 BC-AD 330)

Mani was inhabited for the entire period. Written sources are available from the 8th Century BC onwards. Towns and villages and cemeteries around them were spread around the region; stone paved paths and terraces on the slopes often date back to ancient times.

There are more remains from the historic period but the problems are similar to those discussed above for the Helladic era. Materials were reused and marble fragments, inscriptions, columns and capitals were often built into walls of churches or houses. Some examples are Agios Taxiarchis in Kouloumi (Woodward 1905-6: 241), Agios Dimitrios, Agios Soter, Koimisis and Agios Taxiarchis in Itilon (Forster 1903-4: 160-1) and Agios Spiridon in Vathia (Woodward 1905-6: 246-247). Churches are mostly built on foundations of ancient Greek temples or other buildings, as in the case of the Byzantine and Post-Byzantine churches of Kyparissos, in ancient Caenepolis. In the later Byzantine and Ottoman periods, many marble materials were destroyed by calcinations (Moschou & Moschou 1981: 14). Extensive looting has unfortunately been another destructive practice in the area (Moschou 1975: 173). In addition, besides information from written sources, such as Homeric Messa being identified as modern Mezapo (Woodward 1905-6: 243) and abundant material evidence of the Greek and Roman past, limited archaeological surveys and excavations have been undertaken.

As can be seen in the list below, extended remains of ancient towns and sanctuaries have survived, such as those in Gythion (Forster 1905-6), Kionia (Woodward 1905-6: 253-4; Moschou & Moschou 1978-9; ibid. 1988) and Tainaron (Woodward 1905-6: 249-253; Moschou 1975), but they are often abandoned or badly maintained. There is an emerging need to protect these sites and it would be interesting to explore Mani's archaeology further.

Eastern Roman or Byzantine and Ottoman Periods (AD 330-1830)

Megaw wrote as an introduction to his article on *Byzantine Architecture in Mani:*

> '*The field is richer than this article suggests. The buildings are in a state of preservation unique for Greece, they cover a long period and their study may be expected to elucidate the history of the peninsula in the Middle Ages*' (Megaw 1932-3: 137).

The excavation of Early Christian basilicas in Gythion, Tigani and Alika-Kyparissos proved that the people of Mani were converted to Christianity early on and Christian communities already existed, at least on the coast (Drandakis 1995: 19). These basilicas have not all been entirely excavated. They are typically Early Christian, and often include marble fragments, which had in many cases been removed from ancient Greek or Roman buildings. Agios Petros in Kyparissos is most probably pre-6th Century AD and, as can be seen in the list below, the others probably date to the 6th-8th Centuries AD (Drandakis 1995: 21).

All Greek-Byzantine periods are represented in Mani.

There are churches dating to the Middle, the Late and the Post Byzantine period. Many of these are decorated with paintings of high quality. Most of the church buildings are of small proportions and have provincial masonry, which is either megalithic with dry-stone walls or utilising lime mortar between joints. These are mainly single apsidal chambers (Megaw 1932-3: 138). The more complicated architectonic types are mostly built with the '*plinthoperikleisto*' system of masonry, which means carefully squared stones with bricks in both vertical and horizontal joints (Traquair 1905-6: 181). All variations of barrel-vaulted and cruciform domed churches appear in Mani, as in other Greek-Byzantine territories too: simple two-columned or *distyles*, simple four-columned or *tetrastyles*, semi-complex *tetrastyles* and transverse barrel-vaulted churches, some of them with galleries, lateral chapels or bell-towers. There are exceptions: Agios Petros in Glezou is a free cross with a dome and Agios Leos in Briki is a vaulted dome church (Drandakis 1995: 21-22).

As was mentioned, many of the churches are often decorated with high quality paintings. The earliest are to be found in Agios Prokopios close to Episkopi, which date back to the 9th Century AD (Drandakis 1995: 21). Several churches were decorated between the 10th and 13th Centuries AD. In the 14th Century AD Mani has fewer examples of painted churches and in the 15th Century AD they almost disappear (Ibid. 1995: 23). After the 16th Century AD and until the 19th, paintings multiply again.

As has been shown, all periods of Byzantine art history are represented in Mani, both in architecture and painting. The quantity of churches, estimated approximately at 1500, along with approximately 90 monasteries, makes the area unique; a perfect region to become familiar with or specialize in Byzantine art (Saitas 2002: 6). The churches are in various conditions. Some are very well preserved, whereas others are in extreme danger as they are gradually disappearing. Churches mentioned in former sources are sometimes hard to identify and in extreme cases have completely disappeared. Strong winds, heavy rainfalls and salts are the main natural causes of their decay. Calcinations in the past, abandonment and looting are the main human causes of their decay.

In 197l, the Ministry of Development registered 170 villages out of approximately 250 as 'traditional', i.e. having historical and archaeological value (Saitas 2002: 6). In these villages hundreds of houses-towers were built in the 18th and 19th Centuries AD. These belong to Mani's cultural heritage and often form a picture of its landscape. Many of these have been restored in the last 30 years, as they were private properties and are now used year-round or as holiday homes, or guesthouses. Still, many remain in danger of falling apart.

It is of great importance to mention that apart from the traditional villages there are also the abandoned remains of an entire Byzantine town. Very close to the modern village of Drosopigi, on a hill which is not easily accessible, lie the ruins of Karyoupolis. A castle, the churches of Agios Nikolaos and Agios Georgios, towers and many buildings, preserved in fairly good condition, form what is left of this once flourishing city (Etzeoglou 1988: 3-47).

A number of castles are preserved in Mani as well. Dating and condition vary from case to case and are not systematically researched or published. Apart from certain exceptions, information is only to be found in local bibliography. Some of the castles, like in Tigani, are difficult to reach, as the surrounding area is completely abandoned and paths have disappeared. All of them need conservation work and most of them require urgent attention. Threshing floors, wells, storage rooms, stables, paths, terraced slopes and other similar buildings or constructions are part of Mani's historic landscape as well.

There is a large number of villages, which Moschou calls '*Paleomaniatika*', i.e. 'old settlements of Mani' (Moschou & Moschou 198: 3-29: Ibid. 1982: 261-271). Other researchers have called these settlements or isolated buildings 'megalithic' and have compared them with other megalithic constructions in Malta, the Balearics or Pantelleria (Moutsopoulos & Dimitrokalis 1976: 146-161). Saitas also calls them 'megalithic' and recognises analogies with western European megalithic architecture of the past, but argues that this type of architecture is of local origin (Saitas 1992: 20). In his 1983 article, Saitas describes four 'stonehenges' he has located in southern Mani and asks whether these are related to 'megalithic' architecture of houses, churches and cisterns (Saitas 1983: 151-168).

To conclude, Mani's continuous habitation is an excellent example of a historic landscape's evolution. Further research and conservation work is needed.

Acknowledgements

I would like to express my appreciation to Leda Moschou and Kathy Tubb, who checked this paper repeatedly, both from a scientific and a linguistic point of view.

Bibliography

Alexakis E.P. 1998. *Μάνη, γένη και οικογένεια.* Athens.

Avramea A. 1997. *Le Peloponnese du IV au VIII siecle, changements et persistences.* Paris.

Baladie R. 1980. *Le Peloponnese de Strabon: Etude de geographie historique.* Paris.

Bon A. 1951. *Le Peloponnese Byzantin jusqu'en 1204.* Paris.

Buchon A. J. 1842. *La Grece continentale et la Moree.* Paris.

Cartledge P. 2002. *Sparta and Laconia.* London.

Cockerell S.P. 1903. *Travels in Southern Europe and the Levant 1810-1817 – The journal of C.R. Cockerell R.A.* London.

Darlas A. 1999. *Η Παλαιολιθική Μάνη*. Athens.

Darlas A. 2002. "Η ανασκαφή του παλαιολιθικού σπηλαίου Καλαμάκια, τα νεώτερα δεδομένα" *Πρακτικά του Α τοπικού Συνεδρίου Λακωνικών Σπουδών*: 179-209.

Dickins G. 1904-5. "Laconia. III. Thalamae. Excavations" *British School of Athens* 11: 124-130.

Drakopoulou E. 1993. "Travelers in the 18th century", in G Saitas (ed) *Travelers in the Mani 15th-19th c.* Athens: 42-54.

Drandakis N. 1958. "Σκαφικαί έρευναι εν Κυπαρίσσω Μάνης" *Πρακτικά Αρχαιολογικής Εταιρείας*: 199-219.

Drandakis N. 1964. "Ανασκαφή εις το Τηγάνι της Μάνης" *Πρακτικά Αρχαιολογικής Εταιρείας*: 121-135.

Drandakis N. 1964. *Βυζαντιναί τοιχογραφίαι της Μέσα Μάνης*. Athens.

Drandakis N. 1974. "Έρευναι εις την Μάνην" *Πρακτικά Αρχαιολογικής Εταιρείας:* 110-196.

Drandakis N. 1975. "Έρευναι εις την Μάνην" *Πρακτικά Αρχαιολογικής Εταιρείας*: 184-196.

Drandakis N. 1976. "Έρευναι εις την Μεσσηνιακή Μάνην" *Πρακτικά Αρχαιολογικής Εταιρείας*: 213-252.

Drandakis N. 1977. "Έρευναι εις την Μάνην" *Πρακτικά Αρχαιολογικής Εταιρείας*: 200-228.

Drandakis N. 1988. "Παρατηρήσεις στις τοιχογραφίες του 13ου αιώνα που σώζονται στη Μάνη" *Λακωνικές Σπουδές Θ:* 70-75.

Drandakis N. 1993. *Έρευνα στην Κάτω Μάνη*. Athens.

Drandakis N. 1995. *Οι βυζαντινές τοιχογραφίες της Μέσα Μάνης*. Athens.

Drandakis N. & Gioles N. 1980. "Ανασκαφή εις το Τηγάνι της Μάνης" *Πρακτικά Αρχαιολογικής Εταιρείας*: 247-258.

Drandakis N., Gioles N. & Konstantinidi Ch. 1978. "Ανασκαφή εις το Τηγάνι της Μάνης" *Πρακτικά Αρχαιολογικής Εταιρείας*: 183-191.

Drandakis N., Gioles N. & Konstantinidi Ch. 1979. "Σκαφικές έρευνες στη Μάνη" *Πρακτικά Αρχαιολογικής Εταιρείας*: 215-225.

Drandakis N., Gioles N. & Konstantinidi Ch. 1981. "Ανασκαφή εις το Τηγάνι της Μάνης" *Πρακτικά Αρχαιολογικής Εταιρείας:* 241-253.

Drandakis N., Gioles N. & Konstantinidi Ch. 1981. "Έρευνα στη Λακωνική Μάνη" *Πρακτικά Αρχαιολογικής Εταιρείας*: 254-268.

Drandakis N., Dori N., Kalopisi E. & Panagiotidi V. 1978. "Έρευναι εις την Μάνην" *Πρακτικά Αρχαιολογικής Εταιρείας*: 135-182.

Drandakis N., Dori N., Kalopisi E., Kepetzi S. & Panagiotidi V. 1981. "Έρευναι εις την Μάνην" *Πρακτικά Αρχαιολογικής Εταιρείας*: 449-578.

Drandakis N., Kalopisi E. & Panagiotidi V. 1979. "Έρευναι εις την Μάνην" *Πρακτικά Αρχαιολογικής Εταιρείας*: 156-214.

Drandakis N., Kalopisi E. & Panagiotidi V. 1980. "Έρευναι εις την Μεσσηνιακή Μάνην" *Πρακτικά Αρχαιολογικής Εταιρείας*: 188-244.

Drosogianni F.A. 1982. *Σχόλια εις τοιχογραφίες του Αγίου Ιωάννου του Προδρόμου Καστανιάς*. Athens.

Ducelier A. 1990. *Byzanz*. Frankfurt a. M.

Eliopoulou Rogan D. 1973. *Mani: history and monuments*. Athens.

Etzeoglou R. 1988. "Καρυούπολις, μια ερειπωμένη βυζαντινή πόλη" *Λακωνικές Σπουδές Θ:* 3-47.

Forrest W.G. 2000. *A History of Sparta*. Bristol.

Forster E. 1903-4. "South-Western Laconia: sites" *British School of Athens* 10: 158-166.

Forster E. 1905-6. "Laconia. II. Topography" *British School of Athens* 13: 219-238.

Giannopoulos V. 1993-4. "Συμβολή στη μελέτη του καρστικού φαινομένου της Λακωνικής Μάνης" *Δελτίο Ελληνικής Σπηλαιολογικής Εταιρείας* 21: 72-87.

Gibbons B. 1994. *A guide to the national parks and other wild places of Britain and Europe*. London.

Gioles N. 1991. *Παλαιοχριστανική τέχνη, μνημειακή ζωγραφική*. Athens.

Guillet A.G. 1676. *Lacedemone ancienne et nouvelle*. Paris.

Hooker J.T. 1980. *The ancient Spartans*. London.

Hope Simpson R. 1957. "Identifying a Mycenaean state" *British School of Athens* 52: 231-259.

Hope Simpson R. 1966. "The seven cities offered by Agamemnon to Achilles" *British School of Athens* 61: 113-131.

Jones A.H.M. 1968. *Sparta*. Oxford.

Kalliga H. 1974. "The evolution of settlements in Mani", in B. Doumani & P. Oliver (eds) *Shelter in Greece*. Athens.

Karpodini-Dimitriadi E. 1993. *Κάστρα της Πελοποννήσου*. Athens.

Kathimerini 08.08.2002. *"Εκπαιδευτική ανασκαφή στην Αρεόπολη"*: 15.

Katsafados P. 1992. *Τα κάστρα της Μαίνης*. Athens.

Katsafados P. 1994. *Μάνη, Μέζαπος ή ομηρική Μέσση*. Athens.

Kennell N. 1999. "From periiki to poleis: The Laconian cities in the late Hellenistic period", in S. Hodkinson & A. Powell (eds) *Sparta: New perspectives*. London.

Komis K. 1995. *Πληθυσμός και οικισμοί της Μάνης*. Ioannina.

Kotsakis T. 1980-1. "Ο Evliya Celebi στην Πελοπόννησο" *Πελοποννησιακά* 14: 297.

Koutsilieris A. 1993. *Ιστορία της Μάνης*, Athens.

Kyratsoulis T. 1996. *Towards sustainable coastal tourism development in the Mani Peninsula, Greece*. Unpublished MA dissertation, University of Warwick.

Leake W.M. 1830. *Travels in the Morea*, London.

Liwieratos K. 1994. *Die deutschen Reiseschriftsteller im 18. und 19. Jh. und ihr Beitrag zur Wiederentdeckung Griechenlands*. Unpublished dissertation at the University of Athens.

Madouvalos G. 1984. *Το οδοιπορικό των Μπουλαριών, Διπόρου, Μέσα Μάνης.*Athens.
Megaw H. 1932-3. "Byzantine architecture in Mani" *British School of Athens* 33: 137-162.
Merdenisianos K. 1985. "Σπήλαιο Βύθακας Διρού Μάνης" *Αρχαιολογία* 15: 26-33.
Meyer E. 1974. *Pausanias, eine Beschreibung Griechenlands*. Zuerich..
Morritt J.B.S. 1820. *Account of a journey through the district of Maina, in the Morea, travel in various countries of the East*. London.
Moschou L. 1975. "Τοπογραφικά Μάνης" *Αρχαιολογικά Ανάλεκτα* Αθηνών 8: 160-177.
Moschou L. 1976. "Topographie du Magne – A propos de la region du Tenare Κιvστέρνες – Άγιοι Ασώματοι" *Πρακτικά του Α Διεθνούς Συνεδρίου Πελοποννησιακών Σπουδών* Β: 45-54.
Moschou L. 1982. "Μιά αγροτική περιοχή της Ανατολικής Λακωνικής Μάνης στον κώδικα Ambrosianus – Trotti 373" *Πρακτικά ΙΣΤ Διεθνούς Συνεδρίου Βυζαντινών Σπουδών: Jahhrbuch der Oesterreichischen Byzantinistik* 32/4: 639-656.
Moschou L. 1983a. "Οικισμένος χώρος, παραδοσιολογία και ιστορικά φαινόμενα" *Ανθρωπολογικά* 4: 13-36.
Moschou L. 1983b. "Εικόνες μιάς κοινωνίας πριν από την άλωση" *Αντί* 243: 39-41.
Moschou L. 1988. "Οικιστικά συστήματα των περιοικίδων και των Ελευθερολακωνικών πόλεων στην χερσόνησο του Ταινάρου" *Πρακτικά του XII Διεθνούς Συνεδρίου Κλασσικής Αρχαιολογίας*: 148-154.
Moschou L. 1997. "Πόλεις και μνημεία της Αρχαιότητας στη Λακωνική Μάνη" *Αδούλωτη Μάνη* 3: 17-32.
Moschou L. 1999. "Μάνη: Επαρχίες Γυθείου και Οιτύλου" *Αρχαιολογικό Δελτίο* 54.
Moschou L. 1999-2000. "Lakonian Mani" *AR:* 40-41.
Moschou L. 2003. "Για την ιστορία δύο πόλεων και την αρχαιολογία πολλών άλλων" *Αρχαιολογικά Ανάλεκτα Αθηνών.*
Moschou L & Moschou T. 1978-1979. "Κιόνια Α" *Πελοποννησιακά* ΙΓ: 72-114.
Moschou L. & Moschou T. 1981a. "Palaiomaniatika. Οι βυζαντινοί αγροτικοί οικισμοί της Λακωνικής Μάνης" *Αρχαιολογικά Ανάλεκτα* Αθηνών 14: 3-28.
Moschou L. & Moschou T. 1981a. "Το αρχαίο φρούριο της Τευθρώνης" *Αρχαιολογική Εφημερίδα*: 10-22.
Moschou L. & Moschou T. 1982. "The Paleomaniatika: the transition from ancient polis to Byzantine chora" *Ekistics* 295: 261-271.
Moschou L. & Moschou T. 1988. "Κιόνια Β" *Πρακτικά του XII Διεθνούς Συνεδρίου Κλασσικής Αρχαιολογίας*: 140-147.
Moschou L., Raftopoulou S. & Chatzitheodorou T. 1998. "Τα αρχαία λατομεία στον Προφήτη Ηλία Δημαρίστικων Μάνης και η περιοχή τους" *Αρχαιολογικό Δελτίο* 53: 267-288.
Mouriki D. 1975. *The frescoes of the church of St. Nicolas at Platsa in the Mani.* Athens.
Moutsopoulos N.K. & Dimitrokalis G. "Τα μεγαλιθικά μνημεία της Μάνης" *Πρακτικά Α' Διεθνούς Συνεδρίου Πελοποννησιακών Σπουδών*: 135-169.
Oliva P. 1971. *Sparta and her social problems.* Amsterdam.
Papahantzis N. 1976. *Παυσανίου Ελλάδος Περιήγησις: Κορινθιακά-Λακωνικά.* Athens.
Papathanasopoulos G. 1971. "Σπήλαια Διρού: Αι ανασκαφαί του 1970-71" *Αρχαιολογικά Ανάλεκτα Αθηνών*: 289-304.
Papathanasopoulos G. 1983. "Ο ναυτικός χαρακτήρας του νεολιθικού Διρού" *Πρακτικά Πρώτου Συνεδρίου Λακωνικών Μελετών*: 221-224.
Paradisis A. 1973. *Φρούρια και κάστρα της Ελλάδος.* Athens.
Petrocheilou A. 1992. *Τα σπήλαια της Ελλάδος.* Athens.
Philippson A. & Kirsten E. 1959. *Die griechische Landschaften: der Peloponnes.* Frankfurt a.M.
Pitsios T.K. 1983. "Παλαιοανθρωπολογικά ευρήματα της Μέσα Μάνης ΙΙ" *Πρακτικά Πρώτου Συνεδρίου Λακωνικών Μελετών*: 79-82.
Pitsios T.K. 1985. "Παλαιοανθρωπολογικές έρευνες στη θέση 'Απήδημα' της Μέσα Μάνης ΙΙ" *Αρχαιολογία* 15: 26-33.
Plehn C. -. *Kreuzritterburgen auf dem Peloponnes.* Muenchen-Zuerich.
Polikandrioti R. 1993. "Travelers in the 17th and 19th century", in G. Saitas (ed) *Travelers in the Mani 15th-19th c.* Athens: Ministry of Culture: 27-41, 55-80.
Pouqueville F.C.H.L. 1813. *Travels in the Morea, Albania and other parts of the Ottoman empire.* London.
Sabbatini R. 1910. *Ciriaco di Ancona e la sua descrizione autografa del Peloponneso, trasmessa da Leonardo Botta.* Milano.
Saitas G. 1983. "Ορθόλιθοι στη Μέσα Μάνη" *Πρακτικά Πρώτου Συνεδρίου Λακωνικών Μελετών*: 151-168.
Saitas G. 1992. *Μάνη*. Athens.
Saitas G. (ed) 1993. *Travelers in the Mani 15th-19th c.* Athens.
Saitas G. 1995. Τουρισμός και ανάδειξη της πολιτιστικής κληρονομιάς της Μάνης. *Tehnika Chronika* 5/95: 96-103.
Saitas G. 2002. Κίνδυνοι και προοπτικές για τη Μάνη. *Μάνη* 5: 6-7.
Shipley G. 1992. "Perioikos: the discovery of Classical Lakonia" in J. M. Sanders (ed) *Φιλολάκων: Lakonian studies in honor of Hector Catling.* London.
Shipley G. 1997. "The other Lakedaimonians': the dependent Perioiki poleis of Laconia and Messenia", in M. H. Hansen (ed) *The polis as an urban centre and as a political community.* Copenhagen.
Simopoulos K. 1994a. *Foreign travelers in Greece,* Athens.
Simopoulos K. 1991b. *Foreign travelers in Greece,* Athens.
Simopoulos K. 1992c1. *Foreign travelers in Greece,* Athens.
Simopoulos K. 1992c2. *Foreign travelers in Greece,* Athens.

Stackelberg O.M. von 1834. *La Grece, vues pittoresques et topographiques dessinees par O.M. baron de Stackelberg*. Paris.

Stephanopoli D. and N. 1974. *Ταξίδι στην Ελλάδα κατά τα χρόνια 1797 και 1798*. Athens.

Sykka G. 2002. "Η παλαιολιθική Μάνη αποκαλύπτει τα μυστικά της" *Καθημερινή* 31/8: 14.

Theoharis D. 1971. "Η Παλαιολιθική και η Μεσολιθική εποχή στην Ελλάδα", in *Ιστορία του Ελληνικού Έθνους*, vol. A: 38-47. Athens.

Thommen L. 1996. *Lakedaimonion Politeia.* Stuttgart.

Traquair R. 1905-6. "Laconia. Medieval fortresses" *British School of Athens* 12: 259-276.

Traquair R. 1908-9. "Laconia. III. The churches of Western Mani" *British School of Athens* 15: 177-213.

Vingopoulou I. 1993. "Travelers in the 15th-16th century", in G. Saitas (ed) *Travelers in the Mani 15th-19th c.*: 18-26. Athens.

Vouyouklakis K. 1997. *Η Μάνη ανά τους αιώνες*. Athens.

Wagstaff J.M. 1991. "Further observations on the location of Grande Magne" *Dumbarton Oaks Papers* 45: 141-148.

Waterhouse H. & Hope Simpson R. 1961. "Prehistoric Laconia II" *British School of Athens* 56: 114-130.

Wilcken U. 1976. *Αρχαία Ελληνική Ιστορία*. Athens.

Wollny Poppota E. 1982. *Lakonisches Reisebuch.* Munich/Zuerich.

Woodward A.M 1905-6. "Laconia. II. Topography" *British School of Athens* 13: 238-267.

Zias N. 1985. "Mani", in: E Melas (ed) *Alte Kirchen und Kloester Griechenlands.* Koeln.

The Hominid Dispersal into Mediterranean Europe during the Early to Middle Pleistocene: the Sabre-toothed cat connection

Lisa Marlow - University of Cambridge

Introduction

The dispersal of hominids into Mediterranean Europe occurred between one million and half a million years B.P. The first tentative signs of hominid presence occur in a few scattered sites in southern Europe soon after a million years ago, but the earliest evidence that hominids were not confined to Africa comes much earlier; we know they were in the Levant, the Caucasus, and probably Pakistan and Indonesia by about 1.8 million years ago. Why then, is it almost another million years before we have the same kind of evidence for them in Europe? Why, once here, were they restricted to Mediterranean zones for so long? If we can determine why the delay occurred and how, eventually, they became established across Europe, we will have revealed key ecological and perhaps behavioural traits of these early hominids.

This paper investigates a recent suggestion that the first Mediterranean hominids may have scavenged from the kills of sabre-toothed cats and that this relationship may have offered a crucial mechanism by which they were first able to remain in Europe's difficult biotopes. The proposal that certain predators pioneered hominid incursion into new territories, like southern Europe, by providing meaty carcasses for hominids to scavenge (Martínez-Navarro & Palmqvist 1995, 1996; Arribas & Palmqvist 1999), although interesting in its approach, is unsubstantiated by faunal evidence from sites in Mediterranean Europe.

The 'special relationship' considered here stems from suggestions that early hominids may have occupied an obligate scavenging niche (Blumenschine 1988; Lewis 1997; Turner 1992), and the search for a suitable donor partner has since led to serious consideration of the role of sabre-toothed cats in the carnivoran food webs of the Plio-Pleistocene (Marean 1989; Lewis 1997). Salient features of the cats' dentition implied that they must have left very meaty carcasses behind them (Ewer 1954 cited in Turner 1992; Marean 1989; *but see* Marean and Ehrhardt 1995; Arribas & Palmqvist 1999).

In the first instance the utility of the proposed relationship to the debate regarding hominid dispersal into Europe was based on Turner's (1992) prediction that the earliest evidence for hominid sites should be congruent with the distribution of major carcass producers, and his observation (following Kurtén 1968) that *Megantereon cultridens* seems to have had a remarkably similar distribution across Europe to the first hominids.

Martínez-Navarro and Palmqvist (1996) took this further, implying a more dedicated relationship and going so far as to suggest that archaeologists should look very closely at sites where *Megantereon* has been found, since their presence should be a marker for hominid activity in the region.

If this relationship were shown to be reliable, it would be a remarkable and revealing advance in our understanding of hominid ecology in the Early Pleistocene. It is a provocative argument, so it is discussed in more detail below, before turning to how my research sheds light on the subject.

The sabre-toothed cats

Arribas and Palmqvist (1999: 579-581) detail the morphological characteristics that indicate the likely hunting and food processing capacities of the two European sabre-toothed cats *Homotherium* and *Megantereon*. The greatly enlarged upper canines were relatively delicate and it is thought that their size would have prevented the cats from being able to access flesh attached to bones, because to do so would have risked damage to the teeth. Instead, it seems that the canines' main employment was at the kill: a much wider gape than modern felids and large, differently positioned muscles in the head and neck enabled the penetration of canines into the body of large prey animals. Furthermore, their carnassial teeth were extremely specialised for rapid slicing of flesh, to the exclusion of the bone crushing morphology that often occurs in carnivores (including some other cat species).

There is no consensus regarding *Homotherium's* hunting technique, since it possessed limb proportions midway between those of modern cursorial and ambush predators, but like modern cursorial hunters, it displayed a decreased ability to use forelimbs to grapple with prey once caught. It was of similar size to modern lions (150 – 200 kg) but it had elongated forelimbs, short hind limbs and, hence, a sloping back. An increase in the size of the optic nerve indicates an emphasis on vision and provides support for the thesis that it was a cursorial hunter of open habitats (*but see* Marean 1989). In contrast, *Megantereon's* shorter limbs display the build of an ambush hunter with slower top speeds than Homotherium, and a preference for closed habitats. Well-developed olfactory lobes support this argument. However, their forelimbs were powerful, suggesting good grappling capacities. Analogy with modern jaguar morphology suggests long distance dragging and tree caching behaviours (*but see* Lewis 1997).

Site	*Period*	*Country*	*Notes*
Atapuerca TD3	Early Pleistocene	Spain	Cave site
Atapuerca TD5	Early Pleistocene	Spain	Cave site
Atapuerca TD6	Early Pleistocene	Spain	Cave site. Hominid find spot
Fuente Nueva 3a	Early Pleistocene	Spain	Orce Basin, lake site.
Colle Marino	Early Pleistocene / Early Middle Pleistocene	Italy	Open site
Petralona (Petralonan levels)	Early Middle Pleistocene	Greece	Cave site. Hominid find spot
Petralona (Crenian levels)	Early Pleistocene	Greece	Cave site
Grotte de l'Escale	Early Middle Pleistocene	France	Cave site, possible intentional fire

Table 18.1. Table of European sites before O.I.S. 13 with archaeological evidence (i.e. sites with evidence that Hominids were present at that locality)

Site	*Period*	*Country*	*Notes*
Cueva Victoria	Early Pleistocene	Spain	Cave site
Fuente Nueva 3a	Early Pleistocene	Spain	Archaeological evidence
Venta Micena	Early Pleistocene	Spain	Lake site
Olivola	Plio-Pleistocene	Italy	River site
Upper Valdarno (Tasso Sands)	Early Pleistocene	Italy	Lake/river site
Upper Valdarno (Figline beds)	Early Pleistocene	Italy	Lake/river site
Colle Pardo	Late Pliocene	Italy	
Fontana Acetosa	Early Pleistocene	Italy	
Tiber Basin	Early Pleistocene	Italy	
Pirro Nord	Early Pleistocene	Italy	
Apollonia 1	Early Pleistocene	Greece	Cave site
Dacic Basin	Early Pleistocene	Romania	
Untermassfeld	Early Pleistocene	Germany	River site

Table 18.2. Table of European sites with Megantereon present

In other words, European sabre-toothed cats were hunters capable of successfully predating animals much larger than themselves, while at the same time being unable to access considerable amounts of the flesh and *none* of the nutrients held within the bones of the carcasses they procured. These outcomes appear to provide favourable conditions for opportunists like hyaenas and hominids. However, it is worth introducing a note of caution raised by research into Friesenhahn Cave (Texas, U.S.), a carnivore den attributable to *Homotherium* (Marean & Ehrhardt 1995). Analysis of the faunal assemblage here showed that, contrary to expectations outlined above, predators at this site not only disarticulated and transported parts of carcasses back to their den, but once there, bone damage indicated that they were capable of effectively defleshing the bones with their incisors (though they did not access the marrow). The authors note that the amount of scavengeable material left by these carnivores is less than previously advertised, however *Homotherium* at this cave appeared to specialise in predating juvenile mammoths and the within-bone nutrients available to a scavenger from one of these kills would have been substantial, if access to the carcass was possible. The question of the degree of sociality in *Homotherium* remained unresolved, but this behavioural trait would have an enormous impact upon the availability of prey carcasses to a scavenger (Blumenschine 1988; Marean & Ehrhardt 1995).

Scavenging parties

Of the two candidates then, it would seem that *Megantereon* was the likeliest to have pioneered hominid dispersal into Europe, partly because its smaller size, habitat preferences and probable solitary habits rendered it a more productive and less threatening target for carcass thieving hominids, and partly due to its geographic distribution (and perhaps also its chronological range), which almost match those of early European hominids. *Homotherium* provides less explanatory power as a pioneer carnivore, because it displays a much broader geographic and chronological range than the first European hominids.

There is disagreement over the taxonomy of *Megantereon* in Europe. *M. cultridens* colonised Eurasia from North America in the Pliocene, but *Megantereon whitei*, said by some (e.g. Martinez-Navarro and Palmqvist 1996; Arribas and Palmqvist 1999) to have replaced *cultridens* in Europe at the Plio-Pleistocene boundary, and considered to be an African descendant of *M. cultridens*. Earlier research based on the dimensions of the lower carnassial failed to distinguish two species, and Turner (1987) concluded that *M. cultridens* was the only valid species for this genus. Martinez-Navarro and Palmqvist (1996) argue that the M^1 shows great conservatism and is not a suitable character to base such a study on, and further, find that a discriminant function analysis that includes the premolar P_4, distinguishes African individuals from [Eurasian + North American] specimens, but places Early Pleistocene specimens from Venta Micena (Spain), Dmanisi (Georgia) and Apollonia (Greece) within the African (*M. whitei*) range. Thus, they suggest that the presence of *Megantereon whitei* may have been a necessary ecological component of the earliest hominid colonisation of Mediterranean Europe. Nevertheless, other researchers continue to regard European Pleistocene *Megantereon* as belonging to *M. cultridens*, or an advanced form of that species (e.g. Gliozzi *et al.* 1997). Therefore, I choose to include *all* European Pleistocene *Megantereon* as my subjects for this paper.

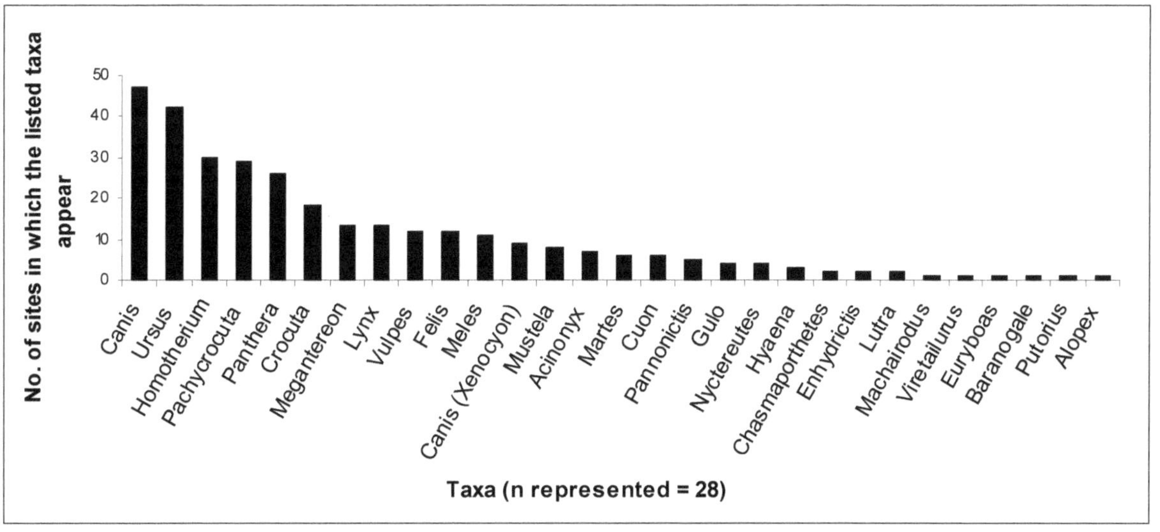

Figure 18.1. Number of carnivore taxa represented at European sites before the widespread occurrence of hominids after O.I.S. 13 (i.e. before 524 Ka BP). (N. sites = 76)

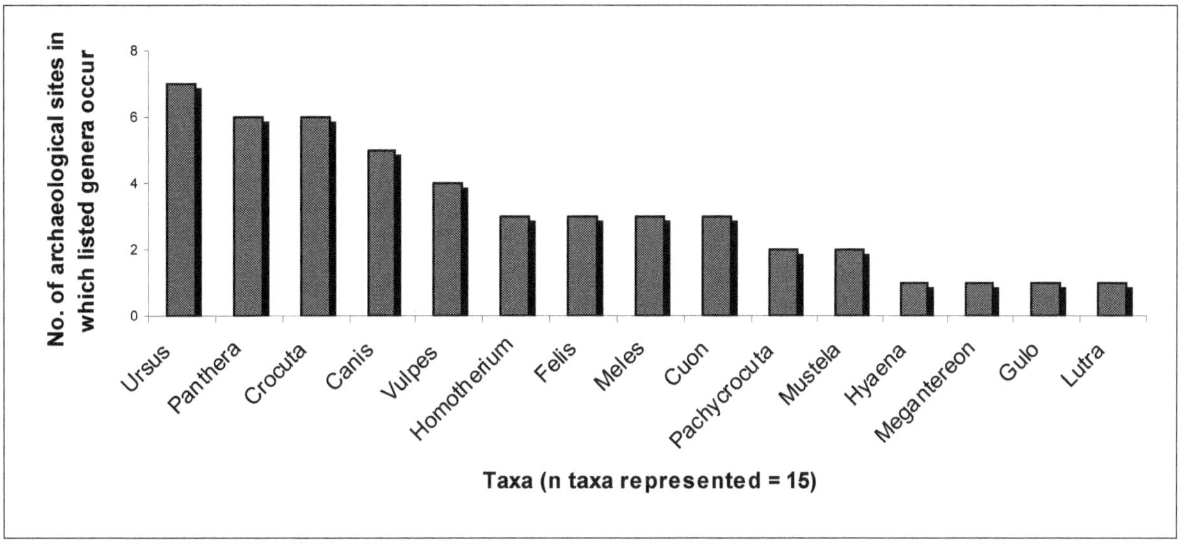

Figure 18.2. Frequency of co-occurrence of carnivores with evidence for hominid activity in European sites dating to before O.I.S. 13. (N. sites = 8).

Arribas and Palmqvist (1999: 578-579) also suggest that there is a close ecological relationship between *Megantereon whitei* and *Pachycrocuta brevirostris,* the giant short-faced hyaena. Not only do the two carnivores arrive in Europe simultaneously at the end of the Pliocene, but the decline of the machairodont cats approximately half a million years ago also coincides with, and is implicated in, the local extinction of the giant hyaena (Turner 1990, 1992). The Late Pliocene arrival in Europe of *M. whitei* "may have played a very significant role in facilitating the dispersal of *Pachycrocuta* and *Homo* outside Africa, since it was an ambush predator with great killing capability in relation to its flesh requirements, and presumably left large amounts of carrion for both hyaenas and hominids" (Arribas and Palmqvist 1999: 579, citing Martínez-Navarro and Palmqvist 1996); while their decline "implied the loss of an important source of partly-consumed carcasses, and thus a change in the interactions between flesh-eating and bone-consuming species of the carnivore guild" (Arribas and Palmqvist 1999: 578, citing Turner and Antón 1996).

This suggests to me a dangerously competitive relationship between hominids and the lion sized *Pachycrocuta,* rather than the avoidance that the huge size of the hyaenas might suggest. In fact, in his 1992 paper on this very topic, Turner suggested that it would have been the *disappearance* of these two large carnivores, as part of a major restructuring at around half a million years ago, which left the European carnivore guild closely resembling that of Plio-Pleistocene Africa. He proposed that this would have opened up a 'hominid-shaped' niche, a suggestion that appears to be supported by the quantitative and qualitative jump in the evidence for hominid presence after half a million years ago. Nevertheless, despite a distinct difference in emphasis, both sets of authors think that sabre toothed cats may have provided a key resource for the very earliest hominids to disperse into Europe.

Carnivore associations

The faunal assemblages of the early Pleistocene sites themselves are used to test the idea that hominids and sabre-toothed cats had an important ecological connection. The proposal under investigation suggests a number of outcomes:

- That hominids should follow *Megantereon* into the same regions – and that hominids should *not* be in an area where there are no big cats.
- That they shared the same landscapes (they should be found in the same faunal assemblages more often than random sampling would suggest).
- That even if they are not in the same assemblages they should be found with similar assemblages of animals (implying that hominids interacted with similar taxa to those which *Megantereon* encountered).

If these outcomes are not supported by the patterns in the data, then the proposal must be reconsidered. The data presented here are preliminary – and take a relatively simplistic approach to the issue. However, I have been unable to find the patterns that one would expect to find if the proposed relationships operated.

Of the hundreds of sites for which faunal data is recorded, so far there are only fourteen sites claimed to represent hominid activity before approximately half a million years ago and not all of these contain carnivoran remains. Of the eight that do (*Table 18.1*), most are in Spain with one each in Italy and France, and two layers in the cave site of Petralona in Greece. Atapuerca TD6 and Petralona also boast skeletal remains of hominids.

A similarly small number of European sites record the presence of *Megantereon* (*Table 18.2*). Three are in Spain, seven in Italy and one each in Greece, Romania and Germany. All predate the Middle Pleistocene. Only one of these is also an archaeological site, so there is nothing to suggest an unusual attraction between hominids and *Megantereon.*

Figure 18.1 presents data showing the number of sites at which European carnivore taxa occur, for 76 Plio-Pleistocene to Early Middle Pleistocene European sites for which faunal data has been collected. In the bar charts presented, the height of the bar indicates the number of sites in which each taxon is represented. *Canis* and *Ursus* appear in the greatest number of sites. *Homotherium* is also important, as are the big cats of the genus *Panthera,* along with the hyaenas *Pachycrocuta* and *Crocuta.* This is not intended as a reliable indication of relative abundance, since there are clearly taphonomic and reporting factors in play, but it does give some indication of the character of the palaeontological record for this period.

The utility of looking at carnivore presence, and crucially carnivore *co*-presence, at these sites is that with this information the predatory guild, and their relationships with one another, can be investigated, and testable hypotheses about how early European hominids fit into these interactions can be built. *Figures 18.2* and *18.3* indicate the differences in the 'shape' of the assemblages of carnivorans associated with the activities of hominids and those who met their deaths in the same locales as did *Megantereon.*

As already noted, *Megantereon* appears in only one of the eight archaeological sites. The popularity of *Ursus* in *Figure 18.2* is likely to be a feature of the number of cave sites in the sample, rather than any special relationship between humans and bears, but the high proportion of sites with the spotted hyaena, *Crocuta* and big cats of the genus *Panthera* may be worth further investigation. *Canis* and

Panthera are both fairly well represented in *Figure 18.2*, though, as we have seen from *Figure 18.1*, both are ubiquitous in this period.

Figure 18.3, showing the frequency with which Early Pleistocene carnivores share a find-site with *Megantereon*, presents a rather different picture. Naturally, *Megantereon* is the most prominent, but *Pachycrocuta* is also important, particularly compared to its position in archaeological sites.

Finally, *Figure 18.4* presents a comparison of the assemblages of carnivoran taxa in archaeological sites, and those occurring in sites with *Megantereon,* against the data for all early sites presented in *Table 18.2*. Much of the variation from the background data (represented by the 'all-sites' data in solid black) may be down to small sample sizes in the 'archaeological' and '*Megantereon*' assemblages (*Tables 18.1 & 18.2*). Yet, it is interesting to note that *Pachycrocuta* appears in a large proportion of the sites containing *Megantereon*, as suggested by the possible flesh eating / bone-consuming link made by Turner (1992), and repeated by Arribas and Palmqvist (1999). By the same token, *Pachycrocuta* appears in a slightly lower proportion of *archaeological* sites (25%) than it does in the background faunal population (38%) – whether this corresponds to the 'avoidance' relationship suggested above is unclear, but it certainly contrasts with the positive relationship suggested by its occurrence in *Megantereon* sites. Indeed, the proportion of archaeological sites presenting *Crocuta* appears the most out of line with the background data.

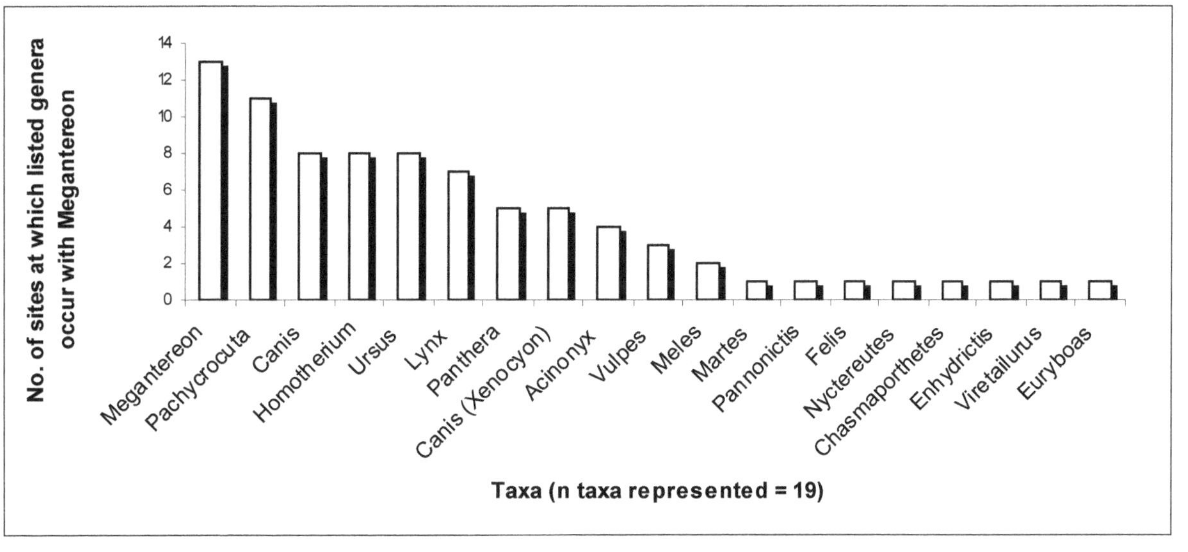

Figure 18.3. Frequency of co-occurrence of carnivoran taxa and Megantereon in European sites dating to before O.I.S. 3. (N. sites = 13).

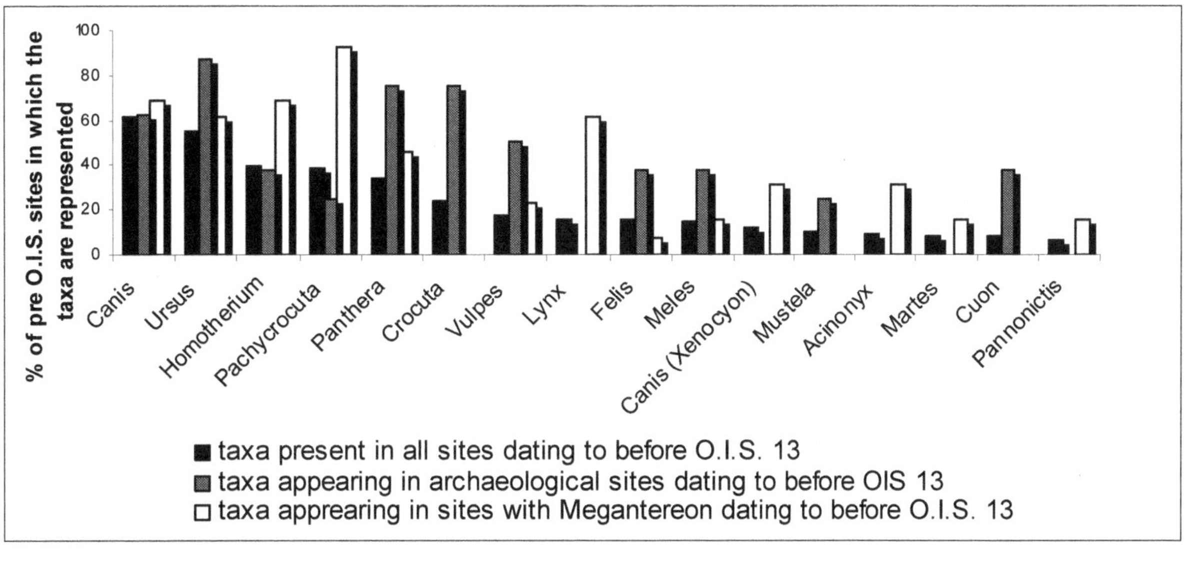

Figure 18.4. Proportion of sites dating to before O.I.S. 13 in which carnivoran taxa are present (N. sites = 76), compared with the proportion they share with hominids (N. sites = 8) and the proportion they share with Megantereon (N. sites = 13). (Megantereon not included in the display).

A Chi-squared test of the goodness of fit[1] between the presence of *Megantereon* and archaeological evidence at sites finds that the pattern of their co-occurrences and absences from sites is in the proportions one would expect, given their relative frequencies. However, the same test carried out with archaeological evidence and *Crocuta* reveals a possible relationship; they seem not only to co-occur at sites, but to be also 'co-absent' from sites more often than expected (X^2 = 10.443, $p<0.05$). Unfortunately, this interpretation is not secure due to the small sample size; the expected cell count for the co-occurrence of archaeological evidence with *Crocuta* was <5. A stronger result for the relationship between *Megantereon* and *Pachycrocuta* (X^2 = 13.421, $p<0.05$) confirms the observations above; they too display positive covariation. These results confirm the statistical significance of the intuitive interpretations of data presented in *Figures 18.1-18.3*, discussed above. More in-depth investigations into relationships, such as these, will increase our understanding of the character of the large carnivore guild that early Mediterranean hominids had to negotiate.

Conclusion

The preliminary investigation of the actual record of large carnivore fauna at archaeological and non-archaeological sites in Mediterranean Europe fails to support propositions about an ecological connection between hominid and *Megantereon*, apart from the one easiest to make: that both *Megantereon* and hominids occur in Spain Italy and Greece at roughly the same time. They do not appear in the same sites any more frequently than one would expect given their relative frequencies. Nor, however, does the evidence support an avoidance relationship. That this approach to the data is capable of revealing such relationships where they exist, is indicated by the results for the covariation between *Megantereon* and *Pachycrocuta* (which previous authors have drawn attention to), and the likelihood of a similar relationship between hominid activity and *Crocuta* (which they have not). The statistics provide the patterns, but the capacity to explain them will only come as the taphonomic, behavioural and ecological factors appertaining to the fauna and sites in question are incorporated into the discussion.

Meanwhile, however provocative and apparently plausible the idea of a hominid – sabre-toothed cat scavenging ecology may appear, it is not so far supported by the data. I think it more likely that a complex interaction of several factors, such as geographical position and biotope, interacted with as yet unidentified permutations of predator and prey species to encourage, or enable hominid incursion into Mediterranean Europe. The aim of my ongoing research is to identify these variables and their interactions, enabling us to expand our understanding of the ecological limitations of Early to Middle Pleistocene hominids.

[1] This test was carried out on a sub-sample consisting of 58 sites for which n taxa >1

Bibliography

Anyonge W. 1996. "Microwear on canines and killing behavior in large carnivores: saber function in *Smilodon fatalis*" *Journal of Mammalogy* 77 (4): 1059-1067.

Arribas A. & Palmqvist P. 1999. "On the Ecological Connection Between Sabre-tooths and Hominids: Faunal Dispersal Events in the Lower Pleistocene and a Review of the Evidence for the First Human Arrival in Europe" *Journal of Archaeological Science* 26: 571-585.

Blumenschine R.J. 1988. "An experimental model of the timing of hominid and carnivore influence on archaeological bone assemblages" *Journal of Archaeological Science* 15: 483-502.

Ewer, R.F. 1954. "Sabre-toothed tigers". *New Biology* 17: 27-40

Gliozzi E., Abbazzi L., Argenti P., Azzaroli A., Caloi L., Carpasso Barbato L., di Stafano G., Esu D., Ficcarelli G., Girotti O., Kotsakis T., Masini F., Mazza P., Mezzabotta C., Palombo M.R., Petronio C., Rook L., Sala B., Sardella R., Zanalda E. & Torre D. 1997 "Biochronology of selected mammals, molluscs and ostracods from the Middle Pliocene to the Late Pleistocene in Italy. The state of the art". *Rivista italiana di paleontologia e stratigrafia.* 103: 369-388.

Kurtén, B. 1968. *Pleistocene Mammals of Europe.* London.

Lewis M.E. 1997. "Carnivoran paleoguilds of Africa: implications for hominid food procurement strategies" *Journal of Human Evolution* 32: 257-288.

Marean C.W. 1989. "Sabertooth cats and their relevance for early hominid diet and evolution" *Journal of Human Evolution* 18: 559-582.

Marean, C.W. & Ehrhardt C.L. 1995. "Paleoanthropological and palaeoecological implications of the taphonomy of a sabetooth's den" *Journal of Human Evolution* 29: 515-547.

Martínez-Navarro B & Palmqvist P. 1995. "Presence of the African machairodont *Megantereon whitei* (Broom 1937) (Felidae, Carnivora, Mammalia) in the Lower Pleistocene site of Venta Micena (Orce, Granada, Spain), with some considerations of the origin, evolution, and dispersal of the genus" *Journal of Archaeological Science* 22: 569-582.

Martínez-Navarro, B. & Palmqvist P. 1996. "Presence of the African Saber-toothed Felid *Megantereon whitei* (Broom 1937) (Mammalia, Carnivora, Machairodontinae) in Apollonia-1 (Mygdonia Basin, Macedonia, Greece)" *Journal of Archaeological Science* 23: 869-872.

Turner A. 1987. "*Megantereon - cultridens* (Cuvier) (Mammalia, Felidae, Machairodontinae) from Plio-Pleistocene deposits in Africa and Eurasia, with comments on dispersal and the possibility of a New World origin" *Journal of Paleontology* 61 (6): 1256-1268.

Turner A. 1990. "The evolution of the guild of larger terrestrial carnivores in the Plio-Pleistocene of Africa" *Geobios-lyon.* 23: 349-368.

Turner A. 1992. "Large Carnivores and earliest European hominids: changing determinants of resource availability during the Lower and Middle Pleistocene" *Journal of Human Evolution* 22: 109-126.

Turner A. & Antón M. 1996 "The giant hyaena, *Pachycrocuta brevirostris* (Mammalia, Carnivora, Hyaenidae)" *Geobios* 29 (4): 455-468.

Turner A. & Antón M. 1997. *The big cats and their fossil relatives: an illustrated guide to their evolution and natural history.* New York.

Numismatics, Hellenism, and the Enemies of Alexander Jannaeus

Killian McAleese - Trinity College, Dublin

In this paper, I will examine some aspects of Judean coinage from the end of the second and the beginning of the first centuries BCE. I would like to point out some principal iconographic elements and legends, and observe their interpretation by modern commentators, particularly in reference to their use as a "numismatic index" (Madden 1967: 69-70) to events described in historical sources. It is my intention, upon doing this, to highlight a potentially successful way to establish a broader picture in which to interpret the events that I will briefly describe.

To begin, the historian Tacitus refers to our period and region of study:

> "*Later on, since the power on Macedon had waned, the Parthians were not yet come to their strength, and the Romans were far away, the Jews selected their own kings*".[1]

He refers here to the "kings" of the Hasmonaean dynasty, whose predecessors, from the second half of the second century BCE, gradually asserted Judean independence and territorial expansion in the wake of the increasingly troubled Seleucid kingdom.[2] Jonathan, brother and successor of the famous Judah Maccabeus, had approximately doubled the territorial extent of Judea by his death in 143 BCE.[3] Jonathan's brother Simon continued to assert Jewish independence, apparently with popular support.[4] We are, for example, told in the first book of *Maccabees* that he was granted Seleucid royal approval to mint his own copper coins,[5] although it is generally accepted that he did not produce any.

John Hyrcanus, Simon's son and successor, significantly added to the Hasmonaeans' territories during his reign from 133 to 104, with the incorporation of areas including Samaria to the north and Idumaea to the South of Judea. Following Hyrcanus' death, his son Judah Aristobulus reigned for approximately one year between 104 and 103, incorporating Galilee, north of Samaria, and taking the significant steps of turning the Hasmonaean dynasty into a monarchy, and, according to the author Josephus, being the first in his line to call himself "king".[6]

The main focus of my interest, here, is Aristobulus' brother, Alexander Jannaeus, who succeeded him in 103 BCE and became the longest serving monarch of his dynasty, ruling until his death in 76. His military gains spread the Hasmonaean kingdom to its furthest territorial extent. Indeed, according to Lester Grabbe (1994: 102), territorial acquisition is one of the two main characteristics of Josephus' descriptions of his reign in the works *The Jewish War* and *Antiquities of the Jews*.[7] The other is internal opposition.[8] If Jannaeus' conquests were particularly aggressive, as the sources suggest,[9] internal resistance to his rule seems to have been equally vigorous. There are two separate instances of sedition described by Josephus. One takes place at a festival in Jerusalem, where Jannaeus was pelted with citrons while sacrificing in his capacity as High Priest.[10] An apparently more significant six-year revolt is described in a separate, longer, section. These six years of civil war, according to Josephus, caused some fifty thousand deaths. During its course, the Seleucid dynast Demetrius III Eukairos was invited to Judea to aid Jannaeus' enemies. It ended with the reported mass crucifixion of eight hundred of these enemies, and, apparently, another eight hundred fleeing the kingdom.[11] These events, according to James S. McLaren (1991: 56), seem to have taken place some time between 96/5 BCE and 86/5 BCE.[12]

The two main questions that have been put to our available evidence are: *who* instigated this rebellion, and *why* did they do so, apparently so vigorously and over such a sustained period of time? The obvious narrative sources are the two aforementioned works of Josephus.[13]

---

1. *Hist* V 8:4-5. Trans. Moore, C. H. Cited by Habicht (1989: 352).
2. See Habicht 1989 for Seleucid decline.
3. According to *1 Maccabees*, he won control of the coastal cities of Joppa, Ascelon and Gazara (*1 Maccabees* 15-16) and three districts of southern Samaria. See Bickerman (1962: 137-143) who writes that from the reign of Jonathan, "Judea becomes a Hellenistic principality" (1962: 143).
4. *1 Maccabees* 14:25-34. All biblical citations are the New Revised Standard Version translation.
5. A "letter" from Antiochus VII (*1 Maccabees* 15:1-9) includes a list of concessions made to Simon by the Seleucid government.
6. *AJ* XIII 301. See note 8 for information on this reference.
7. The Greek version of *The Jewish War*, according to John Bartlett (1985: 79), was probably first published some time between 75 and 79 CE (1985: 76) and *Antiquities of the Jews* in 93/4 CE.
8. *The Jewish War* I 88-89, 90-98; *Antiquities of the Jews* XIII 372-73, 376-81. The term *stasiasdo* is used by Josephus (*BJ* I 88), which can be translated as "quarrel" or "dispute", as well as "revolt".
9. See Kasher on the destruction of Hellenistic cities such as Gaza (1990: 145-149) and Gadara (Kasher 1990: 152-153) and the so-called policy of *Judaization* (1990: 142 and 156-158). Archaeology, according to Kasher, contradicts some of Josephus' reports, although most of the evidence for this argument comes from Jerusalem, the Jordan Valley, and the Judean Desert (1990: 162).
10. *Antiquities of the Jews* XIII 372-73. Josephus' description, here, constitutes a separate narrative, but the incident described may well have been one of many that occurred as result of a greater civil unrest.
11. *Antiquities of the Jews* XIII 380. 4QpNah (Commentary on Nahum) from the Qumran library may also refer to the mass crucifixion. See Halperin 1981.
12. Kasher, however, believes that the end of this "internal crisis" was gradual, happening between 83 and 80 BCE, when the confidence of the people was restored (1988: 99).
13. The books of 1 and 2 *Maccabees*, the Qumran library (1QpNah for example), and later rabbinic sources have also been employed to gain information about the ideology and social conditions of the time. See Mendels 1987 on *1 Maccabees* and ideology during the Hasmonaean period, and Vanderkam 1994 on the Qumran library.

Many, however, have looked to other sources including Jannaeus' coinage in order to answer more ideological questions, and this shall be my primary focus here.

For about a decade after the publication of Ya'akov Meshorer's work on Ancient Jewish coinage[14] the scholarly consensus was that Alexander Jannaeus was the first Hasmonaean to mint coins. It has since been convincingly argued that John Hyrcanus had previously minted his own issue.[15] The obverse of one of these bears the Hebrew inscription *yhwhnn hkhn hgdwl whbr yhwdym* ("Yehohanan the High Priest and the Council/Assembly of the Jews") surrounded by a wreath. A number of coins bear the variant "Yehohanan the High Priest, *head* of the Assembly of the Jews"). The reverse bears double cornucopiae and a pomegranate. Scholars including Eric Gruen have suggested that the inscriptional variation may imply a shift towards greater personal power on the part of Hyrcanus in the latter period of his tenure, which Gruen suggests may have been enabled by the consolidation and passification of his conquered territories (1998: 36).[16] The legend, as I have said, is in Hebrew, and the text is what Birnbaum terms "Neo Palaeo-Hebrew" (1971: 77). This alphabet form seems to have been largely out of use, at least functionally, for two hundred years or more, since the striking of "Yehud" obols in Persian period Jerusalem circa 350 BCE. The Palaeo-Hebrew script can be seen on the reverse, to the right of a falcon. The lily flower on the obverse reappears on later Seleucid and Hasmonaean coinage.[17] Both Hyrcanus' and Jannaeus' issues are confined to small denomination bronze coins.

These coins set a certain precedent for those of later Hasmonaeans, and very similar symbols and legends are employed. Hyrcanus' successor, Judah Aristobulus I, does not seem to have minted coins.[18] Alexander Jannaeus, however, minted several issues, and, interestingly, employed three different languages for his coin legends: Hebrew, Aramaic, and Greek. One series is very similar to that of John Hyrcanus, again, with Hebrew legend and Jannaeus identified as High Priest. Important for our purposes, however, are two groups that are exclusive to Jannaeus' reign. One, which I will call Group 1, bears a lily flower with the Hebrew legend "Yehonatan the King" on the obverse, and on the reverse, an anchor surrounded by a knotted diadem with the Greek inscription "Alexander the King". Another, Group 2, bears the Greek inscription "Alexander the King" surrounding an anchor on the obverse. A star and knotted diadem appear on the reverse, and contained within is the inscription "Yehonathan the King the Priest" in Hebrew. Jannaeus, therefore, is the first Hasmonaean whose coins bear a legend in the Greek language, and the first to call himself king on this particular medium, in both Greek and Hebrew. (According to Josephus, his predecessor Judah Aristobulus was the first to adopt the title.)[19]

The reign of Jannaeus is complicated from a numismatic point of view, but here, at least, it is clear that he is expressing his authority in a very Hellenistic way. The star symbol, here, is a borrowing of the "Macedonian starburst" from Seleucid coins that, according to Avi-Yonah (1967: xxv), stands for the radiate crown particularly associated with royal portraiture, such as on the coins minted in Jerusalem under the Seleucid ruler Antiochus VIII.[20] The diadem is also a prominent symbol of Hellenistic kingship associated with portraiture (Mendels 1992: 57). We can see, then, that the combination of these two symbols appears in place of a portrait of Jannaeus. An actual portrait does not appear on Jewish-minted coins until the time of Philip, son of Herod the Great, almost one hundred years later, but we see here, at this earlier stage, that there was certainly a will to *represent* oneself. While the anchor may allude to the Hasmonaean family tomb which is described in the book of *1 Maccabees* as bearing carved ships,[21] it is clearly another prominent symbol borrowed from Seleucid coinage, appearing on several preceding series of coins, including those of Antiochus VII, minted in Jerusalem around 130 BCE.[22]

Here, then, we can see that Alexander Jannaeus is overtly announcing his kingship through a set of international Hellenistic symbols. Rajak emphasizes this change in symbolism from the coins of Hyrcanus to those of Alexander Jannaeus. She argues that while we can see from the deliberately archaic script of Hyrcanus that the Hasmonaean dynasty at this stage did not wish to be perceived as just another Hellenistic state, Jannaeus is clearly representing himself in more international terms (Rajak 1990: 142-144). It must be noted, however, that this does not mark a complete change in outlook in the dynasty. Jannaeus clearly continues to mint coins in Hyrcanus' style, and these, after all, also bear Hellenistic

14 Meshorer Y. 1982. *Ancient Jewish Coinage-Vol 1: Persian through Hasmonaeans*. New York.

15 Gruen (1998: 36) highlights this, as does Grabbe (1994: 244), who refers to two unpublished collections of the time, one in the ownership of the Hebrew University, and the other from Mt. Gerizim.

16 See also Jeselsohn 1980: 11-17, and Sperber 1965: 85-93 on the possible significance of the use of the term *hbr*.

17 Meshorer highlights the development of this motif through coins issued by Antiochus VII and VIII, Alexander Jannaeus, and Hyrcanus II (1982: 62).

18 Bickerman (1962: 154) assumes that at least some of the coins bearing the name *yhwdh* are to be dated to 104/3BCE. Grabbe, however, points out that there are too many die variants in these issues for them to have been issued in just one year, and that they are more likely to be dated, as a whole, to the time of Judah Asistobulus II (1994: 244).

19 *Antiquities of the Jews* XIII 301.

20 See Meshorer 1982: 40. The star is also seen as the international symbol of good-luck or chance, and other scholars, including Goldstein, have preferred to emphasize its allusion to the "star out of Jacob" of Numbers 24:17 (Goldstein 1989: 336). Antiochus VIII ruled 126/5-96 BCE.

21 *1 Maccabees* 13:29. See Rajak 1990: 271.

22 Gruen 1998: 36. Meshorer (1982: 39) discusses the implications of these issues for Hasmonaean coinage, although his conclusions may be in doubt since his 1982 thesis on the dating of the origin of Hasmonaean coins is now disputed.

symbols such as the cornucopia: another Seleucid numismatic motif with precedents on other coins (Gruen 1998: 36).

One interesting feature of Hasmonaean coinage is evidence for re-striking. We have examples of Group 1, the lily-anchor coins, having clearly been over-struck on the obverse with the legend "Yonatan the High Priest and the Assembly of the Jews", and also on the reverse with double cornucopiae. Goldstein (1989: 334) states that "plentiful and incontrovertible numismatic evidence shows that he [Jannaeus] relinquished the royal title or renounced the wearing of a diadem".[23] I will return to this point.

It has been suggested that these developments provide a numismatic index in the following way. The main protagonists of the six-year revolt have often been identified as the Pharisees.[24] This is largely due to Josephus' description in *Antiquities of the Jews* of Alexander Jannaeus' deathbed speech to his wife, Salome Alexandra,[25] in which he advises her to befriend the Pharisees and yield power to them in order to gain popular support. There are two problems with drawing this conclusion: the first is that at no point is the direct connection made with the Pharisees and those revolting against Jannaeus. Indeed, at no point in the actual narrative description of the events of the revolt are they mentioned. His enemies are, however, called, "the nation" or "the nation of the Jews". The second problem is that by the end of the revolt, which Kasher believes to have been about 83 BCE, Jannaeus' "most powerful" enemies are supposedly all either dead or in exile. The Pharisees, in the speech, are said to reside in Jerusalem at this time.

Numismatic evidence records, according to Madden (1967: 69-70) for example, and Goldstein (1989: 334), the adoption of the title king, followed by the denouncing of this title, and then its re-adoption in a more "watered-down" form. Several scholars suggest that this represents Jannaeus adopting a public image that was seen as too Hellenistic for the Pharisees, and one in which he insensitively represented himself as a Greek "king". Then, in order to make reparations to the Pharisees, and the nation, he renounced the title. Certain coins were over-struck, and on others, dated to the end of Jannaeus' reign, the diadem, a symbol of Hellenistic kingship, was replaced by a solid circle, according to Goldstein (1989: 335-336).

Meshorer (1982: 61), however, concludes that the solid circle is simply a schematic form of the diadem, and it must be observed that the overall iconographic effect remains unchanged.[26] Thus we have examples of type 2 anchor-star coins with "year twenty" and "year twenty-five" dates on them. The latter date two years before Jannaeus' death, giving us a strong indication that the title and diadem were not denounced. We must also observe that the star symbol is probably a *more* explicit reference to kingship anyway, and this symbol remains constant.

There is clearly ideology at work with the Hasmonaean dynasty's use of both Hebrew and Greek, the former giving a sense of longevity or, perhaps, "national" identity,[27] and the latter reflecting a more international outlook.[28] Neither of which, importantly, was the colloquial of Palestine in the first century BCE. Indeed, as we have seen, both languages appear together on coins. This is a manifestation of what Gruen (1998: 9) calls the "complex pattern of reciprocal relations and mutual dependency that undermines the concept of fundamental antagonism" between Hellenistic and Jewish culture. We must observe the wider context in trying to explain why Alexander Jannaeus' reign was, in many ways, such a traumatic one, and be aware, in that context, that he was also the most aggressive Hasmonaean militarily, and that he expanded the kingdom to its furthest limits. His territories suffered up to five violent incursions during his reign by outside forces, one, apparently, immediately preceded the six years of civil war and three during it. When we look at this disturbance in the context of what was going on in Syria, a bigger picture begins to emerge.

After the assassination of Antiochus VIII in 96 BCE, the year of the beginning of our revolt according to McLaren, an extensive ten-year period of bitter dynastic conflict began. This is the wider context in which we must view Alexander Jannaeus' control problems, especially in light of the fact that we know that one of those warring Seleucid dynasts, Demetrius III, actually participated in the war against him. Jannaeus' independence and territorial acquisitions must be seen in light of Seleucid decline, but that decline was not peaceful.[29] The archaeological record provides a mixed picture of the economic conditions of the time in Palestine, with Gamala, in the furthest limits of the kingdom, apparently reaching its economic apex during Jannaeus' reign, while

[23] He bases his theory regarding the diadem partly upon later issues from Jannaeus' reign on which he believes the diadem is replaced by a thick circle. Meshorer, however, prefers to view this as a "schematic form" of the earlier diadem (1982: 61).

[24] This view is widespread. Rabin (1956: 10), however, registers doubt, due to the Pharisees residing in Jerusalem and McLaren (1991: 58 n.1) also acknowledges this.

[25] *Antiquities of the Jews* XIII 4004-402. See also BT *Sotah* 22a.

[26] Goldstein remarks that these examples are "wretchedly carved" (1989: 335). This may indirectly support the idea that the circles are simply schematic forms, as may be expected from lesser quality workmanship.

[27] Birnbaum (1971: 78) discusses the Hebrew language and text on Hasmonaean coins, but does not support the idea of "nationalism" in this context, rather a following of Persian period precedent. See Mendels 1992: 63 on the palaeo-Hebrew script and the idea of a Jewish "national" king.

[28] For a general study on the use of Greek in the ancient Jewish context see Barr 1989: 98-114.

[29] See Shipley (2000: 322) for a detailed description of Seleucid wars and dynastic struggles during Jannaeus' reign. Appian's *Syriake* describes the period and region in question.

other sites reflect destruction (Meshorer 1982: 46).[30] The course of investigation I believe would be appropriate is a further examination into the economic conditions of the time, from an archaeological point of view, because the literary sources, in my opinion, reflect an atmosphere of uncertainty and social and economic stress.

If we are to use the coinage of Alexander Jannaeus as a "numismatic index" we must, of course, bear in mind the ideological impact that these widely distributed small denomination *prutot* may have had upon the people at the time. But we must be aware that there are several categories of evidence available, which strongly reflect some of the same problems of the time in adjacent regions. One category of evidence, or indeed one region, must not merely be considered in isolation.

Bibliography

Avi-Yonah M. 1967 "Prolegomenon", in F.W. Madden, *History of Jewish Coinage, and of Money in the Old and New Testament.* New York.

Barr J. 1989. "Hebrew, Aramaic and Greek in the Hellenistic Age", *The Cambridge History of Judaism*, Vol 2: *The Hellenistic Age*, Cambridge: 79-114

Bickerman E. J. 1962. *From Ezra to the Last of the Maccabees: Foundations of Post-Biblical Judaism.* New York.

Birnbaum S. A. 1971. *The Hebrew Scripts.* Part 1, *The Text.* Leiden.

Goldstein J.A. 1989. "The Hasmonean Revolt and the Hasmonean Dynasty", In Davies, W. D. and Finkelstein, L. (eds.), *The Cambridge History of Judaism, Vol. 2: The Hellenistic Age.* Cambridge: 292-351.

Grabbe L.L. 1994. *Judaism from Cyrus to Hadrian* (2nd edition). London.

Gruen E.S. 1998. *Heritage and Hellenism: The Rejuvenation of Jewish Tradition.* Berkeley, London.

Habicht C. 1989. "The Seleucids and their Rivals" *The Cambridge Ancient History* Vol. III: *Rome and the Mediterranean to 133 B.C.* Cambridge: 324-387.

Halperin D.J. 1981. "Crucifixion, the Nahum Pesher, and the Rabbinic Penalty of Strangulation", *Journal of Jewish Studies* XXXII No. 1.

Jeselsohn. D. 1980. "Hever Yehudim – A new Jewish Coin", *Palestine Exploration Quarterly* 112: 85-93.

Kasher A. 1988. *Jews and Hellenistic Cities in Eretz-Israel: Relations of the Jews in Eretz-Israel with the Hellenistic Cities during the Second Temple Period (332 BCE-70 CE).* Tübingen.

Kasher A. 1990 *Jews, Idumaeans, and Ancient Arabs: Relations between the Jews in Eretz-Israel with the Nations of the Frontier and the Desert during the Hellenistic and Roman era (332 BCE-70 CE).* Tübingen.

Madden F. W. 1967. *History of Jewish Coinage, and of Money in the Old and New Testament*, New York.

McLaren J.S. 1991. *Power and Politics in Palestine: the Jews and the Governing of their Land, 100 BC-AD 70.* Sheffield.

Mendels D. 1987. *The Land of Israel as a Political Concept in Hasmonean Literature: Recourse to History in Second Century BC Claims to the Holy Land.* Tübingen.

Mendels D. 1992. *The Rise and Fall of Jewish Nationalism.* New York.

Meshorer Y. 1982. *Ancient Jewish Coinage*-Vol 1: *Persian through Hasmonaeans*, New York.

Rabin C. 1956. "Alexander Jannaeus and the Pharisees" *Journal of Jewish Studies* VII: 3-11.

Rajak T. 1990. "The Hasmoneans and the uses of Hellenism" in, P. R. Davies & R. T. White (eds.), *A Tribute to Geza Vermes: Essays on Jewish and Christian Literature and History.* Sheffield: 261-280.

Schwartz S. 1991 "Israel and the Nations Roundabout: 1 Maccabees and the Hasmonean Expansion" *Journal of Jewish Studies* XLII: 16-38.

Shipley G. 2000. *The Greek World after Alexander, 323-30 B.C.*, London.

Sperber D. 1965. "A Note on the Hasmonean Coin Legends Heber and Rosh Heber", *Palestine Exploration Quarterly* 97: 85-93.

Vanderkam J.C. 1994. *The Dead Sea Scrolls Today.* Grand Rapids, London.

30 Certain sites, such as Ein Gedi and Beth Sahur, also reveal coin hoards, although the purposes of these are not entirely clear, and could have been associated with good-luck charms (Meshorer 1982: 46 n. 30), especially considering the low value of the coins contained therein.

Gendering figurines, engendering people in early Aegean prehistory

Maria Mina, University College London

Anthropomorphic figurines have always attracted the attention of archaeologists, anthropologists and art historians. The interest is even greater, however, when they come out of prehistoric contexts when symbolic artifacts have been used to provide insights into past societies in the absence of other cultural indicators. In the case of Aegean early prehistory the study of anthropomorphic figurines has not followed a uniform strategy of study. Instead, there is a distinct dichotomy between the way in which Neolithic and Early Bronze Age figurines have been approached and interpreted. In both cases, the methodological route and the obvious biases have hindered the understanding of past societies and their people as attested by anthropomorphic figurines. In this paper I propose theorectical and methodological avenues that can offer a more holistic interpretation of Aegean prehistory through the uniform study of anthropomorphic figurines. *Figure 20.1* summaries how a gender approach can provide us with insights into the minds of prehistoric people which in turn can lead to a fuller understanding of Aegean prehistory.

Even though little interest has been expressed specifically in the study of Aegean figurines and only a few systematic works have been conducted, I will give you a brief review of the main lines of thought as a way of placing their study in the wider academic context.

The first wave of works relating to anthropomorphic figurines dates mainly to the period of the fifties and sixties and all have argued for the existence of the Mother-Goddess cult in a matrilineal society, with the exception of Ucko (1968). In these models Neolithic culture was equated with the earliest stages of development, while later patriarchal societies were viewed as a product of more complex formations. Also in the sixties with the advent of feminism a number of archaeologists following a gynocentric argument attempted to exclude men from the past and argued for a natural biological superiority of women over men. In the 1970s and 1980s a new wave of androcentric interpretations counteracted the previous theories by relating female figurines to men's sexual gratification. What they all have in common, however, is that their main weaknesses lie in the fact that the evidence was used selectively, as a way of structuring the arguments, according to individual archaeologists' biases and expectations from the archaeological record.

Moving on to the Early Bronze Age, the attention that especially Cycladic figurines have attracted is far greater than that for the Neolithic ones. As Gill & Chippindale (1993) have pointed out, most of the studies (see Getz-Preziosi 1987a, 1987b; Renfrew 1991, Thimme 1977) have concentrated on establishing typologies, tracing Masters and compiling glossy museum catalogues which resulted in elevating figurines to a high aesthetic status. With the high demand for Cycladic figurines in the art market, we have reasons to believe that figurines without provenance may be products of forgery.

Academics, however, have conducted research on such unprovenanced museum pieces despite the invalidity that such actions cast on their work. Moreover, the majority of these studies impose modern perceptions concerning the artist's place in his/her society, as well as an invented set of rules that the artist was believed to have followed as in the Classical period (Gill & Chippindale 1993). Apart from the aesthetic appreciation expressed in many of the works, figurines have occasionally been linked to the later Greek pantheon and some androcentric ideas about women's place in a patriarchal society. It is evident, therefore, that on this side of the fence as well, works which do not relate in any way to Neolithic figurines have also followed the same argument that the Neolithic expresses the animus, while the more complex Early Bronze Age provided the basis for the later patriarchal, rational social order.

Recent works mainly on Neolithic figurines of South-East Europe show a clear separation from the earlier interpretations (see Bailey 1994a, 1994b, 1996; Hamilton 2000; Hitchcock 1997; Kokkinidou & Nikolaidou 1997; Langdon 1999; Lee 2000; Talalay 1987, 1991, 1993, 2000). On the other hand, systematic and extensive works on Aegean Early Bronze Age figurines and their meaning are missing. Taking into account the studies carried out so far on early Aegean anthropomorphic figurines there is a clear distinction between those of the Neolithic and those of the Early Bronze Age. Even though in both cases the same class of evidence is analysed, there has been no attempt to contrast and compare the use and meaning of figurines throughout early Aegean prehistory. There is an artificial boundary that is maintained and keeps apart the studies of Neolithic and Early Bronze Age figurines. The existence of such a dichotomy perpetuates the separation in the way figurines have been and continue to be approached. Concerning the Neolithic, in the context of new figurine studies conducted on Balkan and Anatolian anthropomorphic figurines, there has also been a turn in the way they are approached in the Aegean. With the case of Cycladic figurines, however, there is an unwillingness to challenge existing studies and methods of approach.

I will now turn to my area of research that focuses on the study of Neolithic and Early Bronze Age Aegean figurines from a gender perspective, which I consider useful and fruitful in a number of ways.

My decision to apply a gender approach to the study of anthropomorphic figurines serves my research at two levels. On one level, I need to be explicit about my own research agenda and interests which have influenced my

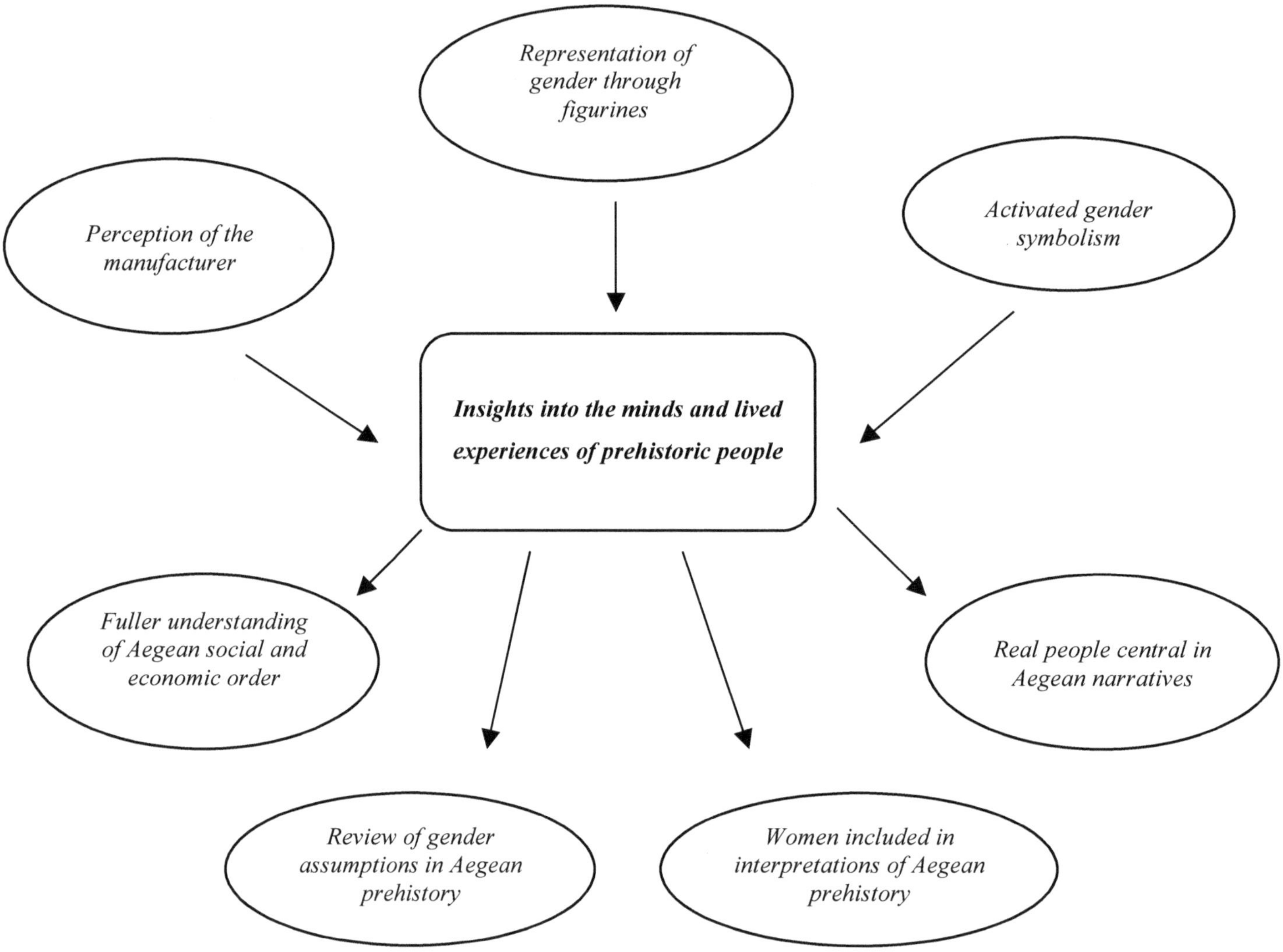

Figure 20.1 Gender approach to anthropomorphic figurines and the implications for our interpretation of Aegean prehistory

decision to approach my data from this perspective, because I believe that being aware of the political interests which form our interpretations, and more importantly making them known to our audience, can lead to fewer uncritical hypotheses. My opinion, therefore, is that the past is often interpreted by imposing our own social preconceptions on how prehistoric societies may have been organised and thus trapping gendered actors in our own social models.

In the case of early Aegean prehistory what soon became apparent to me is that what is lacking from many interpretations of Aegean society, and what has in turn shaped my theoretical concerns, is that we need to make not only women visible in the archaeological record, but also to include in our interpretations a number of gender and age-related experiences lived by prehistoric individuals. Through a gender approach, men can also be potentially viewed beyond the long-held stereotypes which trap them exclusively in the roles of hunters, warriors, or exploiters and prevent us from seeing the cultural roles that men would have played in their societies and in relation to other genders.

On a different level, by applying a clear theoretical framework on both sets of data we take the first step in ensuring that no conscious preconceptions colour our interpretations. The perspective of gender archaeology provides us with a uniform theoretical context in which we can begin to approach anthropomorphic figurines as the same class of material which is in continuous use in early Aegean prehistory. Moreover, a gender approach with the specific questions it raises gives shape to the issues that need to be discussed and moulds a very clear set of methodological rules. The methodology resulting from a gender perspective guards against assumptions that are not based on a systematic study of anthropomorphic figurines. Bearing in mind that most interpretations have been based on biases concerning gender, following a methodology tailored for gender theory allows a critical view on previous works and exposes any preconceptions and unfounded assumptions.

The construction of a methodology resulting from gender archaeology, therefore, focuses on a number of formal attributes which provide us with a list of points against which we can begin to base our understanding of the object represented in the form of an anthropomorphic figurine.

Specific to my research, I have compiled a detailed database in which I record the presence or absence of anatomical characteristics, use of colour and material, detailed break-down of decorative motifs, posture, as well as a list of categories referring to the chronology, context and site of these figurines of both periods. When all these categories are then filtered electronically, patterns begin to emerge which link the represented gender in association with the cultural, economic and chronological context. Such analysis in turn provides us with a systematically sound basis on which we can start to build the next stage of interpretation.

Applying a uniform gender perspective allows me also to bridge the artificial gap between Neolithic and Early Bronze Age Aegean figurines and to detect changes and continuing patterns in their representational form through early prehistory. As the first results of my analysis begin to show, the patterns of figurine use which characterise the Early Bronze Age are already emerging at the end of the Late Neolithic and through the Final Neolithic, which suggests that we are not justified in studying them separately. Moreover, the typological form of Early Bronze Age figurines does not appear abruptly; instead, we need to trace the first signs of change in the transitional period and also include all those figurines which do not follow the common Cycladic repertoire. The anthropocentric character of a gender approach, therefore, and its uniform application allows us to focus on how the human body is rendered through time in terms of its anatomy and cultural identity.

Gender archaeology also concentrates our attention on the symbolism of the represented human body. Concentrating, therefore, on specifically selected attributes means that my study of Cycladic figurines moves away from previous typological schemes and an aesthetically loaded terminology. Such systematic approaches treat figurines as symbolic objects the meaning of which can be found in aspects that surpass aesthetic evaluation and art history canons.

In addition, gender archaeology is an anthropocentric perspective which is ideal for the study of a class of material in the form of human, sexed bodies. A gender approach will not only aid me in rectifying previous gynocentric and androcentric interpretations, but will also allow me to bring into discussion the important parameter of gender which played a major role in the construction of social identities. Starting from the obvious point that Aegean anthropomorphic figurines are deliberately and clearly marked in order to distinguish mainly male from female figurines, we can then realize that a gender approach is not arbitrarily imposed but instead is ideally suited to such type of data in a number of ways.

The most obvious point is that the physical body is represented in the shape of a figurine and the area of research on the body is discussed explicitly in the framework of gender archaeology. The physical body is tightly linked to gender theory because it is involved in the construction and performance of gender with foundations on both biological experiences and cultural factors. Since the majority of anthropological studies have shown that the physical body is the main template on which gender categories are built, we can then begin to realize just how vital is the study of the represented body as expressed through figurines for the understanding of gender and social identity in general.

Moreover, the study of figurines through the gender spectrum also allows us to understand how physical and social identity is perceived and experienced by a member of a prehistoric society as a social actor but also as the manufacturer of such type of artifacts. The deliberate presence or absence of anatomical features on the figurines is one indicator that the manufacturer was making a conscious choice on the way he or she wished to communicate gender through the physical body and its decoration, clothing or hair style. Considering that symbolic objects can be actively manipulated as a way of perpetuating or challenging the social order, the choices of the manufacturer, as they change through time in terms of the represented subject matter, can offer an insight into the social process of early Aegean prehistory.

Finally, a more holistic understanding of the meaning of figurines needs to explore the ideas concerning the gender of the figurine manufacturer as a way of addressing the implications which relate to who was using figurines, how they were used and for what reason. A particularly useful approach developed by Sørensen (1996), however, brings to the surface the fact that the answers often lie in the way we ask our research questions. The methodological device she suggests is that by approaching technology and gender by trying a number of alternative models on the basis of artefactual evidence, we will eventually come to the conclusion that some scenarios are more likely and better supported by our data. By enriching our interpretation with ethnographic analogues and by including men and women in the process of figurine production, we can overcome the established preconceptions and understand better the meaning of figurines in terms of their possible uses in their cultural context.

It has become apparent, I hope, that a gender perspective is ideally suited to the study of anthropomorphic figurines. The systematic analysis of figurines creates the precondition for us to start deciphering the meaning of such symbolic objects without the burden and obstacle of art history typologies and aesthetic evaluation. In addition to providing us with a sound theoretical and methodological framework, a gender analysis of figurines guards against unfounded assumptions and brings to life the social actors of the prehistoric Aegean. It also allows us an insight into the minds of the figurine makers and how they perceived gendered individuals through their eyes and in their specific cultural context. Moreover, a gender approach to figurines has great implications for our understanding of Aegean prehistory and may also lead us to review the already existing interpretations of social organization and order.

We often conduct detailed studies on past societies in an abstract way with faceless people that do not live in a physical body which moulded and dictated their social actions. That is why gendering anthropomorphic figurines in the prehistoric Aegean means we can engender real people and social actors with lived bodies, experienced choices through their daily negotiation in their cultural context.

Acknowledgments

I would like to thank both my supervisors, Prof. R. Whitehouse and Dr C. Broodbank, for their useful comments, but also all those who worked hard to organise the successful SOMA 2003 conference in London.

Bibliography

Bailey D. 1994a. "The representation of gender: homology or propaganda" *Journal of European Archaeology* 2: 215-227.

Bailey D. 1994b. "Reading Prehistoric Figurines as Individuals" *World Archaeology* 25: 321-331.

Bailey D. 1996. The Interpretation of Figurines: the Emergence of Illusion and New Ways of Seeing, in N. Hamilton, J. Marcus, D. Bailey, G. & R. Haaland and P. Ucko, Viewpoint: Can We Interpret Figurines? *Cambridge Archaeological Journal* 6: 291-295.

Getz-Preziosi P. 1987a. *Early Cycladic Art in North American Collections.* Richmond, Virginia.

Getz-Preziosi P. 1987b. *Sculptors of the Cyclades : Individual and Tradition in the Third Millennium B.C.* Michigan.

Gill D. W. J. & C. Chippindale 1993. "Material and Intellectual Consequences of Esteem for Cycladic Figures" *American Journal of Archaeology* 97: 601-647.

Hamilton N. 2000. Ungendering Archaeology: Concepts of Sex and Gender in Figurine Studies in Prehistory, in M. Donald & L. Hurrombe (eds.) *Representations of Gender from Prehistory to the Present.* London: 17-30.

Hitchcock L. A. 1997. Engendering domination: a structural and contextual analysis of Minoan Neopalatial bronze figurines, in J. Moore & E. Scott (eds.) *Invisible People and Processes: Writing Gender and Childhood into European Archaeology.* London: 113-130.

Kokkinidou D. & Nikolaidou M. 1997. Body imagery in the Aegean Neolithic: ideological implications of anthropomorphic figurines, in J. Moore & E. Scott (eds.) *Invisible People and Processes: Writing Gender and Childhood into European Archaeology.* London: 88-112.

Langdon S. 1999. Figurines and Social Change: Visualising Gender in Dark Age Greece, in N. L. Wicker & B. Arnold (eds.) *From the Ground Up: Beyond Gender and Archaeology Conference, University of Wisconsin-Milwaukee, October 1998.* England: 23-29.

Lee M. M. 2000. Deciphering Gender in Minoan Dress, in A E Rautman (ed.) *Reading the Body: representations and remains in the archaeological record.* Philadelphia, Pennsylvania: 111-123.

Renfrew C. 1991. *The Cycladic Spirit: Master-pieces from the Nicholas P. Goulandris Collection.* London.

Sørensen M.-L. 1996. Women as/and Metalworkers, in A. Devonshire & B. Wood (eds.) *Women in Industry and Technology from Prehistory to the Present Day: Current Research and the Museum Experience Proceedings from the 1994 WHAM Conference.* London: 45-51.

Talalay L. E. 1987. Rethinking the function of Clay Figurine Legs from Neolithic Greece: An Argument by Analogy *American Journal of Archaeology* 91: 161-169.

Talalay, L. E. 1991. Body Imagery of the Ancient Aegean *Archaeology* 44: 46-49.

Talalay, L. E. 1993. *Deities, Dolls, and Devices: Neolithic Figurines from Franchthi Cave, Greece.* Bloomington & Indianapolis.

Talalay L. 2000. Archaeological Ms.conceptions: Contemplating Gender and Power in the Greek Neolithic, in M. Donald & L. Hurrombe (eds.) *Representations of Gender from Prehistory to the Present.* London: 3-16.

Thimme, J. (ed.) 1977. *Art and Culture of the Cyclades: Handbook of an Ancient Civilisation.* Chicago.

Ucko P. J. 1968. *Anthropomorphic Figurines of Predynastic Egypt and Neolithic Crete with Comparative Material from the Prehistoric Near East and Mainland Greece.* London.

Naue II swords and the collapse of the Aegean Bronze Age

Barry Molloy - Trinity College, Dublin

The most widespread and popular swords of the later Bronze Age in Europe belonged to the family of grip-tongue swords, the progenitor of which was the Naue Type II sword. It is a double-edged sword of cast bronze with hardened cutting edges and a flanged, full tang grip. By the Iron Age this was the foremost sword form being produced in Greece. This paper will examine the military context of the arrival of this sword into the Aegean and its displacement of the native sword tradition of the Bronze Age, with specific reference to the combat functions of the forms of swords in use at this time.

In order to assess the impact of the arrival of this weapon, a clear understanding of the martial context into which it was adopted must be achieved. The first weapons of sword length in the Aegean are the type A swords, which are believed to have been developed on Crete. Two examples come from the palace at Mallia dated to MM III (Middle Minoan III). However, the majority of extant examples come from the mainland with major concentrations at Mycenae and Pylos. The type A sword was soon followed by the shorter type B, which was in contemporary use. The type Di and C swords come next in this sequence, but there is a period of overlap between these four early types. The types A and C swords are very similar in blade form - they are long blades gradually tapering to a point with straight sides - and they have a thin cross section, typically 1-2 mm, strengthened by a midrib which is typically in the region of 10mm thick at the shoulders of the blade. Types A and C are frequently categorised as "rapiers" (Drews 1993: 198; Harding 1984: 153), but as discussed by Peatfield (1999: 67), they are better understood descriptively as 'cut and thrust' swords. It is not satisfactory to leave the meaning of this term implicit. As it applies to most early swords of the Aegean tradition, Peatfield points out that a brief definition is necessary. While the meaning of 'thrust' may be self-explanatory for this discussion, the meaning of 'cut' is more complex as it takes many forms involving a variety of forces. Amberger (1998: 94) specifies three actions that are present in a cutting attack:

1. Penetration by incision and percussion.
2. Penetration by incision and laceration.
3. Percussion.

It is the first two of these that are most significant with regard to Aegean swords. The so-called 'rapiers' have thin cutting edges and are regarded as relatively fragile due to their long, thin design and hafting mechanisms (Drews 1993: 198; Sandars 1963: 117). However, they can execute drawing attacks in the manner of Amberger's number 2 cut without causing significant trauma to the blade on impact. Such cuts may not cleave a limb from the body, but can easily disable or maim through severing tendons and ligaments. As with their renaissance namesakes, they are therefore capable weapons when it comes to both cutting and thrusting.

The type Di sword marks a departure from the Aegean tradition. In particular, while it retains the midrib profile of the other swords, it is on average considerably shorter. This weapon can be used to strike with some percussive force, as well as to execute draw cuts. This type is followed by the Dii sword, which again marks a significant point of departure. This sword has relatively parallel edges and a pointed oval cross-section, removing the need for a midrib, and allowing a percussive blow to cut deeper than the mid-ribbed blades, and to better withstand blows to hard objects, such as armour or other weapons. The first examples of this sword, such as those from Palaikastro and Ialysos, are dated to the late LM / LH IIIA 2 (Late Minoan / Late Helladic IIIA 2), and they set the standard for a number of subsequent sword forms. While the Gi was contemporary with the later type C sword, the Gii and F swords mainly belong to late LM / LH IIIB. Even though the earliest example of a type F sword, from Zapher Papoura (tomb 95) dates to the LM IIIA: 2 (Hatzaki, Pers. Comm. 2003) and survives into sub-Minoan / Mycenaean contexts, the example from Mouliana (tomb A) is from this later period. These are the typological successors of the Dii sword, with a probable influence from the Ei and Eii dirks, which begin to be produced in LM IIIA. The Gii sword has a roughly parallel-edged blade, although a slight leaf-shape may also be present. It also has a roughly pointed oval cross section with slight variations. The type Fi tapers rapidly to a point and is typically very short, while the Fii can be considerably longer and typically has a blade form, which resembles the Dii and Gii swords. Another source of influence on these later Aegean swords of the late LM / LH IIIB period may have been the Naue II sword described above, that evolved in the area of Northern Italy and the Balkans, and its distribution ranges from northern Italy to Egypt and the Levant. In the Aegean it was adopted alongside the swords of the native Minoan / Mycenaean tradition. Examples of local manufacture soon appeared, including some local innovations (such as the pommel spur). In fact, the latter were also adopted in the lands where the weapon first evolved.

However, it must be noted that all the elements present in the Aegean swords of this period have definite precedents and predecessors within the indigenous tradition. While the Naue ii sword achieved a high level of popularity in Greece, it existed alongside the native tradition, and appears to have been adopted into the native combat tradition (which seems to have evolved during LH IIIB). It must also be noted that it arrived in isolation from the martial package within which it was first developed.

With the change to smaller shields, probably during the LM / LH IIIB period, and certainly by the early LH IIIC, the long weapons of the palatial period were replaced by shorter stouter swords. These would have been very deadly in close quarter combat, especially when ranks closed too tightly for effective spear use, or when the spears got damaged. While a longer sword has better striking range if it is uninhibited, its length can impede dextrous usage when it is in a crowded shield wall context. A shorter sword can be used to execute tight-angled cuts and can be recovered very rapidly to a guard stance, or to strike again offensively or defensively. The replacement of the longer swords in the Aegean by short stout swords during the LM / LH IIIB may be seen as reflecting a deliberate change in the use of the swords on the battlefield. In later times, the Spartans are known to have used their short swords with lethal efficiency. Aegean warriors would have had to be mindful of the pay-off that must be made between length, speed and manoeuvrability when deciding on the dimensions and form of sword they were choosing, be it Naue II or Aegean.

There are distinct differences in the form of the Naue II sword and the contemporary Aegean swords. For example, the classic Naue II sword is more robust than the typical Aegean sword. However, the great degree of variability within these two traditions makes it impossible to isolate examples as being paradigmatic of their class. Any generalisations may be truly unrepresentative of the archaeological record. Kilian-Dirlmeier (1997) has isolated a great number of subdivisions which are very useful to the specialist, but are difficult to work into a general discussion on these weapons with regard to their combat use, especially as there are frequently a nominal number of weapons in a given sub-class or 'variant'.

With these variations in mind, it is important to note that the Aegean smiths were capable of making swords in their own tradition. These swords were as long and functional as the Northern swords, with very similar morphological characteristics to the blades (e.g. type Dii from Ialysos, the unprovenanced Dii in the Ashmolean collection, types Gii and Fii from Siteia, type Gii from Perati, type Fii from Mouliana). Claims that the Naue II sword represented a major shift in the mode of warfare are at present in many regards untenable. If this were the case, spears would be the primary battlefield weapons (much like in the later Greek periods), and swords would be auxiliary. The main evidence for this statement comes from several sources, such as from the art of the LH IIIC (most notably on the Warrior Vase and Warrior Stele from Mycenae), the amount of spearheads recorded in the archaeological record (which is greater than the swords), and from historical combat parallels (for example Anglo-Saxon, Viking and Classical Greek combat). In particular, the vast majority of societies that used spear-wielding infantrymen, from Classical Greece to the Medieval period, fought in a shield wall formation. The swords of the Late Bronze Age were close quarters weapons best suited to tight combat environs that one would expect in an organised and disciplined line of battle. An important observation is that the longer swords in use in Bronze Age Greece, were for the most part, approximately the same length as the Roman Gladius, the archetypal short sword, indicative perhaps of their close-quarters nature.

While some of the Naue II swords were indeed extremely long, nearly a meter in length, others were little longer than daggers. It has been argued that some of the Aegean swords were deliberately lengthened to compete with the Naue II sword (Dickinson 1994: 206). Despite that there are many examples of the latter being produced in short sizes comparable to the Aegean ones, an example from Mouliana is only 27.20 cm. long. It is perhaps significant that no hybrid examples have been found to date, with the broadly defined forms of each tradition being maintained and respected by the smiths, despite the longevity of their contemporaneity.

Another important factor, which is rarely discussed, concerning the exclusive adoption of the Naue II sword during the Iron Age, may have been the manufacturing process both of the bronze and iron examples. The manufacturing procedure for the Naue II sword involved the use of simple bi-valve moulds, whereby a wooden template was impressed into two strips of clay to create a two-piece mould, or, alternatively, a split stone mould could have been carved. The native Aegean swords frequently required several more pieces to accommodate their more complex form (Northover Pers. Comm.: 2003). These were often incised in further detail, after the clay impressions of the template had dried – the "Woodhouse sword" in the Ashmolean museum was probably produced using this process. Whilst simplistic functional arguments relating to this difference in production technology do not account for the widespread adoption of the Naue II, they may to some extent, explain the preference of manufacturing this sword type, perhaps exclusively by some smiths. Of equal importance may have also been the differences in the acquisition costs of these weapons to the warriors who used them, created either by intrinsic value or by differential accessibility. However when it comes to the swords of the Iron Age, forging a weapon of later Naue II design was considerably easier than the native Aegean swords, and was likely to have been a central factor to their demise alongside the age of bronze.

Since the discovery of the Aegean Bronze Age, people have looked for reasons for its collapse. The appearance of the Naue II sword has on a number of occasions been seen as symptomatic of a change in warfare (Connolly 1986: 31; Drews 1993: 193ff), but its form was not as revolutionarily different to the native sword technology as some would argue. This change in warfare has in turn been seen to represent large-scale social upheaval (Drews 1993: 193ff; Hanson 1999: 33) that was intimately related to the decline of the Mycenaean way of life. This event-based understanding of the collapse of the Bronze Age has parallels in many other collapse theories, such as earthquakes or climatic change (Drews 1993: 198; Harding 1984: 153). While it is today generally understood as a process of collapse, taking part over many decades, elements of this event-based approach are still prevalent. The arrival of the Naue II sword was not such an event; it is better understood as a long-term process, which began with its arrival before the troubled times of the Late Bronze Age, and slowly led to the displacement of the sword production, specifically of Aegean traditions. It appears to have taken more than a century to displace the native Aegean sword tradition, and was itself constantly changing in form. In a combat context, the use of the Naue II may be better seen as a personal preference, possibly due to some significant social connotations, or even prestige, which the user desired. Indeed, a Naue II and a Type F sword were found in the same tomb at Mouliana. Perhaps the various types had a notably different intrinsic value in the financial sense, due to different production technologies. Arguments promoting its inherent superiority as a weapon are too subjective (Dickinson 1994: 206) and not supported by the evidence. The two traditions were in contemporary use for over a century, which was over three generations of warriors, and swords of broadly similar morphology with regard to blade length, width and thickness *could* be produced by smiths from both sword-making traditions.

Bibliography

Amberger J.C. 1998. *The Secret History of the Sword".* California

Connoly P. 1986. *The Ancient Greece of Odysseus.* Oxford

Dickinson O. 1994. *The Aegean Bronze Age.* Cambridge

Drews R. 1993. *The End of the Bronze Age.* Princeton

Hanson V.D. 1999. *The Wars of the Ancient Greeks.* London

Harding A. 1984. *The Mycenaeans and Europe.* London

Kilian-Dirlmeier I. 1997 *Die Schwerter in Griechenland (ausserhalb der Peloponnese), Bulgarian und Albanien. Prähistorisches Bronzefunde.* Abteilung IV, Band 12.

Peatfield A.D. 1999. "The Paradox of Violence: Weaponry and Martial Art in Minoan Crete", in R. Laffineur (ed.), *POLEMOS - Le context guerrieer en Égée à l'âge du Bronze. Aegacum* 19 Liège: 67-74.

Sandars N. 1963. "Later Aegean Bronze Swords" *American Journal of Archaeology* 67: 117-153.

The Royal Armouries, 2000. "The Sword" *Arms in Action* video series. Leeds.

Urban Development and Local Identities: The Case of Gerasa from the Late Republican Period to the Mid-3rd Century AD

Rubina Raja – University of Oxford

Gerasa - historical outline and urban history

The ancient city of Gerasa, modern Jerash, is situated inland in North Jordan, in the Ajlun highlands[1] . The city covered an area of 85 ha, which was enclosed by monumental city walls in the late antique period (Kennedy 1998: 40 for a dating of the walls). It was a medium-sized provincial town and it was not situated centrally within the infrastructure of the region. The monuments that shaped the urban fabric, colonnaded streets, sanctuaries, temples, theatres, various shops and a hippodrome, show that the city was a well-furnished town that boasted a wide range of architectural complexes (*Figures 22.1 & 22.2*).

In the 60s BC, Pompey conquered the region of the so-called Decapolis cities on the Transjordanian plateau, to which Gerasa belonged (Pliny NH V, 18.74 mentions the city *Gelasa* as belonging to the Decapolis)[2]. The cities in this region were subsequently made part of the province of Syria (Millar 1993: 39). The region was situated between the Nabataean kingdom to the East and the Herodian kingdom to the West.

In the period between Augustus' rule and the creation of the province of Arabia in AD 106 the regions of the Near East underwent a number of political and administrative reorganisations. The client kingdoms surrounding the Transjordanian plateau gradually came under direct Roman political and military control.[3]

Until the mid-1st Century AD Gerasa consisted of a minor settlement on the so-called Camp Hill and of an altar-like structure on a hill west of Camp Hill.[4] Urban development at Gerasa increased from the mid-1st Century AD and the city continued to expand throughout the 2nd Century AD. The building activity during the 1st Century AD mainly concentrated on the monumentalisation of the Sanctuary of Zeus, including the construction of the first temple of Zeus on the so-called Lower Terrace (Seigne 1985 for the evidence relating to the development of the sanctuary). The South Theatre immediately west of the Sanctuary of Zeus was constructed during the reign of Domitian (Kraeling 1938: inscriptions 51-52).

The major part of building activity at Gerasa dates to the 2nd Century AD. Some of these developments have been connected to the emperor Hadrian's visit to the city in AD 129/130 (Kraeling 1938: inscriptions 30 and 144). The developments of this period were concentrated in the areas surrounding the North-South orientated main street (*Figure 22.1*), around the Sanctuary of Zeus and in the southern part of the town (*Figure 22.2*).

The main characteristics of the urban development Gerasa experienced in the 2nd Century AD included the creation of a grid plan (Seigne 1992: 331).[5] The expansion of the Sanctuary of Zeus, including the building of a temple dating to the Antonine period on the so-called Upper Terrace, and the construction of the massive Sanctuary of Artemis complex also formed part of the 2nd Century developments. Together with the layout of the Oval Piazza in the Trajanic period, a coherent urban plan came into existence during the second half of the 2nd Century AD. The orientation of the urban plan, which was to characterise Gerasa in future centuries, was established through the layout of the main streets and the associated building complexes, such as the macellum and the nymphaeum.

Urban development and local identities

Gerasa saw its first period of urban expansion during the 1st Century AD at a point in time when Roman rule was firmly established in the Transjordanian region and the political situation in the surrounding regions was stable. Several inscriptions that are concerned with the 1st Century AD expansion of the Sanctuary of Zeus attest to the existence of civic administration and religious organisation at Gerasa at this point in time.[6]

One inscription, in Greek, dates to AD 27/28 and refers to the architect who seems to have been in charge of the

1 Research concerning the history and development of the city starts with Kraeling 1938. Although many of the interpretations in this publication are outdated, it presents a large body of archaeological and epigraphic evidence. For illustrations see Fig.1 taken from the South Theatre overlooking the northern part of Gerasa. In the foreground to the right is the so-called Oval Piazza. From the piazza the main street runs northwards. The Sanctuary of Artemis is located in the background to the left. Fig.2 is taken from the Sanctuary of Artemis looking towards the southern part of the city with the South Theatre and the Sanctuary of Zeus to the right in the picture. Both photos were taken by the author. Further illustrations and plans of the city can be found in Kraeling 1938 and Seigne 1992.

2 It is widely accepted that *Gelasa* is a spelling mistake and that Pliny is referring to Gerasa.

3 For a detailed account of the region's history see Millar (1993: ch.11).

4 See Zayadine (1986: 8-9) for a summary of the evidence from the late Hellenistic period as well as further references. The so-called Camp Hill is situated east of the Sanctuary of Zeus (*Figure. 21.2*).

5 The discussions surrounding the dating of the street grid are numerous and a consensus has not been reached. The archaeological evidence, however, does not allow for a definite dating of the street grid to the 1st Century AD. It is likely that the street grid was created during the 2nd Century AD, when most of the monumental building projects were undertaken.

6 Kraeling (1938: inscriptions 2-8) concern the temple and cult of Zeus Olympius and date to the 1st Century AD. See Seigne (1985: 289-290) for the inscription from the lower temenos of the Sanctuary of Zeus discovered in 1985.

expansion of the lower terrace (Seigne 1985: 289-290). Another Greek inscription mentions both the *demos* of Gerasa and a priest of Tiberius. These inscriptions reveal that Gerasa was an organised urban community with the urban institutions which one would expect to find in a city of the Roman empire in the 1st Century AD, such as a city council and imperial cult worship.

The building projects that were undertaken in this period were particularly concerned with the expansion of the Sanctuary of Zeus and the building of the first temple of Zeus. They confirm the impression gained from the epigraphic evidence, namely that the community was determined to express its local identity, also through a built environment.

In the Hellenistic period the altar-like structure overlooked the surrounding region and was not oriented towards the Hellenistic settlement. Thus the importance of the sanctuary to the surrounding region was emphasised, whereas the orientation of the sanctuary in this period points to a local identity already being expressed by the community. Later developments changed the orientation of the sanctuary and it now faced the public centres of the city.[7] The 1st Century AD expansion of the Sanctuary of Zeus was one way of expressing a local identity, which, in this case, was focused around the continuous religious importance of the site and the earlier history of Gerasa.

The 1st Century AD developments at Gerasa were aimed at creating an urban setting and a city plan that incorporated the architectural manifestations of civic and religious institutions. These institutions were embodied in the developments of the Sanctuary of Zeus and the construction of the South Theatre. Both were complexes where civic and religious actions took place. The town, which already had an organised civic administration, was now developing a way of representing a local public identity through architecture as well as through inscriptions.

In the 2nd Century AD Gerasa saw the largest increase in urban development. The city received imperial attention when Hadrian visited the town in AD 129/130.[8] The so-called Triumphal Arch was constructed as a dedication to the emperor and was paid for by the community of Gerasa.[9] Other developments of the Hadrianic period included the South Gate and the Hippodrome, both located in the southern part of the city. However, no further developments took place in this southern area and when the city walls were built in late antiquity, the area between the South Gate and the Triumphal Arch was not included. At this point the Hippodrome had partly fallen out of use and had been converted back to its earlier function as a cemetery (Ostrasz 1995).

The developments of the Hadrianic period give an impression of a scattered development that was not tied into the existing parts of the town. However, as nothing is known about the location, layout and development of the residential areas in this period, it is not possible to fully contextualise the developments that are known to us.[10]

The Hadrianic developments were expressions of the public local identity of Gerasa no matter how disintegrated they seem to the modern viewer. They have to be seen together with the urban development that the site had undergone in earlier periods. The community of Gerasa explicitly expressed their relationship with the new ruling power from an early point during the Roman imperial period, through adopting 'the epigraphic habit' of the Roman empire, as well as incorporating urban elements which represented the society's understanding of itself within a local and imperial context.[11]

The practice of emperor worship, including the introduction of priesthoods connected to the celebration of the emperor, is one example of this. Worship of the emperor was one way in which the community of Gerasa expressed their sense of belonging to the Roman empire after the region had come under Roman rule.

In the Antonine period, the major developments were focused on the two main sanctuaries of the city, the Sanctuary of Zeus and the Sanctuary of Artemis. The Sanctuary of Zeus received its final shape through the construction of the upper terrace and the large temple. The Sanctuary of Artemis situated North of the Zeus sanctuary was, as a result of its grand scale, a prominent component within the urban fabric.[12] Both sanctuaries were situated on hills that overlooked the residential areas of the town. The two sanctuaries were features that influenced (and to a large extent defined) the orientation and layout of the urban space from this period onwards.

---

7 See *Figures.22.1* and *22.2* for the 1st and 2nd Century AD orientation of the Sanctuary of Zeus.

8 Kraeling (1938: inscriptions 30, 58, and 143-145) attest this visit. Welles concludes that Hadrian possibly spent parts of the winter in Gerasa (Kraeling 1938: 425 commentary to inscription 144).

9 Kraeling (1938: inscription 58) is the dedicatory inscription from the arch.

10 Kraeling (1938: 51) is of the opinion that the Hadrianic developments at Gerasa recall the developments that Hadrian undertook in Athens. However, it is unlikely that there is any direct parallel between the developments in Athens, where Hadrian was called the founder of a new and large quarter of the city in an inscription from an honorary arch, and the Hadrianic developments in Gerasa. There is no evidence that the emperor was personally involved in any of the building projects in Gerasa as he was in the case of the Athenian developments.

11 The term 'the epigraphic habit' is borrowed from MacMullen (1982). As all over the Roman Near East, the official inscriptions from Gerasa are predominantly in Greek. A few inscriptions in Latin and indigenous languages have come to light. It is not surprising that Greek was the official language in the Roman period also, as it is a region where Hellenistic culture had had a profound impact on the structure and development of urban life.

12 It is assumed that the Sanctuary of Artemis was never completed. Certain decorative elements of the temple were never finished. However, this did not influence the impact that the scale of the sanctuary had on the urban landscape.

Figure 22.1: North facing Gerasa

Figure 22.2: South facing Gerasa

The 2nd Century urban developments were reflections of the long tradition of civic organisation in Gerasa. The introduction of the Sanctuary of Artemis into the urban fabric has been interpreted as a socio-political decision, following the visit of the emperor Hadrian, in order to influence the economy of the city (Seigne 1992: 338). Although this theory is interesting, there is no evidence that this was in fact the case. The introduction of the cult of Artemis on such a scale as the size of the sanctuary suggests is intriguing. Moreover, the introduction of Artemis as Tyche of the town on coins from the Hadrianic period onwards deserves an explanation. Nonetheless the articulation of the cult of Artemis through the sanctuary and on the coins can also be seen as a representation of a growing sense of a local public identity.[13] There is plenty of epigraphic evidence from Gerasa which attests to the existence of other cults within the city in earlier periods (Kraeling 1938: inscriptions 15-26 and 38-44). Admittedly, the two major cults seem to have been those of Artemis and Zeus, but the community also worshipped a number of other deities, as archaeological and epigraphic evidence testify. That the city could boast two major sanctuaries only adds to the theory that the community had a strong sense of their local identity and used a variety of urban components to express this identity.

From the mid-2nd to the mid-3rd Century AD the developments that took place concentrated on elaborating the existing urban layout through the introduction of a number of arches and gates. Furthermore, public buildings such as the odeion and the West Baths were constructed. The developments of this period show that extensive negotiations of the local identity of Gerasa within its regional context, but also as part of the Roman empire, had taken place. At this point Gerasa's community was familiar with the use and meaning of a wide range of monuments and architectural complexes, which were considered as appropriate representations of a Roman provincial city's public identity. It has also been demonstrated, however, that local interpretations and expressions of a public identity took place in Gerasa, both through the construction of the built environment and the process of urban development.

Bibliography

Kennedy D. 1998. "The identity of Roman Gerasa" *Mediterranean Archaeology* 11: 39-69.

Kraeling C.H. 1938. *Gerasa, city of the Decapolis*. New Haven.

MacMullen R. 1982. "The epigraphic habit in the Roman Empire" *American Journal of Philology* 103: 233-246.

Millar F. 1993. *The Roman Near East, 31 BC-AD 337*. Cambridge.

Ostrasz A.A. 1995. "The hippodrome of Gerasa: A case of the dichotomy of art and building technology" *Studies in the History and Archaeology of Jordan V*: 183-192.

Seigne J. 1985. "Le sanctuaire de Zeus à Jérash: éléments de chronologie" *Syria* LXII: 287-295.

Seigne J. 1992. "Jerash romaine et byzantine: développement urbain d'une ville provinciale orientale" *Studies in the History and Archaeology of Jordan* IV: 331-341.

Zayadine F. (ed.) 1986. *Jerash Archaeological Project 1981-1983, I*. Amman.

13 As mentioned above, there is evidence for the existence of the cult of Artemis prior to the construction of the sanctuary.

Burial customs and social change in Argos from the Protogeometric to the Late Roman Period (1100 BC - 500 AD)

Francesca Ramondetti - Leiden University, Netherlands

The funeral was an important religious event in the life of the ancient Greeks. It was an occasion giving symbolic recognition to a selection of the social identities of the deceased, with the burial context providing material evidence for the representation of social structure. There is a long archaeological tradition of studying the relationship between mortuary practices and social structure. In particular, a theoretical basis formulated by Western archaeologists over many years has been the underlying belief that social structure is reflected in burial practices (Binford 1971; Saxe 1970; Trigger 1989: 289-263).

I will discuss here the case of the Argos cemetery and its development over time and space from the Protogeometric to the Roman Period (c.1100 BC-500 AD). By looking at chronology, typology, context of deposition and literary evidence I will examine the emergence of major political and social changes. Attention will be given to the question of whether this cemetery and others fit into general models of burial customs for Greek and Roman society and whether they reveal certain trends. Greece should be not as seen a monolithic entity, especially in the case of burial customs. In fact, the regional differences in terms of local traditions and social organization are seen as significant.

Since the beginning of the last century systematic excavations of the ancient town of Argos and its cemeteries have been carried out by the French School of Archaeology in Athens. Many studies have been carried out on the Geometric and Protogeometric periods. R. Hägg in *Die Gräber der Argolis* analysed burials from the Submyceanean to the Geometric periods. P. Courbin in *Les tombes géométriques d' Argos I,* discussed in detail and listed all the geometric graves of the Argolid found by the French School from 1952 to 1958. For the Archaic, Classical, Hellenistic and Roman periods however, few graves have been published.

As in other parts of the Greek world, developments over the two centuries of the Geometric period (900-700 BC) include changes in the city of Argos. The city increased in size and the population became increasingly numerous. The society had already become stratified by the end of the 8th Century BC, and the well-furnished tombs of the late 8th Century BC clearly mirror this. Burials of the preceding Protogeometric period (1100-900 BC) at Argos present little variety and are generally poorly furnished. Metal and weapons are extremely rare. Except for a few tombs, the investment in expenditure accorded to each individual does not vary very much: two pots, a cup and a jug form a sort of standard set (Foley 1988: 35).

The dominant grave type is the stone cist. In addition there is a small percentage of burials in simple earth-cut graves (*Table 23.1*). In the Geometric Period, pot burials are the innovation of the time: pithoi were used for adults, whereas amphorae or kraters were generally used for babies or young children.

While in cist tombs the number of grave goods increased and the offerings became richer in quality and quantity over these periods, the pot graves do not show a noticeable increase in the number of offerings. Pottery and metal finds are very rare and a third of the pithoi are without grave goods of any kind (See *Figures 23.1-23.2*). It seems therefore, that aristocratic families used cist tombs, and poorer people tended to bury their dead in pithoi.

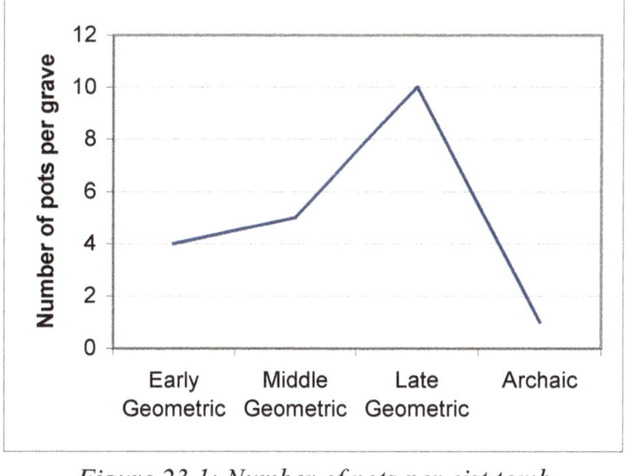

Figure 23.1: Number of pots per cist tomb at Argos (900-700 BC)

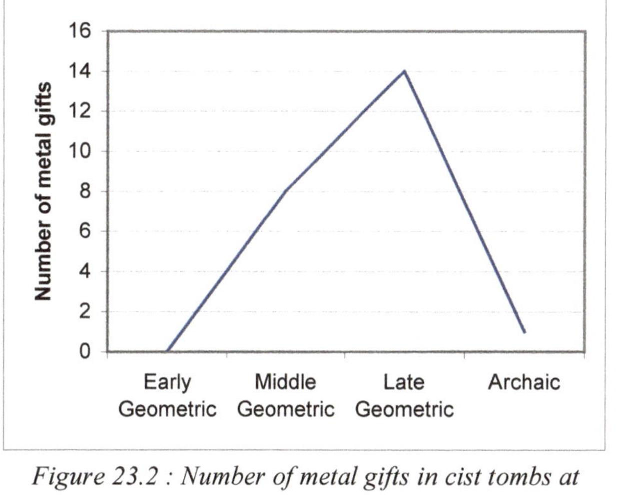

Figure 23.2 : Number of metal gifts in cist tombs at Argos (900-700 BC)

Period	*Sub-mycenaean*	*Proto Geometric*	*Early Geometric*	*Middle Geometric*	*Late Geometric*	
Dates	(SM) 1200-1100	(PG) 1100-900	(EG) 860-850	(MG) 850-800	(LG) 800-700	Total
Cist	11	37	21	30	29	128
Pithoi	1	1	1	2	14	19
Pots	10	1	2	1	9	23
Pits	1	7	4	6	4	22
Total	23	46	28	39	56	192

Table 23.1. An indication of the numbers of graves from the Submycenaean to the Late Geometric period at Argos. All dates BC (After Foley 1988).

If Morris' analysis of Athens holds true at Argos, the real lower class would be absent entirely from the formal cemetery until the Late Geometric period (800-700 BC), and it would therefore be dominated by middle class graves prior to that period (Morris 1989: 72-109).

An interesting phenomenon is the existence of numerous, very large cist tombs furnished with rich equipment. Courbin and Hägg have each pointed out that this phenomena might be the result of a combination of two new trends: firstly, an increase in the quantity of grave goods; and secondly, the enabling of multiple use for later inhumations of the cists as a sort of family grave containing up to 7 people (Hägg 1983: 28-29; Courbin 1974).

Snodgrass saw this as the strengthening of family and kinship ties, especially in a dominant ruling class with a military character (1971: 194-196). In fact, the cist tombs were furnished with weapons, armour and obeloi - iron spits. This latter class of objects was used as currency in early times (i.e. in the 9th-7th Centuries BC). All the jewellery of the Late Geometric Period is confined to cist tombs.

These spacious cists with multiple burials including warrior equipment are completely missing from the contemporary provincial centres of the Argolid, such as Tiryns and Nauplion, Mycenae and Asine. In fact, very few weapons found at these sites can be dated to the Geometric period. As Foley suggested, this may show that Argos had already emerged as the leading community of the region, and that this type of cist not only reflects a social hierarchy within the community of the city, but that the absence of this cist type outside of Argos defines a settlement hierarchy within the Argolid (Foley 1988: 40). This might be significant, as Argos was the only city in the region to develop into a polis, probably with some aristocratic families joining the growing community there (Hägg 1983: 31).

The conclusion reached so far receives some confirmation from independent consideration of the archaeological evidence of the period. Despite great regional variation in these customs, there are some common trends that reveal the underlying social attitudes to the dead. Wealth of grave goods may be a consistent indicator throughout Late Geometric in Argos, since we find the same patterns in Athens.

Another consideration has to be made for population increase. *Table 23.1* shows that the number of graves of the Late Geometric period is higher than that of the Protogeometric period; it seems that Argos enjoyed quite a sharp rise in population - as a consequence of the rise of prosperity throughout the Geometric Period (Snodgrass 1977: 11; Foley 1988: 46). Morris argued that it would be a mistake to consider this numerical peak only as a result of rapid population growth and he suggested that the 8th Century BC increase in burials was caused by changes in the representation of child burials and of the poor. In fact, child burials in Athens are very rare and under-represented down to c. 725 BC, but thereafter the number rises from 8% to 59% of all burials in Athens, and in Argos from 12% to 43% (Morris 1992: 79).

In the 7th Century BC (the Archaic period) the situation changes dramatically. The buried population decreased, nd also the practice of furnishing the dead with grave goods stopped abruptly. Almost all the graves (34 in total, 27 out of them are of uncertain date) are pithos burials and a few are jar burials. Adult graves moved to burial grounds outside the settlement, and had no offerings, while children were buried within the settlement and continued to receive some grave goods - only 4 out of 7 have grave goods (Morris, 1992: 27).

Foley suggested the decrease in population and wealth was the result of general dissatisfaction over the political situation under the reigning king (Pheidon). But the situation seems to be mostly the same throughout the entire region and in other cities in Greece, such as Knossos and Corinth. The total number of graves is 13, mainly without grave goods, and come from the sites of Nauplion, Mycenae, Tiryns and Porto Cheli. At Asine no graves of this period have been found (Foley 1988: 51-52).

By the 6th Century BC and moving into the Classical period (476-323 BC), the hereditary monarchy had ceased to be of political significance and the State was controlled by a closed oligarchic system. Argos was more democratic but the power was still held by elite families. All kinds of burial types have been found in this period, including cist, tile and simple pit graves. The sarcophagus must have been the most expensive choice for inhumation, and a few are found in Argos. Almost every burial contains grave goods and their variability is high. Only 14 out of 88 graves are found empty. The majority of grave goods are ceramic (see *Figure 23.3*). Metal offerings are few (17 out of 88 have metal grave goods – labelled MET in *Figure 23.3*). Children's burials (in cist graves with adults, in pots and clay larnakes) contained many gifts, often more than those of adults (Barakari-Gléni 1998: 524; Morris 1992: 27). The cemeteries are organized in grave enclosures as family plots and it seems that each enclosure contained about an equal number of grave goods.

In the Hellenistic period (323-146 BC) many political changes took place; the gradual interference of external forces (at first Macedonian and later Roman) encouraged the elite to legitimize their social position and to rebuild networks of power in the city. In the year 336 BC, after Philip's murder, fluctuations in policy in the City State resulted from the existence in Argos of individual politicians who tried to take advantage of changes in public opinion, in order to create better opportunities for their own careers.

Bar two large cemeteries, it is interesting to note that a few isolated burials have been found in the centre of the city (Barakari-Gléni & Pariente 1998: 175). This is very curious because from the Archaic period onward, no cemeteries were allowed inside the city. Does this follow a trend, or does this indicate a special ritual function for the intramural burials?

Cists, tile graves and simple sarcophagi continued as the main forms utilized, and were grouped in enclosures as family plots (See *Table 23.2*). An important innovation was the monumental tomb, a burial type originating in Macedonia (three have been found in different areas of the city) for wealthy individuals and their families. Inhumation is the main burial form, and this persisted throughout the entire Hellenistic Period (only 24 out of a total of 329 graves were cremation burials).

A general increase in grave goods and a more obvious display of wealth is evident in the Hellenistic period (compare *Figures 23.2* and *23.4*). This picture fits into the general model of the period: with the crisis and collapse of the democracy, the individual became the centre of all interests. A leading wealthy class emerges, as indicated by the presence of Macedonian tombs and sarcophagi, which were accessible to a restricted number of citizens. Only a few graves do not have grave goods at all, and probably belong to the lower class, however some contain a very high number of pots and metals (see *Figure 23.4*). Some graves have more than 10 pots and a couple up to 30-35 pots per grave.

By 100 BC, the Romans had started to increase their hold in Greece, and Argos was under the general supervision of the Roman governor in Macedonia. In Argos, Roman tombs of different types have been found almost everywhere in the modern city. Cist and tile tombs are predominant (75 out of 176), and sarcophagi are few. The burials of infants or children in pots are common, and there are very few examples within coffins.

The most interesting finds are the monumental tombs of the second Century AD. The grave goods were not particularly rich, but in one tomb jewellery and a gold medallion were found.

The archaeological evidence from the remaining graves is very poor. Apart from a few cases, the offerings consisted mainly of pottery. If we look at the numbers of vessels per grave, a total of one or two vessels make up the most common grave good frequency, although graves with no vessels are just as common. Besides pottery, lamps, coins, miniatures, inscriptions and glass vessels have been found (See *Figures 23.5 – 23.6*). The decreasing number of grave goods compared with the preceding Hellenistic period may suggest that the manner of displaying wealth was changing.

In Argos, the Late Roman graves are mainly located in three nuclei: north (5th-7th Centuries AD), south (4th-6th Centuries AD) and east (4th- 5th Centuries AD). The majority of the tombs were grouped in funerary chambers.

Area	*Type of Grave*	*Pot*	*Lamp*	*Glass*	*Jewellery*	*Tombstone*	*Cross*
NORTH BASILICA	Tile\Cist						
PANAGOPOULOU	Chamber tomb	X	X		X	X	
KOUYA	Tile-Cist	X			X		
YEROULI	Chamber tomb	X	X				
PILINGUIKOU	Tile-Cist	X	X			X	
EFSTATHION	Tile-Cist	X	X		X		
SOUTH	Cist	X	X				
EAST	Tile		X	X			X

Table 23.2. Late Roman – Early Christian Period (400-600 AD). Distribution of grave-goods.

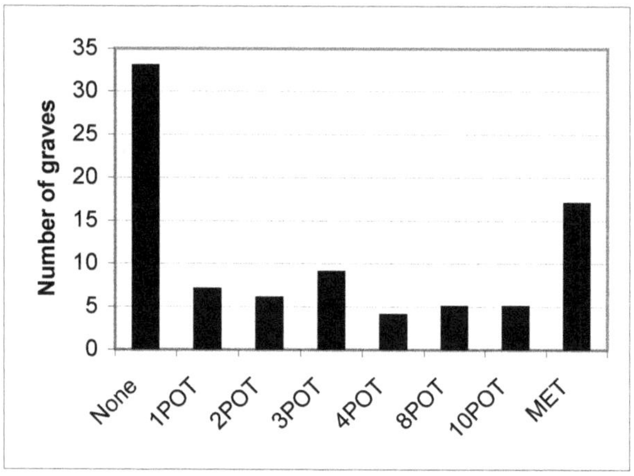

Figure 23.3: Quantity of grave goods per grave at Argos, Classical period (476-323 BC). MET = metal grave goods

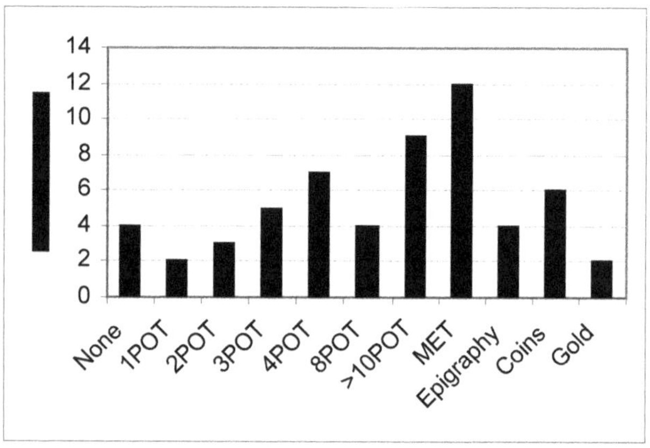

Figure 23.4: Quantity of grave goods per grave at Argos, Hellenistic period (323-146 BC).

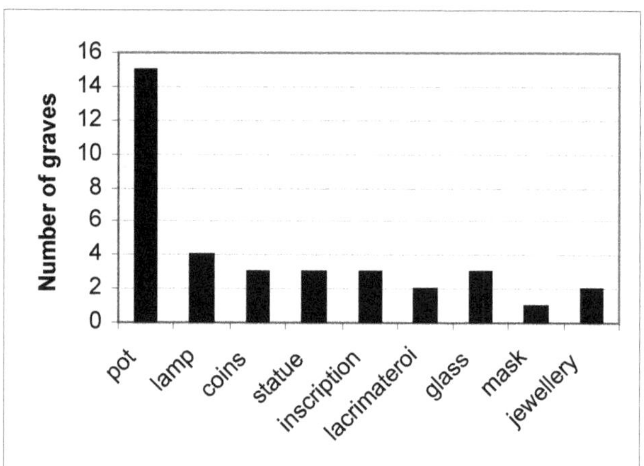

Figure 23.5: Roman Argos. Number of graves containing grave good types.

Although the cemeteries have not been fully excavated it is possible to examine some elements. All the areas were divided into small groups (perhaps family plots?), each of which apparently has differences in grave goods. The plots of the Northern area (Panagopoulou, Kouia, Yerouli, Pilinguikou, Efstathion) seem to have a greater number of rich burials, with gold, jewellery, bronze and marble inscriptions. The character of the grave goods of the areas East and South seems to be more modest with crosses, lamps, glasses and bronze coins (Oikonomou-Laniado 1998: 405-416) (See *Table 23.2*).

It seems that while the upper class was buried in the monumental tombs and the middle class used cist and tile tombs, the very poor tombs may have disappeared due to a lack of grave markings or use of tiles to cover the burial. Alternatively, as in Rome, the poor people used to bury the dead in *puticuli*, a common pit, without any ceremony. Evidence of such a pit has been found in Athens (Baziotopoulou-Valavani 2002: 187-201), but not in Argos.

A decrease in the number of grave goods from the Hellenistic Period onwards is evident. The quantity and the number of types of grave goods are lower, and the number of tombs without any gifts is higher. The general evidence for the Roman Period from cemeteries of the Empire does not show a great quantity of grave goods, but in Argos the situation seems to be rather dramatic, especially for the Early Roman period. The material available is very scarce and further studies will help to arrive at a better understanding of the situation.

It is difficult to talk about a process of Romanization in the context of death and burial at Argos. Quite the contrary, it appears that the Romans adopted Greek customs, for example inhumation, which was still rare in Italy. The burial data shows that ethnic and cultural integration, if at all detectable, mainly concerned the living and not the dead. Costume, status symbols and social structure surely changed. According to the archaeological evidence, the afterlife beliefs did not remain constant.

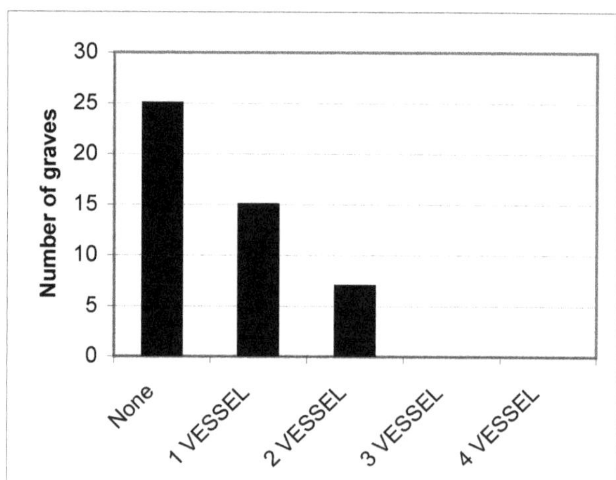

Figure 23.6: Roman Argos. Number of vessels per grave

The character of the graves has been a useful source for observing new, and changing political and social realties. In fact gradual changes are visible, corresponding to the evolution of society. Despite its own particular regional developments, the changes within the cemetery of Argos over the centuries that this article examines follow general developments across the whole of Greece.

Bibliography

Baraka A. 1998. "Ρομαικα ταφιχα μνμεια του Αργους", in A. Pariente & G. Touchais (eds.) *Argos et l'Argolide: topographie et urbanisme. Actes de la table ronde internationale.* Athènes-Argos, 1990 : 385-390.

Barakari-Gléni K 1998. "Οι νεκροπολεισ τον Αρχαιον και Κλασικον Χρονον στην Αραια πολη τοί Αρ γοσ", *Πρακτικα*, Τομο B, Athens : 509-533.

Barakari-Gléni K. & Pariente A. 1998. "Argos du VII au II siècle av. J.C.: synthèse des données archéologiques", in A. Pariente & G. Touchais (eds.) *Argos et l'Argolide: topographie et urbanisme. Actes de la table ronde internationale.* Athènes-Argos, 1990 : 165-176.

Baziotopoulou-Valavani E. 2002. "A mass burial from the cemetery of Kerameikos", in M. Stamatopoulou & M. Yeroulanou (eds). *Excavating Classical Culture. Recent archaeological discoveries in Greece.* Oxford: BAR International Series 1031: 187-201.

Binford L.R. 1971. "Mortuary Practices: their study and their potential" in. J.A. Brown (ed.) Approaches to the Social Dimensions of Mortuary Practices. *Memoirs for the Society of American Archaeologists* 25: 6-29.

Courbin P. 1974. *Tombes géométriques d'Argos. I (1952-1958),* EtPelop VII, Paris.

Foley A. 1988. *The Argolid 800-600 B.C. An Archaeological Survey.* Gotheborg.

Hägg R. 1983. "Burial customs and social differentiation in 8th-century B.C.", in R. Hägg (ed.) *The Greek Renaissance of the Eighth Century B.C.: Tradition and Innovation. Proceedings of the Second International Symposium at the Swedish Institute in Athens 1981.* Stockholm: 27-31.

Hägg R. 1990. "Argos and its neighbours: Regional variations in the burial practices in the Protogeometric and Geometric Periods", in A. Pariente & G. Touchais (eds.) *Argos et l'Argolide: topographie et urbanisme. Actes de la table ronde internationale.* Athènes-Argos, 1990: 131-135.

Morris I. 1987. *Burial and ancient society: the rise of the Greek-City State.* Cambridge.

Morris I. 1992. *Death-Ritual and Social Structure in Classical Antiquity.* Cambridge.

Oikonomou-Laniado A. 1998. "Les cimetières Paléochrétiens d'Argos.", in A. Pariente & G. Touchais (eds.) *Argos et l'Argolide: topographie et urbanisme. Actes de la table ronde internationale.* Athènes-Argos, 1990:405-411.

Saxe A.A. 1970. *Social dimension of mortuary practices.* PhD. Thesis, University of Michigan.

Snodgrass A. 1971. *The Dark Age of Greece.* Edinburgh.

Snodgrass A. 1977. *Archaeology and the rise of the Greek-State.* Edinburgh.

Trigger B.G. 1989. A history of Archaeological thought. Cambridge.

Open endings at Osteria dell'Osa (Lazio) : Exploring domestic aspects of funerary contexts in the Early Iron Age of Central Italy.

Erik van Rossenberg - Leiden University, Netherlands

Introduction: funerary evidence and layers of social life

In burial archaeology it is commonplace that funerary practices can tell us as much about the living as about the dead. As the dead cannot simply bury themselves, a dead person is subject to other people's rules and traditions in the social reproduction of collective identities, such as the household or the community. In this paper I will argue that we may have overlooked a strong domestic dimension in funerary contexts of the Early Iron Age in Central Italy.

The social entities of (extended) families or households have been identified in communal cemeteries as burial clusters, but these are often misrepresented as bounded spatial entities. What underlies this misrepresentation is a misconception of the individual. The individual is usually taken in its most literal sense of being an indivisible unit. This is in stark contrast with contemporary notions in social theory that individuals are in fact fragmented bodies. They are fragmented between the several levels on which their experiences of social life take place. In the act of burial the living constitute the deceased as such a fragmented person. We would both overlook the actual actors and reinforce the non-fragmented conception of the individual, if we regard burials as closed contexts.

The many dimensions of burial goods

In other words, we should not regard the funerary assemblage – that is the person buried and the accompanying burial goods - as the reflection of just one individual. For instance, in using the term 'personal belongings' as a substitute for burial goods, we reduce the meaning of these objects to one dimension. Although often the qualitative aspect – the occurrence of so-called luxuries or prestige goods – is taken into consideration as well, the basic interpretive strategy is one of counting. Differences in the number of burial goods are equated with differences in rank and wealth.

However much this sketch may be a caricature, it does highlight the fact that we tend to forget to establish if we are dealing with personal belongings in the first place, before we start counting objects as indicators of social status. In regarding burial as an act, we should remember that the members of the living are the actual actors. The creation of meaning lies in their hands and is for their sake, leaving the dead person almost as a passive body that is worked upon and whose identity can be manipulated. Thus we might expect that some of the burial goods were not only included with the idea of an afterlife in mind, but also for the sake of social reproduction of households and communities.

Another interpretive approach to burial goods is gender attribution. In Italian later prehistory, weapons have generally been attributed to male individuals and spinning and weaving equipment to female individuals (cf. Whitehouse 2001). "Warriors" and "weavers" (actually: spinners) have come to be regarded as the stereotypical male and female identities by which members of the elite expressed their social status in communal cemeteries of the Early Iron Age. As argued above for elite status, in gender attribution there is a risk of conflating the idea of burials as closed contexts with the conception of individuals as non-fragmented. We single out classes of objects to represent a particular gender, disregarding the possibility that an object could have been placed within the grave on behalf of the living rather than for the deceased, and could therefore have been implicated in several layers of meaning simultaneously.

To investigate this possibility, the well-published Early Iron Age cemetery of Osteria dell'Osa, Lazio (Bietti Sestieri 1992a, 1992b) is a valuable starting-point, because it consists of a large sample of some 600 graves and has yielded detailed information on both skeletal remains and the positions of burial goods.

At Osteria, spinning and weaving equipment is taken to represent the female gender (Bietti Sestieri 1992b: 496-498). This assumption is generally corroborated by the skeletal evidence, so that in cases of indeterminate physical identification, funerary assemblages with spindle-whorls and spools are attributed to a female individual (and in their absence to a male). Furthermore, individuals with large quantities of them among their burial goods (note the counting strategy) are ascribed an elite status, representing the female elite identity of the weaver. If the dimension of age is added to the analysis, however, we can find another layer of meaning (*Table 24.1*).

In the burials with only one or two spindle-whorls, these objects seem to have been primarily gender indicators, with a more or less even distribution over all age groups. Among the approximately 140 burials containing only one spindle-whorl, two thirds of the individuals are over 30 years of age. On the other hand, in higher frequencies spindle-whorls and spools gained an age-specific connotation, associated as they are with the age group of under 30 years. Among these latter burials there are two individuals identified through osteological analysis as male adolescents. I would therefore suggest that we

Frequency	*1*	*2*	*3*	*4*	*5*	*6*	*7*	*9*	*10*	*12*	*13*	*16*	*19*	*20*	*28*
< 30 years	47	10	5	7	2	4	6	1	2	3	2	1	1	1	1
> 30 years	90	9	1	3	1	1	1								

Table 24.1. Frequencies of spindle-whorls and spools compared with the age of buried individuals at Osteria dell'Osa (compiled after Bietti Sestieri 1992b).

interpret this particular depositional practice not so much as an expression of elite status or gender, but rather as an expression of the great loss that the death of individuals in these specific age groups must have entailed for the households and community of Osteria.

Domestic aspects: secondary funerary practices and acts of structured deposition

On a general level, the material expression of this domestic dimension has already been recognised in the form of burial clusters as indicated in the layout of Early Iron Age cemeteries in Central Italy (cf. Toms 1998: 160). For instance, at Osteria two main groups and a number of sub-groups have been distinguished (Bietti Sestieri 1992a: chapters 7 & 8). That the living must have had a practical knowledge of who was buried where, can be deduced from the fact that individual graves generally do not intersect one another. Moreover, we tend to interpret situations where graves do intersect as a deliberate act of creating a relationship in terms of kinship (cf. Smith 1996: 82).

By no means the only example at Osteria is the cremation burial of an elderly (presumably female) individual (no. 482), found in the grave of a primary adult male burial (no. 483). In the act of burial the latter has been carefully (re)positioned, more or less sitting against the former (Bietti Sestieri 1992a: 121-125; 1992b: 209-212). The excavators interpret this as a clear case of so-called secondary funerary practices, i.e. the repositioning of bones and presumably burial goods. At Osteria continued concern with the dead, sometimes represented by votive depositions of miniature pottery found in the grave-fill or on top of the grave, seems to have been a generalized practice, as it is evidenced in more than half of the 600 graves.

The paired burial described above (nos. 482-483) also exemplifies that there would have been strict rules (or communal conventions) for acts of deposition within the grave. In the case of grave no. 483 the original position of the head is reconstructed on the basis of preferences in the orientation of the body (with the head towards the east) and in the position of burial goods (the razor near the head). Given such preferences in depositional practice, I would argue that domestic aspects should not only be discernable in terms of structural relationships between graves, but also as acts of structured deposition within burials themselves.

To illustrate this, I will now turn to the analysis of burial good associations, and in the following section, their position with respect to the body or the cremated remains in the urn. For the moment I will focus on the most frequently occurring classes of metalwork, i.e. razors, knives and spearheads. The main observation to be made concerning their association is that they do occur in combination(s) within cremations, whereas in inhumations they are mutually exclusive. This reinforces the interpretation that cremation can be regarded as a special type of burial at Osteria, restricted to a small number of predominantly male adolescents and adults (Bietti Sestieri 1992b: 203-207).

Rather than ascribing elite status to cremated individuals, I would stress the highly symbolic character of this type of burial, which is highlighted by the predominance of miniature burial goods, including domestic pottery forms. The fact that some of the urns took the form of houses, the well-known 'hut urns', also suggests that the symbolism in cremation burials was of a domestic nature. Furthermore, the selection of individuals for cremations and "weaver" burials was complementary at Osteria. Children were excluded and women rarely selected for cremation, let alone for hut urn burials. This suggests that there is a domestic dimension to be found in both "weaver" burials and cremations.

The elaborate domestic symbolism of cremations and their relative scarcity compared with inhumations, should be seen in the light of the fact that they provided the focus for the emergence of the cemetery at Osteria, both spatially and chronologically. Cremations are among the earliest burials and ended up in a position at the centre of burial clusters. Thus they seem to represent the heads of families or households (Bietti Sestieri 1992b: 204; Smith 1996: 64), or perhaps the "founding fathers" of the community of Osteria. Particularly, the domestic symbolism represented by the hut urn assemblages suggests that these burials can be regarded as acts of structured deposition that literally and metaphorically mark the formation of a new community by several households or communities. We might even consider a scenario in which families on the move brought with them the remains of someone who died and was cremated elsewhere, to represent their household in the new community.

Subjecting the dead, positioning the burial goods

To explore the idea of structured deposition, I will now turn to the analysis of burial good positions with respect to the body or the cremated remains in the urn.

As mentioned above, in cremations at Osteria particular

classes of objects co-occur regularly, unlike inhumations. Nevertheless, some preferences seem to emerge in the positioning of (miniature) razors, knives and swords with the cremated remains inside the urn, and (miniature) spearheads and figurines outside the urn (*Table 24.2*). The preferred position of razors fits the pattern in the Lazio region as a whole, where 79% of the razors have been found inside the urn. At Osteria the exceptions can be explained in terms of a change in the sequence of depositional events in the combination of razors with spearheads and figurines. Generally, the latter two classes were the first to be placed, at the bottom of the large ceramic vessel which usually contained the cremation burial, followed by the urn – with the other metalwork inside - and the associated pottery.

	Inside	*Outside*
Razors	8 (73%)	3 (27%)
Razors Lazio (PBF)	26 (79%)	7 (21%)
Knives	4 (67%)	2 (33%)
Swords	2 (67%)	1 (33%)
Spearheads	5 (36%)	9 (64%)
Arrowhead		1 (100%)
Tweezers		1 (100%)
Figurines		2 (100%)

Table 24.2. Position of objects with respect to the urn in cremations at Osteria dell'Osa (compiled after Bietti Sestieri 1992b and Bianco Peroni 1979).

In four cases at Osteria the urn and burial goods had been placed in a grave normally used for inhumation and were probably involved in an act of structured (re)deposition. Although this makes the sequence of events less clear, we can observe that the positioning of the associated pottery - in the western part of the grave - recalls that of inhumations rather than proper cremations. Thus, these particular burials suggest a general similarity between the concept of cremation and inhumation. This is confirmed, for instance, by shared preferences in the orientation of the body and the urn (*Figure 22.1*). If we accept the case made above for the domestic character of cremations, the cross-over burials represent experiments in transferring domestic symbolism to inhumations.

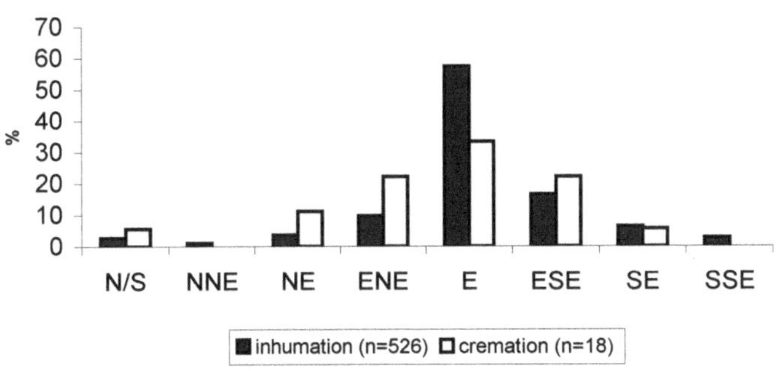

Figure 24.1. Orientations at Osteria dell'Osa (compiled after Bietti Sestieri 1992b).

Within inhumations at Osteria which include spindle-whorls, again we can distinguish between those burials with lower and higher frequencies of objects. Whereas in the latter, these objects are found dispersed over the full length of the body, in the former the preferred position is near the head or generally the upper part of the body (*Table 24.3*). At a later stage in the life of the cemetery this pattern broke down and spindle-whorls could be found at the feet as well. Compared with classes of metalwork in inhumations at Osteria, the initially preferred position of spindle-whorls recalls that of razors – invariably near the head – and, at a later stage, spearheads – with spears apparently placed in a functional position alongside the body – but not knives, which were preferably placed at the feet or generally near the lower part of the body (*Table 24.3*). This puts knives among the objects related to (ritual) food consumption on the part of the household, such as spits and ceramic vessels – sometimes with the remains of foodstuffs - which were more numerous at the feet than at the head.

	Upper part	*Centre*	*Lower part*
Spindle-whorl (n=1)	126 (86%)	4 (3%)	17 (11%)
Spindle-whorls (n=2)	24 (67%)	4 (11%)	8 (22%)
Spindle-whorls (n=3+)	126 (51%)	18 (7%)	104 (42%)
Spools	20 (23%)		67 (77%)
Razors	13 (93%)	1 (7%)	
Razors Lazio (PBF)	18 (75%)	5 (21%)	1 (4%)
Tweezers	3 (75%)		1 (25%)
Spearheads	15 (79%)		4 (21%)
Swords	7 (88%)	1 (12%)	
Pins	14 (74%)	4 (21%)	1 (5%)
Knives	5 (28%)		13 (72%)
Spits	3 (25%)		9 (75%)

Table 24.3. Position of objects with respect to the body in inhumations at Osteria dell'Osa (compiled after Bietti Sestieri 1992b and Bianco Peroni 1979).

	Upper part	*Centre*	*Lower part*
Razors		5 (42%)	7 (58%)
Swords	1 (7%)	8 (53%)	6 (40%)
Knives	1 (8%)	5 (42%)	6 (50%)
Spearheads	4 (29%)		10 (71%)

Table 24.4. Positions of objects with respect to the body in inhumations at Fossa, Abruzzo (compiled after Cosentino et al. 2001).

The normative position of razors near the head in inhumations at Osteria, fits the pattern for the Lazio region as a whole, where 75% of the razors have been found at the upper part of the body (*Table 24.3*). By comparison, at the cemetery of Fossa in the adjacent Abruzzo region, they were invariably placed near the legs or at the feet (*Table 24.4*). Usually it is assumed that razors were implicated in a male warrior ideology or simply indicated male gender (cf. Toms 1998: 163, 166, 174), as a counterpart to the spindle-whorl. However, its clear-cut association with the head could link the razor to secondary funerary practices which involved repositioning of the skull at Osteria (cf. Bietti Sestieri

1992a: 121). This might also explain the curious position of swords near the head at a later stage, again as opposed to their preferred position near the lower part of the body, for instance, at Fossa (*Table 24.4*).

This shows that funerary contexts can be regarded as a locale where the living subjected the dead to their rules in the positioning of burial goods. Further analysis along these lines will attempt to represent both primary and secondary burial practices as embedded in the social reproduction of households and communities.

Acknowledgements

First I would like to thank both the organizers of SOMA 2003 and the editors of its proceedings, who did a great job in improving this contribution. I'd also like to thank the Dutch Institute at Rome (NIR) for awarding me a two-month scholarship during which the ideas presented here could develop, and my supervisors John Bintliff, Harry Fokkens and Peter van Dommelen for their continuing support.

Bibliography

Bianco Peroni V. 1979. *I rasoi nell'Italia continentale*. Prähistorische Bronzefunde VIII (2). München.

Bietti Sestieri A.M. 1992a. *The Iron Age community of Osteria dell'Osa. A study of socio-political development in central Tyrrhenian Italy*. Cambridge.

Bietti Sestieri A.M. (ed.) 1992b. *La necropoli laziale di Osteria dell'Osa*. Roma.

Cosentino S., V. D'Ercole & G. Mieli 2001. *La necropoli di Fossa I. Le testimonianze più antiche*. Pescara.

Smith C.J. 1996. *Early Rome and Latium. Economy and society c.1000 to 500 BC*. Oxford.

Toms J. 1998. "The construction of gender in Early Iron Age Etruria", in R.D. Whitehouse (ed.) *Gender and Italian archaeology. challenging the stereotypes*. Accordia Specialist Studies on Italy 7. LondonL 157-179.

Whitehouse R.D. 2001. "Exploring gender in prehistoric Italy" *Papers of the British School at Rome* 69: 49-96.

A Scale of Identity in the Mycenaean Argolid

Daniel Sahlén – University College London

Introduction

In their article *Making Mycenaeans* (1999), Davis & Bennet discuss the political meaning of the Battle Frieze fresco in the palace of Pylos (Lang 1969, fresco 22 H 64) and its connections to the establishment of the Mycenaean state of Pylos during LHIIIA-B. In her publication of the frescoes at Pylos, Lang described the Battle Frieze as a depiction of a battle between Mycenaeans (Greeks) and barbarians (Lang 1969: 71-72). Davis & Bennet (1999: 111-12) emphasise that this interpretation is part of the contemporary view of Mycenaeans as an ethnic group in earlier studies, but sees the division of Mycenaeans and barbarians as an artistic expression of *we* and the *others*. According to Bennet & Davis (1999: 114), this manifestation of identity was part of the development of Mycenaean state ideology.

According to Lang (1969: 71, and plate L) and Davis & Bennet (1999: 109), the Battle Frieze shows two types of warriors, based on a fragment that shows two almost complete figures, "the Mycenaeans" with Mycenaean armour (most significantly the boar's-tusk helmet), and the "others" dressed in animal skins. I question the interpretation of this painting, since the differences in armour that Lang and Davis & Bennet describe are not that obvious on the fragments Lang published (Davis & Bennet 1999: 109; see also plate L in Lang 1969). It should also be mentioned that the two combatants on the surviving piece both carry a typically "Mycenaean" dagger. Davis & Bennet argue that this frieze and, more importantly, the symbolic meaning of *we* the Mycenaeans and the *others*, was an important part of the state formation at Pylos.

That Mycenaean Greece was neither a unified society nor had a homogeneous material culture has recently been argued in several regional studies (e.g. Cavanagh 1998; Papadimitriou 2001; Voutsakis 1998), but how unified were these Mycenaean "regions?"

The concept of identity has been fundamental for archaeological research to understand differences in material culture and stylistic variation (Shennan 1989: 20-21), but in recent decades questions concerning multiple identities have been raised. Wiessner (1983; 1989) built upon the theories about social identity by Tajfel (1978; 1981) in her studies of the San projectile points. Tajfel's theories were focused on social processes of stereotyping and categorising and argued that this was an important cognitive mechanism for individuals and groups to identify themselves and others. Tajfel stressed that *"he [or she] is member of numerous groups"* (Tajfel 1978: 61), so instead of one homogeneous cultural identity we may anticipate that there will be levels and subgroups of identities. Hodder (1979; 1982) studied the cultural expression in the Baringo district and noted boundaries between and within social groups in the districts. Hodder argued that stylistic variation could be explained not only as different *"archaeological "entities""* (Hodder 1979: 452), but also as relationships between individuals and subgroups.

It should be pointed out that there is a clear difference between Tajfel's social identity and the identity Hodder discussed. The former is a result of an active expression of *me/we* and *you/them*, while the latter is formed by inactive processes without the aim of marking social boundaries. I therefore stress a difference between expressed and formed identity. These types of identity can be seen in different materials and expressed in different ways in archaeological contexts. Expressed identity is more likely to be stressed in a socially important context, for example in religious or political activity, or where social interactions occur (cf. Wiessner 1983). Formed identity can instead be seen in a domestic context (Hodder 1982). The Battle Frieze is a clear example of expressed identity, but it should be said that manifested identity can be expressed in a much less dramatic way.

Age and gender are of course important aspects of social identity, but I will focus on forms of identity relevant to the spatial and social classification and organisation of individuals.

Identities in the Mycenaean Argolid

The social structure in the Argolid during the Mycenaean period has been widely discussed (cf. Bennet & Galaty 1997). In my study, I will use this discussion as a starting point for the consideration of levels of social identity, but I emphasise that social structure is not the same as social identity. Social structure is a political and economic framework, but social identity is a cultural division and indicates how people identify themselves and how they mark this. In previous discussions about social structure in the Argolid, four different levels have been in focus: Mycenaean culture, the region, intra-regional division, and kinship/family. I see from these levels, together with the individual, five potential levels of social identity.

In the introduction I have stressed two types of identity, expressed and formed identity, and that these can be seen in different archaeological materials. We have no other wall painting in the Argolid that can be compared with the Battle Frieze at Pylos as an expression of identity, but we can use alternative materials. The most relevant material derives from mortuary practices, firstly because grave goods constitute the largest group of material from the Bronze Age Argolid, but also because they are placed in the grave for a reason, and should be seen as a social action, typical or characteristic for the group that constructed the grave and left the goods. Formed identity can be seen in material of secular use and everyday activities. Perhaps the most useful evidence here is pottery, since the large quantity of this material and its production (cf. Hodder 1982) is the result of both internal (within the settlement or the family group) and external relations (Mycenaean pottery from the Argolid has been found extensively in the Aegean and the eastern Mediterranean (Mountjoy 1999)). I will now discuss the different levels of identities in the archaeological material from the Argolid. The aim of this discussion will mainly be to stress the potential rather than discuss the evidence of identities in this material.

Cultural identity

By cultural identity, I mean a cultural expression that is shared by a larger group of people. This does not have to be homogeneous, but should be based on some shared ideas and beliefs expressed in the material culture. This does not have to be linked to ethnicity but this is often the case. In the Mycenaean material we can see an assemblage of artefacts, the Mycenaean *koine*, which to some degree is similar throughout the Mycenaean areas. This *koine* is a mixture of earlier traditional MH objects, influenced by Minoan objects. In graves, some materials have been argued to have had a significant ideological meaning, for example: drinking vessels, bronze weapons, jewellery, clay figurines, cloth and body ornaments. These artefacts dominated the material culture in the Argolid during LHIII (Voutsakis 1995; Papadimitriou 2001). This suggests that the people on the Greek mainland became more defined as Mycenaean, but also that the expression of a Mycenaean identity became more important to them.

Regional identity

Regional identities are subgroups of cultural identity. They share the same cultural expression but they have developed or adopted regional characteristics. This is often due to regional circumstances, for example different natural settings or external contacts.

In the case of Mycenaean Greece this is interesting because of the traditional division of Mainland Greece. Bronze Age Greece is usually divided up after the classical *poleis* or the modern *nomoi*, but they *"are artificial divisions which do not reflect prehistoric settlement patterns"* (Mountjoy 1999: 15). In his classic work about the origins of the Mycenaean civilisation, Dickinson (1977: 87) used larger territories to address this, which *"have often produced archaeological evidence for a distinct local 'character'"*.

With an awareness of the artificial nature of these regions several scholars have made regional studies of dynamics in different regions. In several studies of the spatial distribution of graves, Mee & Cavanagh (e.g. 1990; 1998) have stressed the regional significance of the Argolid, a conclusion that is drawn by Voutsakis also (Voutsakis 1998: 56). As discussed above, if the "Mycenaean" koine increased in clarity of definition during LHIII, this would mean that regional differences decreased or regional identity became less important.

Intra-regional identity

With intra-regional identity I mean an expression of unity that is made inside the region and which marks opposition and/or independence. This can be a consequence of tradition or be actively created because of a conflict of interests, and the differences are likely to be emphasised in the times of increased stress. The division of Crete and Messenia into politically dependent sub-regions have been extensively discussed in the comparison of Linear B texts and archaeological remains (Ventris & Chadwick 1973; Bennet 1990; Bennet 1995; Driessen 2000).

The Argive Plain has usually been divided up into centres around the palaces, with some sense of clustering in the later part of LH. Some argue that these centres were linked together in a large Mycenaean kingdom (Bintliff 1977), and others that palaces were the centres of smaller independent states (Kilian 1988), possibly linked together as peer-polities (Renfrew 1986; Halstead & O'Shea 1986). Cherry & Davis (2001), in a resent study drawing heavily on results from the Nemea Valley Archaeological Project (Cherry n. d.), have discussed the status of the Nemea valley as a hinterland to Mycenae. Elsewhere, I have discussed the possible division of the Argive plain into three independent districts (Sahlén 2003), potentially associated with areas outside the defined region of the Argolid itself (cf. Cherry & Davis 2001).

Kinship identity

In anthropology, kinship is often seen as a "pre-political" level of organization of power, but kinship can be seen as relationships on different levels. It can be seen as the basis for interaction between families in pottery exchange (Hodder 1982), and as the political structure of a tribe-based society (Sahlins 1971; for a recent discussion in a Mycenaean context see Wright 1995). I address the issue of kinship identity because kinship has to some degree been discussed in the Aegean Bronze Age, though mostly in later periods (Cavanagh & Mee 1998; Morris 1987; also Tsountas & Manatt for an early account). Kinship identities can in some cases explain intra-regional divisions (see Cavanagh & Mee 1990), but kinship can also be important on a much lower scale of one or more family groups. Wright (1995) has argued that kinship was the basis for social structure in the Argolid, and similar conclusions have been proposed based on the evidence at Asine (Nordquist 1987). Kinship can also be related to issues of status and class. Voutsakis has argued that the gap between the elite and the common people increased during late Mycenaean times (Voutsakis 1998: 62).

The major argument for a political structure based on kinship has been the division of groups of graves, for example at Mycenae (Alden 2000) and at Prosymna (Blegen & Blegen 1937), but actual kinship between the buried has not been proved. Few studies have been made comparing the grave goods in single graves or groups of graves, but recent studies at Argos have shown that there are significant differences between groups of graves.

Individual identity

Individual identity is the foundation of social identity. Tajfel emphasised the importance of self-image in the categorisation of others (1978: 62). In archaeological research the study of the individual can increase our understanding of groups in the society, since the individual as a category reflects a group of individuals, rather than a specific person. The individual has had a remote role in the study of Mycenaean society; the categories of individuals that have been mostly discussed are the wanax/king (e.g. Kilian 1988), the warrior (e.g. Kilian-Drilmeier 1985) and, lately, individual pot makers (e.g. Lindblom 2001).

1: Cultural Identity
Shared expression

2: Regional Identity
Differences due to regional circumstances

3: Intra-Regional Identity
Expression of opposition or independence

4: Kinship/Family Identity
Differences indicating kin relations

5: Individual Identity
Fundamental for our understanding of social groups

Figure 25.1: Potential scale of identity in the Mycenaean Argolid

Conclusion

Social psychology and post-processual archaeology have acknowledged social identity as an important social process in the formation of human groups. I have pointed at five potential levels of identity in the Argolid: cultural, regional, intra-regional, kinship and individual (*Figure 25.1*). These can be seen through different types of materials, but the context of these materials has to be considered when we read the meaning of these identities. There is a difference between an expressed and a formed identity, and this can be seen in the archaeological material. The expressed identity has been marked with an aim to stress difference or group belonging, while a formed identity is an archaeological construction of patterns, which can be read as differences between social groups. Obviously, when the purpose is to mark social belonging, this is formed through materials that have a social function. In the Argolid we can see how funeral data is well suited to studies of expressed identity, since this material is strongly represented in the archaeological record of the Argolid and it is clearly the result of a social action. Domestic pottery, on the other hand, is useful for investigations of formed identity, since this is formed after generations of craft production and internal and external contacts.

Acknowledgements

I would like to thank the SOMA committee for a well organized conference and the publication of the papers presented, and my supervisor Dr Todd Whitelaw for discussing the issues presented in this paper, and helping me with the English in earlier drafts of this paper (any errors are of course my own). I would also like to thank Dr Cyprian Broodbank for his seminars during my year at UCL.

Bibliography

Alden M. J. 2000. *The prehistoric cemetery: Pre-Mycenaean and early Mycenaean graves*. Oxford.

Bennet J. 1990. "Knossos in context: Comparative perspectives on the Linear B administration of LMII-III Crete" *American Journal of Archaeology* 94: 193-211.

Bennet J. 1995. "Space through time: Diachronic perspectives on the spatial organization of the Pylian state", in R. Laffineur & W.-D. Niemeier (eds.) *Politeia: Society and state in the Aegean Bronze Age, Aegaeum 12*. Liège: 587-602.

Bennet J. & Galaty M. 1997. "Ancient Greece: Recent developments in Aegean archaeology and regional studies" *Journal of Archaeological Research* 5 (1): 75-121.

Blegen C.W. & Blegen E.D.P. 1937. *Prosymna, the Helladic settlement preceding the Argive Heraeum*. Cambridge.

Cavanagh W. 1998. "Innovation, conservatism and variation in Mycenaean funerary ritual", in K. Branigan (ed.) *Cemetery and society in the Aegean Bronze Age. Studies in Aegean Archaeology 1*. Sheffield: 103-08.

Cherry J.F. "The Nemea Valley Archaeological Project archaeological survey"
http://classics.lsa.umich.edu/NVAP.html

Davis J.L. & Bennet J. 1999. "Making Mycenaeans: Warfare, territorial expansion and the representation of the other in the Pylian kingdom", in R. Laffineur (ed.) *Polemos le contexte guerrier en Égée á l'âge du Bronze. Actes de la 7e rencontre égéenne internationale Université de Liège, 14-17 avril 1998, Aegaeum 19*. Liège: 105-120.

Dickinson O.T.P.K. 1977. *Origins of the Mycenaean civilisation, Studies in Mediterranean Archaeology 49*. Göteborg.

Driessen J. 2000. *The scribes of the room of the chariot tablets at Knossos: Interdisciplinary approach to the study of a Linear B deposit*. Salamanca.

Halstead P. & O'Shea J. 1982. "A friend in need is a friend indeed: Social storage and the origins of social ranking", in C. Renfrew & S. Shennan (eds.) *Ranking, resource and exchange*. Cambridge: 92-99.

Hodder I. 1979. "Economic and social stress and material cultural patterning" *American Antiquity* 44 (3): 446-454.

Hodder I. 1982. *Symbols in action: Ethnographical studies of material culture*. Cambridge.

Kilian K. 1988. "The emergence of wanax ideology in the Mycenaean palaces" *Oxford Journal of Archaeology* 7: 291-302.

Kilian-Drilmeier I. 1985. "Noch einmal zu den "Kriegergräbern" von Knossos" *Jahrbuch des Römisch-Germanisch Zentralmuseum*. Mainz: 196-214.

Lang M. 1969. *The palace of Nestor in western Messenia, vol. 2: The Frescos*. Princeton.

Lindblom M. 2001. *Marks and markers: Appearance, distribution and function of middle and Late Helladic manufacturers' marks on Aeginetan pottery, Studies in Mediterranean Archaeology 128*. Jonsered.

Mee C. & Cavanagh W. 1990. "The location of Mycenaean chamber tombs in the Argolid", in R. Hägg & G.C. Nordquist (eds.), *Celebrations of death and divinity*. Stockholm: 55-64.

Mee C. & Cavanagh W. 1998. *A private place: Death in pre-historic Greece. Studies in Mediterranean Archaeology 125*. Jonsered.

Morris I. 1987. *Burial and ancient society: The rise of the Greek city-state*. Cambridge.

Mountjoy P. 1999. *Regional Mycenaean decorated pottery*. Rahden, Westfalia.

Nordquist G.C. 1987. *A Middle Helladic village: Asine in the Argolid*. Uppsala.

Renfrew C. 1986. "Introduction", in C. Renfrew & J.F. Cherry (eds.) *Peer polity interaction and socio-political change*. Cambridge: 1-18.

Sahlén E.D. 2003. *An analogical perspective on regional politics in the Argolid during the Mycenaean period*. Unpublished MA dissertation at University College London, Institute of Archaeology.

Shennan S. 1989. "Introduction: Archaeological approaches to cultural identity", in S. Shennan (ed.), *Archaeological approaches to cultural identity*. London: 1-32.

Tajfel H. 1978. *Differentiation between social groups*. London.

Tajfel H. 1981. *Human groups and social categories: Studies in social psychology*. Cambridge.

Tsountas C. & Manatt I.J. 1897. *The Mycenaean age: A study of the monuments and culture of pre-Homeric Greece*. London.

Ventris M. & Chadwick J. 1973. *Documents in Mycenaean Greek; Three hundred selected tablets from Knossos, Pylos, and Mycenae with commentary and vocabulary*. Cambridge.

Voutsakis S. 1998. "Mortuary evidence, symbolic meanings and social change: A comparison between Messenia and the Argolid in the Mycenaean period", in K. Branigan (ed.) *Cemetery and society in the Aegean Bronze Age. Studies in Aegean Archaeology 1*. Sheffield: 41-58.

Voutsakis S. 1995. "Social and political processes in the Mycenaean Argolid: The evidence from the mortuary practice", in W.-D. Niemeier & R. Laffineur (eds.) *Politeia: Society and state in the Aegean Bronze Age, Aegaeum 13*. Liège: 587-602.

Wiessner P. 1983. "Style and social information in Kalahari San projectile points" *American Antiquity* 48 (2): 253-76.

Wiessner P. 1989. "Style and changing relations between the individual and society", in I.

Hodder (ed.) *The meaning of things: Material culture and symbolic expression*. London: 56-63.

Wright J. 1995. "From chief to king in Mycenaean Greece", in P. Rehak (ed.) *The role of the ruler in the prehistoric Aegean: Proceedings of a panel discussion presented at the annual meeting of the Archaeological Association of America, Aegaeum 11*. Liege: 63-80.

Expressions of Ethnic and Gender Identities in Egypt during the Early 1st Millennium B.C.E.

Heidi Saleh – University of California, Berkeley

Introduction

Awareness of ethnic identity was pervasive among the ancient Egyptians, who divided humankind into four basic types: Egyptian, Asiatic, Nubian, and Libyan (Leahy 1995: 226).[1] Ethnic affiliation became increasingly important as Egypt started to expand its "empire," and as the Egyptians came into closer contact with "others" during the New Kingdom (c. 1550-1069 B.C.E.).[2] However, Nubians, Asiatics, and Libyans are known to have lived in Egypt, and ancient Egypt was far from a homogenous population.

The Libyans are one of the least understood of the foreign groups residing in Egypt because few archaeological traces of native Libyan culture have survived in either Libya or Egypt (O'Connor 1990:45). This minority group was able to reach the highest echelons of Egyptian society. During the Third Intermediate Period (c. 1069-664 B.C.E.), the Egyptian state weakened and leaders of Libyan descent became Pharaoh, inaugurating the so-called Libyan Dynasties 22-24 (c. 945-715 B.C.E.) (Shaw 2000: 481-482).

Prior to the Third Intermediate Period (henceforth T.I.P.), Libyans were depicted in Egyptian art as captive enemies or evil forces losing on the battlefield. Those Libyans settling in Egypt during the New Kingdom portrayed themselves, with few exceptions, as Egyptian, suggesting that they had been assimilated (Ward 1994: 62). Stereotypically, however, New Kingdom references to or depictions of Libyans show a rebellious ethnic group that commonly threatened and caused problems for the Egyptian state.

This paper examines T.I.P. Libyan influence on traditional Egyptian conceptualizations of ethnic and gender identities by focusing on changes in the depictions of Libyans and in the representation of women in Egypt.[3]

Methodology

My primary data consists of T.I.P. stelae. Stelae serve as a rich medium for investigating issues of self-representation because, for the Egyptians, they visually present a person's quintessential identity. As icons of the ideal, stelae portray the owner's most desired identity and serve as useful indices that can help explore how people wanted to be remembered. I analyze two genres of stelae (donation and funerary) in order to trace changes in the depiction of Libyans and women and to see whether these changes occur in both private and public contexts.

Donation stelae record a gift of land bestowed by an individual to a temple or its personnel (Schulman 1966: 33). They are mostly stone and evidently were erected in open fields for all to see (Meeks 1979: 608). Funerary stelae, on the other hand, depict the deceased making an offering to deities in return for assistance in the Afterlife (Hölzl 2001: 320-322). Libyan Period funerary stelae were largely made of wood and served as private religious objects typically placed in the burial chamber adjacent to the coffin (Aston 1987: 566). Therefore, no living person was intended to have access to them. Because these two genres of stelae function on different visibility levels (donation stelae are public monuments, while funerary stelae are private objects hidden in a tomb), I am investigating whether self-representation on a stela is a function of its visibility. That is, will an individual be more reticent or more expressive about showing certain aspects of his/her identity depending on whether anyone else will see it or not?

Donation Stelae

When Libyans gained governmental control, official donation stelae, which earlier rarely featured identifiable foreigners, appeared illustrating Libyan patrons. Libyans, as new members of the ruling class, seemingly had no reservations about representing themselves as ethnically Libyan. Harnakht's donation stela (*Figure 26.1*), for example, draws attention to Harnakht's Libyan ethnicity and tribal affiliation textually, by identifying himself as the "Great Chief of the [Libyan] Ma" tribe, and iconographically, by wearing a feather on his head, which is a feature of his indigenous costume and a non-Egyptian custom (Fazzini 1988: 16). Similarly, Ti-Taru declared his ethnic identity on his donation stela (*Figure 26.2*) through his foreign name, his title of "Great Chief of the Libu" clan, and his non-Egyptian headdress comprising a single vertical plume (Fazzini 1988: 16; Kitchen 1969-1970: 65).

The depictions of Libyans on T.I.P. donation stelae differ noticeably from previous traditional depictions of imperial Egypt where Libyans, like other foreigners, were presented as uncivilized. During the New Kingdom, Libyans appeared either as exotic barbarians with flamboyant costumes, or they were highly acculturated and almost indistinguishable from Egyptians. The Libyan Period donation stelae, on the other hand, illustrate, on an official and public scale, positive expressions of a non-

---

1 According to texts and depictions in the tomb of Seti I (c. 1294-1279 B.C.E.), there are four "races" of humankind (Hornung 1999: 62).

2 All dates mentioned are taken from Shaw (2000).

3 This work introduces the subject of my dissertation entitled "Changes in Ethnic and Gender Identities under the 'Libyan' Dynasties of Third Intermediate Period Egypt (c. 1069-715 B.C.E.): Analyses of Donation and Funerary Stelae."

Figure 26.1. Brooklyn 67.118. Limestone donation stela of Harnakht from Mendes (T.I.P.; 52.3 X 32.3 cm). Reprinted from Kitchen © 1969-1970, by permission of the Brooklyn Museum of Art.

Egyptian ethnicity, a rare phenomenon in the history of Egypt. T.I.P. Libyans selectively and subtly employed certain ethnic markers to emphasize their own ethnicity. In the T.I.P., the depictions of Libyans are transformed from those of "primitive" foreigners to respectful benefactors making traditional donations to Egyptian temples.

Another notable change on Libyan Period donation stelae is the appearance of women as patrons, in particular women of foreign descent. Karamama, a member of the royal family whose name and descent are Libyan, commissioned her own donation stela. Oddly, she appears to emerge from a box-like container (probably a sarcophagus) holding a roll of papyrus. The meaning of this scene, so far unique in Egyptian iconography, is poorly understood (Fazzini 1988: 16). However, that women commissioned their own donation stelae and were depicted on those stelae as principal benefactor (with unorthodox iconography), signals a period of artistic experimentation and cultural flexibility with regard to women's roles.

Funerary Stelae

Feminine identity finds fuller expression on private Libyan Period women are prominent on funerary stelae, a development that may be linked to the different ethnic background of the ruling class. Libyans may have had different standards for women's roles. Whereas in earlier instances most women were portrayed alongside male relatives (Robins 1995; Roth 1999), in theLibyan Period many women had their own funerary stela.

Figure 26.2. Brooklyn 67.119. Limestone donation stela of Ti-Taru from Kôm Firin (T.I.P.; 39.3 X 18.5 cm). Reprinted from Kitchen © 1969-1970, by permission of the Brooklyn Museum of Art.

Furthermore, a new female body aesthetic was introduced in contrast to the male body type, which maintained mostly the late Ramesside[4] artistic traditions (*Figure 26.3*). Libyan Period women are displayed with a voluptuous body type that contrasts markedly with the earlier slim and streamlined Ramesside women. Art historians have noted that there is "a new ideal for the female figure," yet the explanations and motivations behind this change are still unclear (Robins 1997: 204).

Major changes in the female body aesthetic include: 1) the presence of abdominal rolls of flesh, usually indicated as small rolls or horizontal lines sketched below the navel; 2) a large, pendulous breast drawn in profile 3) ; and a heavy mid-section comprised of full thighs, round buttocks, and a protruding belly (*Figure 26.4*).[5]

---

4 The Ramesside Period consists of Dynasties 19-20 (c. 1295-1069 B.C.E.).

5 The appearance of this new female body aesthetic is not restricted to funerary stelae; it appears on at least one donation stela (JE.40716) (Fazzini 1988: 4, 33), as well as other objects such as coffins, papyri, and monumental reliefs. Only mortal women,

Figure 26.3 M.M.A. 28.3.35 (Rogers Fund, 1928). Wooden funerary stela of Aafenmut from Thebes (T.I.P.; 23 X 18 cm). Reprinted from O'Neil (ed.) © 1987, by permission of the Metropolitan Museum of Art.

Figure 26.4. E 2043. Wooden funerary stela of Ta-sheryt-akhet from Thebes (T.I.P.; 24.7 X 19 cm). Reprinted from Capel and Markoe (eds.) © 1996, by permission of the University of Pennsylvania Museum of Archaeology and Anthropology.

The modification of the female body aesthetic in the Libyan Period most likely reflects a different ethnic taste fostered by the Libyan dynasts, now in a position to dictate and define the ideal body type. Traditionally in Egypt, the king mandated the ideal form; everyone else, regardless of his/her own ethnic preferences, conformed to the ideal set forth by the Pharaoh. As it is unlikely that all the women on these funerary stelae were of Libyan descent, it appears, therefore, that Egyptian elites at this time represented themselves in accordance with the Libyan-introduced fashions.

Another indication of the foreign character of these funerary stelae is the prevalence of non-Egyptian names. Although one cannot assume that a name reflects a specific ethnicity, foreign names that appear on Libyan Period stelae are most likely Libyan (Goudriaan 1988: 91; Ward 1994: 62).

Finally, as with the donation stelae, funerary stelae also exhibit atypical and innovative iconographic and stylistic features. For example, on the stela of an unnamed woman, a date-palm grows in the area usually reserved for an altar (Aston 1987: 567; Munro 1973: 11-12; Scott and van Siclen III et al 2001: 18). Moreover, certain types of scenes, such as the tomb-in-the-desert-cliff scene, which were previously uncommon, gain popularity and occur on the lowest registers of some funerary stelae (Roveri 1991: 66; Saleh and Sourouzian 1987: 243).

The forms and shapes of funerary stelae evolved as well. Double-sided funerary stelae now occur, such as that belonging to lady Thenetenaset, which depicts her as the primary owner on the recto while her husband accompanies *her* on the verso (Fazzini 1988: 36). In addition, at least one funerary stela has a wooden, double-plumed sun-disk affixed to its top (Perez Die 1991: 99-100). Such differences in iconography, style, and form suggest that the Libyan Period was a time of artistic fluidity.

Conclusion

T.I.P. stelae illustrate a break with the traditions of earlier Egyptian stelae. Variations in the portrayal of ethnic minorities and women are seen in two different stela genres, which reflect both public and private spheres, implying that change took place regardless of the object's visibility. The public donation stelae mark an owner's ethnicity by emphasizing certain elements of his/her ethnic background. The Libyan chiefs stressed their identity through their titles and by depicting themselves with a traditional Libyan feather headdress. Donation stelae also signal ethnicity by declaring the foreign names of the patrons. Women donors play a more public role than before on donation stelae, possibly reflecting a re-formed outlook on women's roles brought about by an ethnically different ruling class. In addition, T.I.P.

however, are represented with this new, full figure. Goddesses hold a more sacred and hence traditional role, thereby preserving the earlier feminine ideal.

donation stelae exhibit some experimental and innovative iconography, suggesting that, in these cases, artistic convention remained sensitive to change.

Funerary stelae, on the other hand, are private (usually found in the burial chamber), personal, and mostly perishable (wooden). Although no ethnic costume markers appear on this class of stelae, Libyan names are used to invoke ethnicity. Women are prominent on funerary stelae, exhibiting a new body aesthetic and appearing individually without their male relatives. This modification in women's roles and depictions may reflect the mentality of a different ethnic group. As with the donation stelae, funerary stelae also suggest stylistic and iconographic experimentation and innovation, implying that the T.I.P. was a time when artistic decorum was not enforced rigorously.

It is possible that these changes in the Libyan Period were associated with a contemporary re-structuring of protocols regulating the illustration of ethnic and gender identities. The ethnically mixed atmosphere of the T.I.P. evidently provided opportunities for greater freedom in the expression of self-identity that lead one to ask whether the alteration of one element of identity (i.e. ethnicity) affects other elements of identity (such as gender) and whether there are greater opportunities for women to re-define their social status when there is a re-definition at the ethnic level of society.

Ancient Egyptian culture was selective in adopting innovative objects and ideas, so it is important to examine these stelae comprehensively in order to offer possible explanations for this change in attitude of a culture so bound by custom. The Libyan Period was an era in which an ethnic shift in the royal household re-structured various aspects of the political, religious, social, and artistic realms of ancient Egypt, providing an opportunity for new expressions of ethnicity and gender.

Bibliography

Aston D. 1987. *Tomb Groups from the End of the New Kingdom to the Beginning of the Saite Period,* Ph.D. Dissertation. Birmingham.

Capel A. and G. Markoe (eds.). 1996. *Mistress of the House, Mistress of Heaven: Women in Ancient Egypt.* New York.

Fazzini R. 1988. *Egypt Dynasty XXII-XXV*. Leiden.

Goudriaan K. 1988. *Ethnicity in Ptolemaic Egypt.* Amsterdam.

Hölzl R. 2001. "Stelae," in D. Redford (ed.) *The Oxford Encyclopedia of Ancient Egypt* 3: 319-24: Oxford.

Hornung E. 1999. *The Ancient Egyptian Books of the Afterlife*, trans. D. Lorton. Ithaca.

Kitchen K. 1969-1970. "Two Donation Stelae in the Brooklyn Museum," *Journal of the American Research Center in Egypt* 8: 59-71.

Leahy A. 1995. "Ethnic Diversity in Ancient Egypt," in J. Sasson (ed.), *Civilizations of the Ancient Near East* I-II: 225-234.

Meeks D. 1979. "Les donations aux temples dans l'Égypte du Ier millénaire avant J.-C.,"
in E. Lipiński (ed.) *State and Temple Economy in the Ancient Near East* II: 605-687. Leuven.

Munro P. 1973. *Die spätägyptischen Totenstelen.* Glückstadt.

O'Connor D. 1990. "The Nature of the Tjemhu (Libyan) Society in the Later New Kingdom," in A. Leahy (ed.) *Libya and Egypt c. 1300-750 B.C.* : 29-113: London.

O'Neil J. (ed.). 1987. *The Metropolitan Museum of Art: Egypt and the Ancient Near East.* New York.

Perez Die M. 1991. *Egipto y Proximo Oriente.* Madrid.

Robins G 1995. *Reflections of Women in the New Kingdom: Egyptian Art from the British Museum 4 February – 14 May 1995.* Atlanta.

Robins G. 1997. *The Art of Ancient Egypt.* Cambridge.

Roth A. 1999. "The Absent Spouse: Patterns and Taboos in Egyptian Tomb Decoration," *Journal of the American Research Center in Egypt* 36: 37-53.

Roveri A. 1991. *Museo Egizio Torino.* Turin.

Saleh, M. and H. Sourouzian 1987. *Official Catalogue of the Egyptian Museum Cairo.* Mainz.

Schulman A. 1966, "A Problem of Pedubasts," *Journal of the American Research Center in Egypt* 5: 33-41.

Scott G. and C. van Siclen III. 2001. *Egypt Revealed: Archaeological Treasures from the Museum of Fine Arts, Boston.* San Antonio.

Shaw I. (ed.). 2000. *The Oxford History of Ancient Egypt.* Oxford.

Ward W. 1994. "Foreigners Living in the Village," in L. Lesko (ed.) *Pharaoh's Workers: The Villagers of Deir el-Medina:* 61-85. Ithaca

Altars and cult installations of Punic tradition in Western Sicily

Federica Spagnoli – University of Rome "La Sapienza"

This paper focuses on altars and cult installations found in some of the most important Punic and Greek sanctuaries of three major sites in Western Sicily: Selinunte, Solunto and Monte Adranone.

Selinunte

The sacred areas of Selinunte, a foundation of Megara Hiblaea of 7th century B.C., are in the Acropolis and on the southern flank of the Eastern Hill where lie the Sanctuary of *Malophoros* and the Triolo North Building.

The *Temenos* of *Malophoros* has a rectangular plan; the monumental entrance (*Propylon*) of the sacred area is on the south-eastern side of the enclosure; the main temple is situated on the western terrace. It was founded in 7th century B.C. and fully reconstructed between the 4th and 3rd century B.C. In front of the entrance of the temple there is an altar with a rectangular shape (*Figure 27.1*)[1]: the upper part is decorated by an Egyptian *cyma*, the upper surface is concave ending in two swellings; in the centre there is a rectangular cavity for burnt offerings, as is testified by a circular burnt impression; the altar is carefully coated with white plaster.

Figure 27.1: Temenos of Malophoros, altar with two vertical swellings.

There are two altars inside the *adyton* of the temple: the first is an installation made of limestone blocks, rectangular in shape and rimmed on the top by an Egyptian *cyma*: the upper surface is a stone slab also with traces of burning[2]. Both altars are erected directly on the virgin soil, perhaps because they were related to the cult of a chthonic deity. The second altar (*Figure 27.2*),[3] located just in front of the niche in the *adyton*, is a little limestone cult stand, rectangular in shape and rimmed on the top and on the bottom with a upright *cyma*. A deep rectangular cavity is cut on the upper surface of the altar, perhaps for libations.

Figure 27.2: Temenos of Malophoros, arula inside the adyton.

In the north-eastern part of the *Temenos*, a sacred precinct hosted a small temple consecrated to a male deity, the *Meilichios* (Riotto 1985: 36; Tusa 2000 :10). Two altars (*Figure 27.3*) are located in front of the main façade of the temple; both are made of limestone blocks and have a rectangular shape, standing on a base with steps[4] (Gabrici 1927: 106-107).

Figure 27.3: Precinct of Meilichios, two altars

1 The structure measures 1.20 by 0.58 by 1.00 m and is oriented NW-SE: Gabrici 1927: 112-114, fig. 69-70, pl. XV: 2.

2 In the reconstruction by E. Gabrici two vertical slabs, similar in shape to the concave surface of the previous altar, lay on the upper surface (Gabrici 1927: 115-116, fig. 72).

3 Dimensions: 0.89 by 0.44 by 0.263 m; rectangular cavity: 0.35 by 0.21 by 0.06 m.

4 Three wells were found near the temple. The wells probably provided the water for the sacrifices (Tusa 2000: 12).

In 1969, to the west of the precinct of *Meilichios*, in that part of the *Temenos* between the Temple of *Malophoros* on the south and the northern limit of the sacred area, another enclosure was found (Tusa 1984: 11-15), where there were a number of sacrifice burials equipped with Greek imported pottery and *terracottas* (Tusa 1984: 14; Tusa 2000: 11); several stelae mark the spot of deposition (Famà 2000: 59-89). In the centre of this enclosure stands a monumental altar (*Figure 27.4*)[5] with three upright stone slabs on it. Two dressed blocks, in horizontal position, support the vertical ones; they stand on a rectangular plinth. The southern slab has an Egyptian *cyma* not very pronounced. The sacred building called North Triolo is situated south-east of the *Temenos* of *Malophoros* (Parisi Presicce 1986: 44).

Figure 27.4: Temenos of Malophoros, three-baetyl altar.

Against the end side of the *cella*, there is an altar (*Figure 27.5*) made of three rectangular stones with square hollows of different size on top, where three baetyls, now lost, were fixed; in front of the structure there is a standing platform for the priest[6]. Outside the southern wall of the building, there is another cult installation, consisting of a rectangular pillar with tapering top, and of a stepped altar (*Figure 27.6*) made of dressed limestone blocks. This altar also rests directly on the soil.[7] The latest cult installation in the North Triolo Building, between the second half of the 4th century and the beginning of the 3rd century B.C., was a base (*Figure 27.7*) located in the centre of the *ante-cella* and made of some collapsed blocks from the original building. On the surface, this base was heavily corroded by fire[8] (Tusa 1966: 152-153; Tusa 1971: 59-61).

Solunto

Solunto is one of the most ancient Phoenician sites in Sicily; the settlement is situated in the North, between Palermo and Himera. The sacred area with the altar is situated in the public district of the city, and it consists of two buildings, Building A and Building B (Tusa 1966: 143-153).

Figure 27.5: Triolo North Building, three-baetyls altar.

Figure 27.6: Triolo North Building, altar with pillar.

5 The base is 1.56 by 1.22 by 1.05 m (Gabrici 1927: 103-104).

6 *Prothysis*: the stone is 1.55 m long by 0.59 m wide, by 0.42 m deep (Parisi Presicce 1986, fig. 47).

7 The presence of irregular holes on the eastern corner of the upper surface, though perhaps not original, suggests a cult function (Parisi Presicce 1986: 45, fig. 43).

8 Round the altar were also two cremations, the first one (Tomb 2) put in a big pot *acroma*, the second (Tomb 3) in a burnt jug. An inhumation (Tomb 1) was found in the same room during the excavations of 1982 (S. Tusa 1984: 33-34, fig. 7: 3; Parisi Presicce 1986: 51).

The main altar (*Figure 27.8*) lies at the centre of room "a" of Building A, at the south corner of the structure; the altar is composed of a masonry platform (3.40 by 3.00 by 0.60 m) square in shape, and of a plastered sloping floor; three baetyls are set into the platform on its western side, and eastwards, on the front side, are two projections made of square bricks; the northern side ends with a square basin.[9]

Several different altars and cult installations were found in Building B (west of Building A). Room h, the central and probably the main room of the western complex, has a rectangular plan, and contains a bench and two altars against its east, south and west walls: the first altar leans on the western wall, while the second is in front of the entrance (Famà 1980: 23). In the north-western corner there is a semicircular platform (Famà 1980: 25).

Figure 27.7: Triolo North Building, altar.

Figure 27.8: Solunto, three-baetyl altar.

The first altar (*Figure 27.9*) is rectangular and is made up of three rows of stone blocks arranged in two lines. On three sides there are traces of white plaster; the second altar of room h (*Figure 27.10*), is built of small stone blocks;[10] the upper surface is made of two rectangular stone slabs with different thicknesses. It does not show any trace of plaster.

Monte Adranone

The Sanctuary erected in the centre of the residential area of this site consists in a rectangular building (Fiorentini 1972-73: 241, 243-244;Fiorentini 1982-83: 180). In the north-eastern room, against a long side, there are three sandstone basins; in the larger room two baetyls rest on square stone bases, situated next to the long wall, in front of the door. A square altar (*Figure 27.11*) lies in front of the south-western *baetyl*; it is made of stone blocks, slab-base and vertical projections on the upper surface. This one shows traces of fire. Behind the building there is a reservoir that supplied water to the city and to the Sanctuary for lustral worship.

Typology and functions

The comparative study of cult installations in the Punic sanctuaries of Selinunte, Solunto and Monte Adranone, allows us to distinguish three typologies of built altars connected to different cult activities and rituals.

A.1 SQUARE ALTARS
A.2 SQUARE ALTARS WITH VERTICAL SLABS
A.3 ALTARS WITH THREE BAETYLS, MADE OF STONE AND BRICKS

A.1:
Seven altars belong to this typology: five of them were found on the sacred area west of the Acropolis of Selinunte, two in the Punic sanctuary of Solunto. The typology is documented from the 5th century B.C. to the 2nd century A.D.

1. Square altars are generally situated inside the ante-cella or in front of it, next to the entrance of the temple.
2. The majority of these structures have a cavity, perhaps for libations, and the altars outside the temple of Meilichios (*Figure 27.3*), which, being located nearby sacred wells, may be linked to some ritual purification activity involving the use of water.
3. Square altars are made of stone blocks and can rest directly on the earth; they can also be decorated with Egyptian cymas, stand on a stepped base, and be plastered, like the structures outside the temple of Meilichios.
4. The structure in *Figure 27.6* presents, on the upper surface, a number of holes which may be interpreted as devoted to libation rites.[11]

9 Inside it there were fragments of pottery "sigillata italica" dated to the first century A.D., and bones of birds and rodents.

10 It measures 0.51 by 0.50 by 0.76 m.

11 The altar of the Phoenician sanctuary of Sarepta presents similar arrangements (Pritchard 1978: 136-137; Shaw 1989: 177).

5. As with the latter specimens, the altar in *Figure 27.7* is put together with a triangular stele: it was probably used for the burial of burnt sacrifices, as shown by the traces of fire, the burnt animal bones, the remains of votive objects and the presence of three tombs.
6. Two altars found in the main room of Building B in Solunto (*Figures 27.9-10*) do not have signs of burning: they are connected to a bench that runs along the walls of the room and to a half-rounded platform in the corner; in court "d", next to room "h", there is also a reservoir with a channel connected to it.

Figure 27.9: Solunto, Room h, altar.

Figure 27.10: Solunto, Room h, altar.

A.2:
This typology differs from the previous due to the presence of two raised edges on the upper surface. In the altar in pl.: 1, found in front of the entrance of the Temple of *Malophoros*, these elements are vertical and concave like the horns of the Mycenaean and Cypriote altars (Karageorghis 1976); a depression between the raising elements possibly served for burnt offerings. The altar in pl.: 11 from the Punic Sanctuary of Monte Adranone and the altar in the *adyton* of the Temple of *Malophoros* have straight vertical slabs. These installations also show traces of fire.

A.3:
This type includes structures built of stone with three vertical pillars on the top, which can be considered baetyls or stone slabs. The structure of North Triolo Building, which does not have the three pillars, probably had a support function for stelae or baetyls.

However, since the three pillars do not seem to have a tectonic function, nor a decorative purpose, they perhaps conceal a symbolic religious significance as in several Phoenician and Punic religious compounds. This is also demonstrated by the retrieval of offerings and *ex-voto*. These altars again display traces of fire.

The peculiar morphology of those structures characterized by the presence of fixed vertical elements allows comparisons between these and altars found outside Sicily, in different regions of the Mediterranean and in different cult contexts at the end of the Bronze Age and the Iron Age.

In Minoan frescoes and relief carvings altars are represented with pointed horns placed on the upper surface (Platon 1971, fig. 3); one can find the same typology with few differences also in altars of Cypriote sanctuaries at the end of the Bronze Age and at the beginning of the Iron Age (Du Plat Taylor 1957, pl. V: 6). In Syria and in Palestine, at the beginning of the first millennium BC, small altars with four horns at the corners of the upper surface, had a wide circulation (Gitin 1992: 42-49; May-Engberg 1935, pl. XII:10). The presence of three elements in relation to a cult installation is documented also in the Fosse Temple of Lachish (Tufnell 1940: 38-39): the raised platform has three rectangular projections stretching towards the centre of the room (we are talking about three pedestals for as many deities or three altars for three different types of offerings: incense, meat and water).

In Temple 200 of Tell Qasile (first half of 11th century B.C.), in front of the podium, there are three elements in a line parallel to the benches leaning on the walls. The excavator explained these structures as pedestals for the statues of the deities (Mazar 1980: 27). One can find the same typology in Crete and on the islands of Malta and Motya. The altars are situated at the sanctuary of Kommos (Shaw 1989: 179, fig. 18, 19, 20), outside the sanctuary of Tas Silg and the temple of Cappiddazzu respectively (Tusa 1964: 40). Moreover, these kinds of altars are also represented on three funeral stelae coming from Lilibeo in Sicily, Nora in Sardinia and Sousse in Tunisia (Moscati 1972, fig. 537, 320, 549).

In accordance with these remarks, one can suggest an oriental origin for this typology, employed in Selinunte to satisfy the religious demands of the Punic people.

Figure 27.11: M. Adranone, Punic Sanctuary, cella with altar and two baetyls.

With regard to Selinunte one can suggest that the Punic people restored the sanctuary of *Malophoros* and practised the worship of the main deities of their religion, Tanit and Baal Hammon, probably merging their cult with the Greek one that had been practiced before (Famà 2000: 89; Tusa 2002: 567-570). One can make the same remarks about the altar of the sacred area of Solunto, which shows other features typical of Punic sanctuaries in Sicily, like the bench, its rectangular plan, the presence of animal bones, and materials imported from Greece and pottery of local production (Tusa 1971: 63-64).

Bibliography

Famà M. L. 1980. "L'area sacra con altare a tre betili di Solunto" *Sicilia Archeologica* 42: 7-15.

Famà M. L. 1989."Il Museo G. Whitaker. Arule e louteria", in A. Ciasca et al. *Mozia* (Itinerari IV), Roma: 81-82.

Famà M. L.-Tusa, V. 2000. *Le stele del Meilichios di Selinunte* (Archeologia mediterranea e del Vicino Oriente 1). Padova.

Fanfara G. 1986. "Edificio Triolo Nord: lo scavo" in S. Tusa et al. "Selinunte-Malophoros. Rapporto preliminare sulla II campagna di scavi" (Sicilia Archeologica 60-61): 25-40.

Fiorentini G. 1972-73. *Kokalos* 18-19 :241-244.

Fiorentini G. 1980. "Santuari Punici a Monte Adranone di Sambuca di Sicilia" in G. Bretshneider (ed.), *Φιλίας χάριν, miscellanea di studi classici in onore di E. Manni*, III. Roma: 905-915.

Fiorentini G 1982-83. "M. Adranone nell'età tra i due Dionisi" *Kokalos* 28-29: 180-186.

Fiorentini G. 1995 *Monte Adranone* (Itinerari 16). Roma.

Fourmon M. H. 1981."Sélinonte: fouille dans la région nord-ouest de la rue F" *Sicilia Archeologica* 46-47: 5-27.

Fourmont M. 1982. "Santuari punici in Sicilia" *Kokalos* 28-29: 195-236.

Gabrici E. 1927."Il Santuario della Malophoros a Selinunte" *Monumenti Antichi pubblicati per cura della Regia Accademia Nazionale dei Lincei* XXXII. Milano.

Gitin S. 1992. "New Incense Altars from Ekron, context, typology and function" *Eretz Israel* 23: 43-49.

Karageorghis V. 1976. *Kition, Mycenaean and Phoenician Discoveries of Cyprus*. London.

May H. G.-Engberg R. M. 1935.*Materials Remains of Megiddo Cult* (OIP 41). Chicago.

Mazar A. 1980. *Excavations at Tell Qasile, part one. The Philistine Sanctuary, Architecture and Cult Objects* (Qedem 12). Jerusalem.

Moscati S. 1972. *I Fenici e Cartagine* (Società e Costume 8). Torino.

Parisi Presicce C. 1986. "Edificio Triolo Nord: la struttura" in S. Tusa (ed.) "Selinunte-Malophoros. Rapporto preliminare sulla II campagna di scavi" *Sicilia Archeologica* 60-61: 40-53.

Platon N. 1971. *Zakros. The discovery of a Lost Palace of Ancient Crete*. New York.

Pritchard J. B. 1978.*Recovering Sarepta, a Phoenician City. Excavations at Sarafand, Lebanon 1969-1974, by the University Museum of Pennsylvania*. Princeton.

Riotto M. 1985. "Il Santuario della Malophoros a Selinunte: spunti per una discussione storico-religiosa" *Sicilia Archeologica* 59 : 25-51.

Shaw J. W. 1989. "Phoenicians in Southern Crete" *American Journal of Archaeology* 93: 165-183.

Tufnell O. 1940. *Lachish II. The Fosse Temple*. London.

Tusa S. 1984. "Selinunte-Malophoros. Rapporto preliminare sulla prima campagna di scavi, 1982" *Sicilia Archeologica* 54-55: 17-58.

Tusa S. 1986. "Selinunte-Malophoros. Rapporto preliminare sulla II campagna di scavi" *Sicilia Archeologica* 60-61: 13-22.

Tusa V. 1964. "Il Santuario del Cappiddazzu" in A. Ciasca et al. *Mozia I. Rapporto preliminare della Missione Archeologica della Soprintendenza alle Antichità della Sicilia Occidentale e dell'Università di Roma*. Roma: 21-40.

Tusa V. 1966. "Le aree sacrificali di Selinunte e Solunto" in A. Ciasca et al. *Mozia II. Rapporto preliminare della Missione Archeologica della Soprintendenza alle Antichità della Sicilia Occidentale e dell'Università di Roma*. Roma 143-153, pls. XCI-XCV.

Tusa V. 1967. "Le divinità e i templi di Selinunte" *Kokalos* 13: 186-193.

Tusa V. 1971. "Selinunte Punica" *Rivista di Archeologia e Storia dell'Arte* XVIII: 47-68.

Tusa V. 1984. "Nuovi rinvenimenti nell'area del Santuario della Malophoros a Selinunte" *Sicilia Archeologica* 54-55: 11-15.

Tusa V. 1986. "I Fenici e i Cartaginesi" in G. Pugliese Caratelli (ed.) *Sikanie, Storia e civiltà della Sicilia greca* (Antica Madre 4). Milano.

Tusa V. 2000."L'area sacra al Meilichios" in M. L. Famà-V. Tusa *Le stele del Meilichios di Selinunte* (Archeologia mediterranea e del Vicino Oriente 1). Padova: 9-15.

Tusa V. 2002. "Il Santuario della Malophoros: un santuario pansicano" in M. G. Amadasi Guzzo, M. Liverani, P. Matthiae (eds.) *Da Pyrgi a Mozia. Studi sull'archeologia del Mediterraneo in memoria di Antonia Ciasca*. Roma: 567-570.

Sacred landscape and the construction of identity: Samnium and the Roman World

Tesse D. Stek – University of Amsterdam

Rome conquered the Italian peninsula in the fourth and third centuries BC by a blend of military campaigns and political strategy. This paper examines the consequences of, and responses to, Roman expansion in an area of Central Italy where the Samnites Pentri are believed to have lived in ancient times (modern Abruzzo and Molise). To understand the socio-political and cultural changes that occurred here in the third and second centuries BC, an analysis of the "sacred landscape" of Samnium is proposed.

Romanization

The impact of Roman conquest on non-Roman peoples, with all its cultural corollaries, has long attracted scholarly attention. In the conventional view, Roman cultural traits spread at the expense of local traditions as a logical consequence of the expansion of Roman imperialism. This process was referred to as "Romanization", a concept devised at the end of the 19th century. Since then the meanings and preconceptions attached to it have varied considerably, reflecting historical circumstances and the development of archaeology and ancient history.

The nationalist and colonialist ideology which prevailed when the concept of Romanization was first introduced meant that a Roman perspective dominated. Little room was left for the wishes and active responses of the conquered. In reaction to this, and to explain marked regional differences in the process of Romanization, scholars in the 1980s preferred a "bottom-up" approach and took a native, rather than a Roman, perspective. Regarding Roman imperialism as essentially a *laissez-faire* policy, they posited the initiative at the local level. Local élites were believed to have played a key role in a process of "self-Romanization", actively adopting Roman goods and customs in order to maintain their status within their own communities (e.g. Millett 1990). More recently, scholars have pointed out that these revisionist studies still operate within a colonial framework, albeit inverted. The adoption of Roman goods is still seen as a straightforward process, aimed at political and economic gain, and as the outcome of contact between two peoples and, therefore, cultures (Freeman 1993; Hingley 1996; Woolf 1997).

Philip Freeman in particular has raised doubts about the interpretation of "Roman" material culture. How do we know if certain styles or elements that we habitually call "Roman" were perceived as such by their users? The presence of "Roman" elements does not, he argues, prove a desire to be seen as Roman (Freeman 1993: 444). More generally, from a post-colonial perspective, more room should be left for the possibility of diversity and complexity in responses to Roman pressure. The survival of traditional elements should be given the same significance as change, and the possibility of opposition and resistance by local groupings should not be ignored (Hingley 1996). Moreover, Roman attitudes and intentions will have varied considerably over time and from place to place (Williams 2001).

Without discussing the appropriateness of the term "Romanization" (cf. Mattingly 2002), we can safely say that the issues it covers can be studied as a complex integrative process, in which the formulation and re-formulation of identities play a key-role (cf. e.g. van Dommelen 2001). People express their worldview through the construction of identities, giving themselves a place in communal or contrastive terms. These identities are constantly reformulated in the light of change. Further, people express their identities in symbolic form, which – fortunately for us - can have a material dimension. The archaeologist can attempt to reconstruct the process by studying, on the one hand, the appropriation, instrumentalisation or rejection of new elements and, on the other, the assertion, re-invention or abandonment of "traditional" elements, all within the logic of the cultural system in question. It follows that, in order to study perceptions of a changed world, an holistic approach and a "native perspective" should as far as possible be taken.

If one accepts this view, the study of "Romanization" is about how conquered peoples dealt with the new socio-political order. A wide range of diversity can thus be expected, since different histories, different Roman attitudes and different pre-Roman social and ideological structures will greatly influence the course the process takes. Clearly, identities are also constructed at various "levels", and can differ or even oppose one another depending on the context in which they are formed. This paper looks at the societal and communal levels, which it seeks to illuminate by examining the so-called "sacred landscape".

Sacred landscape

Rituals and religion play an important role in the creation and maintenance of a common identity. Sanctuaries serve communities as religious, political and social gathering places and are thus important arenas in which communal identities are articulated and reproduced. Drawing conclusions about broad cultural processes from individual sanctuaries can be misleading, however. As has been said, the interpretation of material culture *an sich* has proved a major problem in the analysis of cultural change. This paper will focus rather on the spatial and functional significance of sanctuaries in relation to settlements, necropoleis, and the natural landscape. The aim is to provide a context for the interpretation of the material evidence. Following Susan Alcock, this network will be referred to as the "sacred landscape" (Alcock 1993). The underlying assumption is that changes in the function and form of sanctuaries must correspond to changes in the needs of the groups that built, visited and re-built (or finally abandoned) them. In this sense, the sacred landscape, developing over space and time, can be read as a socio-ideological document.

Samnium: the Pentri

This paper will concentrate on the "heartland" of ancient Samnium, believed to have been inhabited by the Pentri, one of five subgroups considered to have made up the "Samnites". In antiquity the mountainous landscape formed one of the most impervious and (at least from a Central-Tyrrhenian perspective) remote areas of Central Italy, hard to access by land and with none of the advantages of the Adriatic coastal area. In the historical sources, the Pentri are notorious from the middle of the fourth century to the Social War in the 90's BC as the major obstacle on Rome's route to absolute power over the Italian peninsula (Salmon 1967; Dench 1995; Tagliamonte 1997). Their geographical position and historical role have helped to create an image of the area as the "core-region" of Samnite culture and resistance to the spread of Roman dominion. The Pentri are also comparatively well known through the material record. Their hill-forts and sanctuaries, which are dotted all over the area, form the most visible elements of the Samnite landscape and have therefore attracted (and dominated) scholarly interest (La Regina 1976; id. 1989; Tagliamonte 1997).

In what follows, general developments from the pre-Roman period to the third and the second centuries BC will be roughly sketched, with the focus on the development of the sacred landscape. Attention will then be paid to the culmination of this process in the second century BC, in the light of the theoretical debate outlined above.

Pre-Roman period

During the fourth and early third centuries BC, which witnessed the first series of violent encounters with Rome (traditionally beginning with the First Samnite War in 343 BC) the Samnite landscape appears to have been dominated by hill-forts, located at the most visible points. The periodisation and the function of these constructions have still to be studied (Oakley 1995).

Sanctuaries do not yet appear in monumentalised form and in this period are modestly located halfway down the hills. At some cult places votive objects and weapons are deposited. Weapons of foreign origin have been found in Pietrabbondante, some of which can be dated as early as the late fifth century and have been interpreted as booty offered and displayed in the sanctuary after battle. At that time Pietrabbondante may have already been serving as a symbolic central place for a larger group, at least in times of war (La Regina 1984). Most cult places in this period, however, seem to have served largely local purposes, as suggested by their direct relationship with the pattern of settlement. Archaeological evidence for the form of these early sanctuaries is lacking, but Livy has provided us with a valuable description of a rite performed by the Samnites at Aquilonia in 293 BC, in the course of the Third Samnite war (Livy 10.38). According to Livy, the time-honoured Samnite ritual held to form the *legio linteata*, the élite soldiers of the Samnite army, took place in a square sacred area measuring 200 by 200 feet, marked out with linen tapes. Altars were erected inside the square, a practice that can be compared with the altars mentioned in the religious text on the second century Agnone tablet.

It seems that in the period before the third century BC, no claim was laid upon the landscape by emphasising the sacred. A sacred landscape was not yet accentuated by monumental architecture, nor by selecting eminent or spectacular positions in the natural landscape.

Third and second centuries

The Samnite Wars ended in 290 BC with a unequal treaty for the Samnites, but the arrival of Pyrrhus in 280 BC triggered renewed Samnite military action as they (presumably including the Pentri) sided with Pyrrhus against Rome. After the Roman victory, the conquered Samnites were punished and were granted only semi-independent status. The pattern of settlement changed dramatically: in 263 BC the Romans placed the Latin colony Aesernia in the middle of Pentrian territory, and established a *praefectura* at Venafrum, the important passage to Campania. Robust houses in stone masonry also appear at this time, while farms and little villages pop up all over the countryside.

It is precisely during this period that the sacred landscape is structured more solidly. At the locations of sanctuaries that had formerly been open-air cult places, such as Pietrabbondante and Campochiaro, temples were built in an eclectic "Hellenistic" style that sometimes have direct parallels in Latium and Campania. The location of these temples seems to be determined by tradition, as is suggested by the presence of votive objects from previous periods. It is in the "traditional" Pentrian regions that these sanctuaries flourish, away from the new colony at Aesernia. Here a typical Roman-style triple cella temple was built, a small copy of the *Capitolium*, symbolising and propagating *Romanitas* as in colonies all over Italy.

The monumentalisation of the Pentrian sanctuaries seems

to have been dictated by the wish or need to express and reinforce ties with Pentrian traditions. This process comes to a climax at the end of the second century BC, when it becomes far more archaeologically visible. Third century sanctuaries are all now embellished or rebuilt, while many more small sanctuaries appear all over Samnite territory. In the latter case, the location is determined by local demands, directly linked to the changed pattern of settlement. The monumentalisation, however, carried out in a uniform style, meets a supralocal need. The combination of location and architectural uniformity seems to reinforce the bond between local systems and a larger entity. The focal point of this entity is the main sanctuary of Pietrabbondante, the supralocal function of which becomes clear from the second-century *safinim*-dedication, which is thought to refer to the Pentri as an ethnic group and may reflect a self-conscious appeal to their Samnite tradition (Ve. 149; Dench 1995: 139).

At the end of the second century the impressive temple-theatre complex at Pietrabbondante is erected, where Hellenistic elements from the Latial and Campanian regions are freely used. The temple is laid out on a triple cella plan, reminiscent of the typical "Roman" temple of the time and exemplified at Aesernia. Unfortunately, too little is known about Samnite religion and architecture to establish whether this feature would then have been regarded as typically "Roman". If that were indeed the case, it would suggest the conscious appropriation of "hostile" elements. On the other hand, when one looks more closely at the central sanctuary of Pietrabbondante and at the "satellite-sanctuaries" dispersed in Pentrian territory, one detects signs of a certain traditionalism. As Adriano La Regina has pointed out, the area occupied by the *templum* of Pietrabbondante measures 200 by 200 feet, thereby recalling the Samnite *area sacra* where the *legio linteata* was formed according to Livy (see above; La Regina 1976).

This may represent more than just an analogy: recently, Livy's Aquilonia has, on other grounds, been identified with Pietrabbondante (Sisani 2001). If this were correct, it would mean that the basic outline of the traditional sanctuary at Aquilonia/Pietrabbondante was never affected by the rebuilding programmes of the third and second centuries. Even more suggestive is the fact that the basic concept of Pietrabbondante (the square plan, with the cella/altar in the centre and porticos at the sides) is echoed in other, smaller second century sanctuaries, such as S. Giovanni in Galdo, and maybe Quadri and Atessa (La Regina 1976; id. 1989).

The Pentri example presents an interesting illustration of the problems surrounding the interpretation of material culture, as outlined in the introduction. Widely-spread Hellenistic cultural forms, and maybe even elements that could have been regarded as "Roman" in this context, are employed to serve local purposes and provided meaning that is at direct variance with any straightforward notion of "Romanization". Although one should be wary in using terms like cultural resistance (cf. Brown 1996), there seem to be strong indications in the Pentri case to support such an approach. One of these is the well-known parallel/opposition between the Roman she-wolf and the Italian calf, whereby in the case of the Pentri the association with the Samnite bull seems to be added. On coins minted just prior to the Social War, the Italian or Samnite bull is depicted as trampling the Roman she-wolf (Dench 1995: 213-215).

The most plausible interpretation of the monumentalisation of the sacred landscape in the third century, and again on the eve of the last insurrection against Rome at the end of the second century, can be found in the theories of the social anthropologist Anthony Cohen (1985). Cohen argues convincingly that as the structural base of a community is threatened, so its symbolic boundaries will be strengthened. In this sense, the adoption of "Hellenistic" (and "Roman") elements are not to be interpreted as "self-Romanization", but as a desperate search for an own, Pentrian identity. Thus impelled, the patrons and architects consciously recall the vetust Samnite tradition in the square lay-out. At the same time, the adoption of newly available cultural forms must have deeply changed the Pentrian world: in re-constructing a Pentrian identity new elements were added, which in turn will have influenced Pentrian thought and cosmology.

Acknowledgements

This paper represents the first outline of an approach to Romanization-processes through the analysis of the 'sacred landscapes' of four different regions in Central Italy. The relatively well-known example of Samnium is to be seen as a point of departure for this larger study, which is to be submitted as PhD-thesis to the University of Amsterdam. This project is undertaken in close cooperation with Jeremia Pelgrom, who studies the sacred landscape of Lucania. I thank the SOMA 2003 committee for their kindness to accept the written version of my paper, as I was not able to present it at the symposium due to an accident opening a bottle of wine. Furthermore, I want to thank Jeremia and Ellen Thiermann for their support, and the Dutch Institute at Rome (NIR) for granting me a scholarship. I also would like to thank the National Health Service for helping me after a night of arterial bleeding in the waiting room, and all that for free.

Bibliography

Alcock S. E. 1993. *Graecia capta: the landscapes of Roman Greece*. Cambridge.

Brown M. F. 1996. "On Resisting Resistance" *American Anthropologist* 98 (4): 729-735.

Cohen A. 1985. *The symbolic construction of community*. London-New York.

Dench E. 1995. From barbarians to new men: Greek, Roman and modern perceptions of peoples of the central Appenines. Oxford.

Freeman P.W.M. 1993. "'Romanisation' and Roman material culture" *Journal of Roman Archaeology* 6: 438-445.

Hingley R. 1996. "The 'legacy' of Rome: the rise, decline and fall of the theory of Romanization", in

J. Webster & N. Cooper (eds.) *Roman Imperialism: post-colonial perspectives*. Leicester: 35-48.

Keay S. J. & Terrenato N. (eds.) 2001. *Italy and the West: comparative issues in Romanization*. Oxford.

La Regina A. 1976. "Il Sannio", in P. Zanker (ed.) *Hellenismus in Mittelitalien*. Göttingen: 219-254.

La Regina A. 1984. "Aspetti istituzionali nel mondo sannitico", in *Sannio: Pentri e Frentani dal VI al I sec. a.C.* (atti del convegno 10-11 novembre 1980) Isernia: 17 ff.

La Regina A. 1989. "I Sanniti", in G. Pugliese Carratelli (ed.) *Italia omnium terrarum parens*. Milano: 301 ff.

Mattingly D. J. 2002. "Vulgar and weak 'Romanization' or time for a paradigm shift?" *Journal of Roman Archaeology* 15: 536-540.

Millett M. 1990. *The Romanization of Britain: an essay in archaeological interpretation*. Cambridge.

Oakley S. P. 1995. *The hill-forts of the Samnites*. London.

Salmon E. T. 1967. *Samnium and the Samnites*. London.

Sisani S. 2001. "Aquilonia: una nuova ipotesi di identificazione" *Eutopia* n.s. I 1-2: 131-147.

Tagliamonte G. 1997. *I Sanniti: Caudini, Irpini, Pentri, Carricini, Frentani*. Milano.

Van Dommelen P. 2001. "Cultural imaginings. Punic tradition and local identity in Roman Republican Sardinia", in Keay & Terrenato 2001: 68-84.

Vetter E. 1953. *Handbuch der italischen Dialekte*. Heidelberg.

Williams J. H. C. 2001. "Roman intentions and Romanization: Republican northern Italy, c. 200-100 BC", in Keay & Terrenato 2001: 91-101.

Woolf G. D. 1997. "Beyond Romans and natives" *World Archaeology* 28: 339-350.

Investigating Colonialism and Post-Colonialism in the Archaeological Museum Space: The Case of the Lebanon and France

Lina G. Tahan – University of Cambridge

Introduction

The objective of this paper is to propose a theoretical model with which to study the history of Lebanese Archaeological Museums. It will present a definition of what colonialism and post-colonialism are according to both colonial and post-colonial theory. Moreover, it will set the background for an investigation of how Lebanese museums are still influenced by colonial ideas in terms of the display of artefacts.

I shall explore the relationship of coloniser versus colonised within archaeological museums, taking as an example the Louvre Museum in France and its exhibition of Lebanese archaeological heritage. As the Lebanon was under the authority of the French between 1918 and 1943, the development of archaeological research and the organisation of museums followed the plans set by the French.

Finally, I will present the theoretical models I intend to apply in my assessment of the role of Lebanese museums within the context of colonialism and post-colonialism. This paper makes reference to the works of Simpson (1996) and Barringer & Flynn (1998), which include a history of museums and collectors, as well as making representations of them in the colonial and post-colonial eras.

Colonialism/Post-Colonialism

The definition of these terms in such a particular context needs to be examined very closely, especially since a great deal of confusion and scepticism surrounds the relatively new field of post-colonial studies (Loomba 1998: xi).

It is very important to point out that colonialism and imperialism are often used interchangeably. The word 'colonialism' comes from the Latin 'colonia' which meant 'farm' or 'settlement', and referred to Romans who settled in other lands but still retained their citizenship (Loomba 1998: 1). According, to the *Oxford English Dictionary* (OED), the term 'colonialism' is defined as follows: *"the policy or practice of acquiring political control over another country, occupying it with settlers, and exploiting it economically"*. It seems that this definition avoids referring to the colonised people. It only refers to the colonisers. Hence, it clears the word 'colonialism' of the notion of an encounter between peoples, but refers to the practice of retaining an authority over another country. One should note that colonialism is not an identical process in all parts of the world. In the case of the Lebanon, it was part of the Ottoman Empire until the latter's defeat in World War I and was then ruled by France as a Mandate protectorate from 1918 till 1943. In a sense, colonialism can be defined as the conquest and control of other people's land and goods (Loomba 1998: 2). Archaeology became part of these goods because the past was appropriated by the colonisers and used for various political reasons.

Therefore, one can say that the history of archaeology in the Lebanon can be attributed to French involvement. In short, it was initiated by the western travellers, the commercial and diplomatic agents that had long been engaged in the looting and selling of antiquities. The travellers and agents who came to the Lebanon ushered in new and different kinds of colonial practices which altered the whole country. Hence, it is argued that colonialism started first with the coming in 1855 of the travellers and agents who started exploring the Lebanon and acquiring archaeological artefacts to ship back to their respective countries.

All these practices contributed to shaping the history of the colonised country and the colonising power. This paper will present some of the artefacts, focussing mainly on the Yehawmilk stele and the sarcophagus of Eshmounazar II, collected by the colonisers (the French) and displayed in their own museum: the Louvre. Such artefacts have been instrumental in forging and maintaining relationships across space and time, between peoples of different cultures (*Figures 29.1 & 29.2*).

As the Lebanon was under the hegemony of the Ottoman Empire, archaeology was under their control and Constantinople was made the centre to house those artefacts in the Ottoman Imperial Museum. During that period, the American University of Beirut Museum was the first museum to be founded in the Lebanon (in 1868). When the French took over, the Lebanon became a Mandate territory and its boundaries were redefined as independent to Greater Syria. It is during this period that archaeology was also redefined, along with the construction of the National Museum of Beirut in 1929 and the establishment of an Antiquities Law in 1933.

The fluctuations in meaning can complicate our comprehension of the term 'post-colonial', a term that is the subject of an ongoing debate which will be discussed later in this paper. Archaeology is a discipline that is strongly influenced by the position that the countries and regions in which it is practised occupy within the modern world-system. The Lebanon, for example, has a rich heritage, and so archaeology has played a significant role in the creation of the Lebanese state and the shaping of its identity.

Figure 29.1: The Yehawmilk Stele

'Colonialist archaeology', according to Trigger (1984: 360) is the archaeology that grew either in countries whose local population was wholly replaced or overwhelmed by European settlement, or in countries where the political and economic hegemony remained under European domination for a considerable period of time. In such countries, Trigger explains that archaeology was practiced by a colonising population that had no historical ties with the locals whose past they were studying and representing in museums.

The achievements of ancient Near Eastern civilisations were appropriated for Western Europe by claiming that western Europeans rather than the people who lived in the Near East today were their true spiritual heirs (Trigger 1984: 365).

In the Lebanon's case as a Near Eastern country whose past was appropriated by France, French archaeologists began their archaeological investigations in the nineteenth century. They stressed that several streams of cultural influences from the Lebanon and elsewhere converged in France.

One needs to be aware that archaeology as a discipline, in a previously colonised country, has its roots founded in colonialism (Gosden 1999: 16). Also, the colonial origins of the museum continue to have a lasting influence over these cultural institutions and over the public perceptions of them (Simpson 1996: 1).

Colonialism has been used in different ways to suit certain theoretical grounds. Two issues have been considered important in archaeology: firstly, the fact that one or more groups of foreigners are located in an area which is a considerable distance from their place of origin i.e. the colonisers; secondly, the fact that the relations between colonisers and colonised were relations of domination and socio-economic exploitation (Rowlands 1998: 328).

Therefore, one can conclude that colonialism is about a set of power relations that involve complex themes such as domination/resistance, coloniser/colonised, civilised/primitive. These relations can be generalized as differences between the self and the other. Rowlands' theory demonstrates that understanding colonialism is not an easy task if one wants to focus on historical change, particularly in the case of the Lebanon.

According to Gosden & Knowles (2001: 5), there are three basic models to understanding colonial relations. Firstly, there is acculturation, which views the local culture as being dominated by the culture of the incoming people. This may be seen as harmless change that can be labelled as 'development' or a less positive one, cultural loss. The second one suggests the lack of change and maintenance of tradition, which was a model used by many anthropologists who stress the ethnographic present. The third model views change as a hybrid whereby new cultural forms emerge out of the encounters of existing cultural logics.

After examining definitions of 'colonialism' and 'colonial archaeology', we should turn to 'post-colonialism' and 'post-colonial archaeology'. The term 'post-colonial studies' is now generally accepted as the name for a field of interdisciplinary studies which encompasses a wide variety of types of analysis. What links them is a concern with the imperial past, with the different varieties of colonialism within the imperial framework, and with the links between the imperial past and the post-colonial present. It is the links concerning museums in the Lebanon which interest us.

Colonialism is not something that occurs outside a country or its people, and does not operate with the collusion of forces inside, but a version of it can be duplicated from within, so that post-colonialism, far from becoming a term that can be indiscriminately applied, appears to be orchestrated around contradictions and qualifications. It is important to think of post-colonialism not as just literally following colonialism and implying its demise, but more flexibly, as the challenge of colonial domination and the legacies of colonialism (Loomba 1998: 12). Such a position would allow us to think of the transfer of archaeological material culture displaced by colonialism, such as the Yehawmilk stele or the sarcophagus of Eshmounazar II for example (displayed in the Louvre Museum), as 'colonial' objects since they cannot be repatriated to the Lebanon.

Figure 29.2: The Sarcophagus of Eshmounazar II (Source: Louvre Museum, Department of Oriental Antiquities)

A country such as the Lebanon may be regarded as post-colonial (in the sense that it got its independence from France in 1943), but is still colonial or rather, 'neo-colonial', in the sense that it has remained culturally dependent on France, or any other Western country. This is reflected in museums of archaeology, which are regarded as institutions of culture. The Lebanon is still dependent on foreign expertise to create museums or to influence particular kinds of displays or to retain the same traditional displays renewing only their appearances, as is the case of the National Museum of Beirut.

According to Gosden (2001: 241) in his article *"Postcolonial Archaeology: issues of Culture, Identity, and knowledge"*, all archaeology, in a previously colonised country, is post-colonial. Moreover, museums should be in practice post-colonial, yet they have retained a colonial discourse and this is the case for the Lebanese archaeological museums and the Louvre Museum. It is in this sense that one should consider the impact of post-colonial theory on archaeology, archaeological museums and on the new legal and intellectual context of archaeology in many parts of the world (Gosden, 2001: 241). An important feature of post-colonial thought is that it emphasises culture and attempts to identify and remove colonial habits of thought within the western intellectual tradition.

The Concept of 'Museum Culture'

Museums are human-made objects. The museum, as the largest object itself, houses the collection of objects. In addition to that, the museum has a past, a particular version of which is that it is part of colonial construction. Through museums one can see directly or indirectly that the relationship of coloniser versus colonised is obvious. These relations always involve material culture (Gosden & Knowles 2001: 6), thus creating a particular museum culture. Consequently, an artefact is best seen as indicative of an ongoing process, rather than a static one. This process is reflected in the museum space, as there is a series of continuous social encounters surrounding the artefact, linking 'the archaeological site' to 'the museum' (Ibid. 2001: 5).

Culture is defined as a set of values that are attached to people, places and things and the processes that create and change such values (Gosden 1999: 204). Such places can be museums. Hence, in order to develop a museum culture one needs to become aware of that concept which is totally ignored in the Lebanese context. The creation of such a culture was purely a Western idea that has not infiltrated Lebanese museums. Particularly important to the shaping of the Lebanese museum in the post-colonial period is the formation of a new nation-state which usually defines a 'bounded' citizenship and identity. Colonialist expansion provided three things: 'material' for display, territories to be governed and the development of 'museological' ways of seeing the world. Hence, the museum is not merely a product of, or a site for displaying the narratives of modern developments, but also a tool through which modernity is shaped. It should also be a place where 'voices' can be heard in order to overcome social differences, exclusions and contradictions (Macdonald 1998: 9).

Adopting a Theory: 'the Archaeological Orient' and the Museum

Colonialism, with its ideologies and power relations, influences the ways in which archaeological artefacts are understood (Barringer & Flynn 1998: 1). One of the aims of this paper is to explore this issue further, by looking at the meaning of the artefact that is inflicted by the context in which it is displayed. The removal of the artefact from a colonial periphery to the imperial centre profoundly alters the ways in which it is introduced. Accordingly, rather than addressing the museum's collections in isolation, the museum will be considered as a cultural formation. As a result, the museum is comprised of objects ordered into taxonomies, whose interpretation is conditioned by labels, panels, guides and catalogues (Barringer 1998: 12). Moreover, it becomes clear that colonialism can raise a range of historical, archaeological as well as interpretative issues essential to museum professionals and their publics in the post-colonial era. Finally, the issue of representation raises important questions, such as the impact of the power relations of colonialism on the interpretation and definition of archaeology in museums and the present possibilities for the display of the appropriated objects.

Colonialism and archaeology can provide timely and substantive insights into issues of more general - indeed of global – interest. Among the most powerful and influential analyses of colonialism's effects on the culture of the coloniser and the colonised is Edward Said's *Culture and Imperialism*. Said's (1993: 5) aim is *"... to connect these different realms* [Culture and Imperialism] *to show the involvements of culture with expanding empires, to make observations about art that preserve its unique endowments and at the same time map its affiliations ... "*.

Over the past two decades a wide range of academic discourses under the label of post-colonialism has been devoted to disentangling and elucidating European colonial processes. It is true that archaeology has been slow to enter this discourse; however, it is my argument that through analysing archaeological exhibits, archaeology will provide a fertile ground for understanding colonialism and associated processes of cultural interaction, in terms of cultural contexts, both recent and past. It is my belief that studying colonialism and archaeological artefacts may not only prove fruitful because it has not actually been systematically attempted before, but it may also provide valuable links to understanding the role that archaeological museums can play in the present.

George Orwell's (1949: 199) insightful statement *"Who controls the past controls the future: who controls the present controls the past"*, perhaps written with reference to Nazi manipulation of the past during World War II, is just as valid today as it was in 1949. Control over the past can often be employed to legitimise actions in the present and thus affect the future. For instance, this can be applied to the archaeology and generally the past of the Lebanon. During the colonial period the past of the Lebanon was exploited and, by doing so, France had control over the present and the future. Even today, it is still evident that the colonised countries rely on the help of the coloniser in terms of museum organisation, design and display. Even though post-colonial Lebanese archaeological museums seem to be independent, they are still controlled in some way or other by an imperial power, since France wants to interpret and research the Lebanese archaeological heritage and owns some of its objects, and this can be seen at many levels.

During the colonial period, vast quantities of cultural material were collected and placed in museums in France. In many instances, the manner in which items were collected resulted in poor documentation, particularly in the early years of collecting. Likewise, the issues surrounding the acquisition and ownership of cultural material are rarely analysed in French museum exhibitions, yet the attitudes and actions of the colonialists and the deleterious effects of expansionist policies are part of the history of the archaeological objects in museum collections, and also part of the history of both colonised and coloniser (Simpson 1996: 25).

Museums have long served to house a diverse heritage. Often beginning with the private collections of elites obtained through conquest and exploitation, museums also came to conserve a national and a cultural heritage, as well as the heritage of the 'other', and to educate the public (Kaplan, 1994: 9). It is the nature of these collections, and the way in which they are used to illuminate and interpret the past that is of interest to us. In the Western world, museum professionals seem to be discovering the impact of colonialism in their respective museums (Kaplan 1994: 9); on the other hand, in the developing world, many museum professionals do not seem to be concerned with the impact of the colonial legacy in their respective museums in order to make new choices.

The Lebanon resembles any other developing country in assigning national roles to its museums. The objective is to present the past and create a coherent picture of the Lebanon's heritage. Thus Lebanese museums aim at creating a national identity. The nature of the Lebanese past has particular qualities since it encompasses the colonial experience, which is often viewed negatively. Lowenthal (1976: 89) has suggested that countries freeing themselves from colonial rule try to erase the colonial experience and to find an autonomous national identity in the pre-colonial past. Whether this can be applied to the Lebanon needs to be researched further.

So far, the interaction between colonisers and colonised has remained largely ignored in museum exhibitions in France and the Lebanon. Archaeological material on display has tended to focus on the material culture of the Lebanon, but has failed to reflect the actions and attitudes of the colonialist era (Simpson 1996: 25). Hence, this 'cultural colonialism' continues to control the representations of Lebanese archaeology and culture.

Acknowledgements

I would like to thank David A. Barrowclough for reading and commenting upon this paper. It would be appropriate to mention the Late Edward W. Said whose contributions laid the foundation of my PhD research project for an enquiry into colonial material culture that has wide reverberations across humanity. This paper is dedicated to the memory of Professor Edward W. Said who will remain an influential figure in post-colonial studies.

Bibliography

Barringer T. & Flynn T. 1998. "Introduction", in T. Barringer & T. Flynn (eds.) *Colonialism and the object: Empire, material culture and the museum.* London: 1-8.

Barringer T. 1998. "The South Kensington museum and the colonial project", in T. Barringer & T. Flynn (eds.) *Colonialism and the object: Empire, material culture and the museum.* London: 11-27.

Gosden C. 1999. *Anthropology and archaeology: A changing relationship.* London.

Gosden C. 2001. "Postcolonial archaeology: Issues of culture, identity, and knowledge", in I. Hodder (ed.) *Archaeological theory today.* Cambridge: 241-61.

Gosden C. & Knowles C. 2001. *Collecting colonialism: Material culture and colonial change.* Oxford.

Kaplan F. S. (ed.) 1994. *Museums and the making of ourselves.* London.

Loomba A. 1998 *Colonialism/ postcolonialism.* London.

Lowenthal D. 1976. "The place of the past in the American landscape", in D. Lowenthal & M. J. Bowden (eds.) *Geographies of the mind: Essays in historical geography in honour of John Kirtland Wright.* New York.

Macdonald S. 1998. "Exhibitions of power and powers of exhibition: An introduction to the politics of display", in S. Macdonald (ed.) *The politics of display: Museums, science, culture*, 1-24. London.

Mitchell T. 1991. *Colonising Egypt.* Berkeley.

Orwell G. 1949. *Nineteen eighty-four.* Harmondsworth.

Pearsall J. 2002. *The concise Oxford English Dictionary.* Oxford. Tenth edition, revised.

Rowlands M. 1998. "The archaeology of colonialism", in M. Rowlands & K. Kristiansen (eds.) *Social transformations in archaeology*, 327-333. London.

Said E. W. 1993. *Culture and imperialism.* London.

Simpson M. G. 1996. *Making representations: Museums in the post-colonial era.* London.

Trigger B. G. 1984. "Alternative archaeologies: Nationalist, colonialist, imperialist" *Man* 19: 355-70.

Ethnic Identity in Archaic Pompeii

Ellen Thiermann – University of Amsterdam

Introduction

The ethnic origins of the city of Pompeii have been the subject of intense debate for more than a century. Numerous efforts have been made to prove either a Greek or an Etruscan initiative behind the founding of the famous city, not least because Pompeii has become a symbol of the antique roots of Western civilisation. None of these efforts has yielded conclusive results and I will argue in this paper that no satisfactory answer is possible, and that the question should be reformulated in the light of recent theoretical developments. Our views on ethnicity and on how we, as archaeologists, should approach it have changed fundamentally in the last decade. But the debate has tended to gravitate towards theory, and the practical applications have only just begun to be explored. The most elaborate case study on the subject is Hall's famous study of Greece (Hall 1997), whereas the Italian Archaic period has remained relatively untouched. To correct this imbalance, a case study of ethnicity in Archaic Pompeii is proposed here. First, the concept of ethnicity will be briefly discussed, along with possible approaches that can be taken towards it. The written and material sources relating to Pompeii in the Archaic period (6th – mid-5th Century BC) will then be analysed within a new theoretical framework.

Theories of Ethnicity

The term "ethnicity" originated in the 1950s. It was added to the list of terms used for classifying human collective existence – concepts such as "race", "people" and "culture" - and partly replaced them (Jones 1997). Since then, the definition of ethnicity as a primordial and unchangeable characteristic of man has changed. In the post-modern social sciences, especially in anthropology and sociology, an ethnic group is defined as *"a self-ascribing social collectivity"* that believes in a common descent and *"constitutes itself in opposition to other groups of a similar order"* (Hall 1997). It follows that an ethnic group does not constitute a stable phenomenon in the course of history, but can be subject to change in reaction to social, economic or political circumstances. Moreover, the importance attached to ethnic identity, as opposed to other forms of social and cultural identity, can vary both individually and spatially. In general, ethnic identity will be more emphasised in times of "stress", i.e. when the integrity of an ethnic group is threatened.

The current debate has centred on the criteria upon which the construction of ethnic identity is based. Amongst the criteria that have been proposed are: a common myth of descent, a (putative) common history, a (putative) original homeland, a specific shared culture, a common language, common religion, common physical appearance, a common name ("ethnonym") or group solidarity (Hall 1997; Renfrew 1996; Smith 1986). Obviously, none of these criteria can form a stable basis for ethnic identity on its own. Rather, as has been frequently argued, every group chooses a specific set of criteria that fits its own circumstances. In my view, however, freedom of choice is often rather restricted. Recent research has pointed out the crucial role played by a common past, be it real or invented (Hall 1997). It is this belief in a common past that seems to represent the essential characteristic of ethnic identity, as distinct from other forms of cultural or social identity. Secondary criteria, like language, religion or physical appearance, may serve as visible markers for the ethnic group. My own suspicion is that their presence or absence depends on their relevance to the group's (putative) common past. The central importance of a common past coincides with sociological theory, which stresses that the construction of identities is far more determined by oral or written discourse than by real action (Bourdieu 1977). This suggests that manifestations of ethnic identity would be much easier to find in the written record than in material culture.

Archaic Pompeii

To subject these theoretical presumptions to critical examination, we will now consider the case of Pompeii in the Archaic period. As has been said, scholars in the first half of the 20th Century discussed who was responsible for the foundation of the city in a highly politicised manner (e.g. Maiuri 1973 (a posthumous edition of earlier articles); Mau 1908). The few elements of Archaic material culture known at that time were taken, via style analysis, as direct indicators of either Greek or Etruscan dominion. Thus scholars were convinced of the absolute "Greekness" of the Doric temple, which could only have been built by Greek colonists. The bucchero sherds from the votive deposit of the temple of Apollo had to prove the presence of the Etruscans. After World War II, for obvious reasons, ethnic attributions ceased to be an openly expressed focus of archaeological research in Pompeii. However, thanks to a recent series of small excavations below the level of the eruption of 79 AD, new evidence has been added to our still very meagre archaeological knowledge of the Archaic period at this otherwise well-studied site. Yet once again, in these very recent studies of pre-Roman Pompeii, the "ethnic issue" is carefully avoided.

Thanks to its geographical position, Campania was a region of intensive cultural and economic interaction in the Archaic period. Most of the small Campanian settlements that were scattered all over the region were inhabited by the indigenous Italic population, though this cannot be presumed to have been an entirely homogenous group. Since the 8th Century BC, Greek colonies had been established along the coast: at Pithecusa on the island of Ischia, and Cuma and Poseidonia. By contrast, a strong Etruscan influence is attested in the centre of Campania, especially evident at Capua. In this dynamic and stimulating climate, a number of settlements developed into urban centres during the 6th Century BC, amongst them Pompeii.

In what follows, the archaeological, historical and epigraphical evidence is approached not according to material categories, but rather via what I have called "thematic contexts". These contexts have been devised on the basis of the criteria used to establish ethnic identity, as described above. Thematically similar criteria have been grouped to form the following contexts: the historic-mythological, the material-cultural, the religious and the linguistic. There are two advantages to operating within these contexts: first, that the criteria used to construct collective identity should become visible; second, that the respective roles of written sources and material culture within this process should become clear.

The historic-mythological context

The first context embodies the historical view and mythological traditions of the Archaic period and therefore relies upon written sources. These are unfortunately limited to Greek writers. Several of these, most of them cited in Strabo's Geography (V 4, 3-13), reported on the ethnic situation in Campania and Pompeii, employing the colonial Greek perception of "the Other". The presence of Greek colonists is described at length, of course, but they are not mentioned for Pompeii. It is the Etruscan element that is emphasised for the whole of Campania, including Pompeii. Here the Etruscans are linked with the *Pelasgoi*, a mythical people who frequently occur in Greek historical writings, especially in relation to urbanisation processes in the past, which Greek authors had difficulty in labelling ethnically. In this context, the term *Pelasgoi* could be seen as shorthand for an ethnically mixed community (Cerchiai 1995: 105). For the centuries preceding and following the Archaic period, various names are used to describe the Italic population of Campania: Opikoi, Osks, Campanians and Samnites. It is probable, however, that these different denominations do not reflect population changes caused by large-scale invasions of foreign peoples, but rather Greek perceptions of the far-reaching social transformations taking place over the centuries within the same indigenous population.

The written sources also inform us about the myth of Herakles, which is closely linked to Pompeii. The cult of Herakles was probably introduced to the region by the Greeks of Cuma. They situated some of his adventures in the Campanian landscape, where the hero became one of the most worshipped divinities. One version of the myth tells that Pompeii got its name from the place where Herakles celebrated his triumph (the *pompa triumphalis*) after completing his twelve tasks. This myth of origin provides the city with mythological legitimation, hereby creating an urban identity. However, ethnic categories are not referred to in this process.

The material-cultural context

The second context concentrates on all the materially expressed aspects of culture preserved in the archaeological record. Most of the small excavations have brought to light modest domestic remains consisting of wall structures of *pappamonte* (soft tufa) and ceramics. Unfortunately, neither the dimensions nor the state of preservation of these remains permit the reconstruction of the layout for any single house, which might have provided a marker for ethnic identity. However, from the distribution of the structures it is evident that the Archaic city already had the same plan and extent as the Roman city, although it was much less densely inhabited (Fulford & Wallace-Hadrill 1999).

In addition to this domestic material, two monumental temples from the Archaic period are known. The sanctuary of Apollo on the Forum was monumentalised at the end of the 6th Century BC (De Caro 1986). The temple stood on a high podium and had wooden columns. It also had a so-called "Campanian" roof, which had been developed in the centres of Cuma and Capua and had become very popular throughout the region. The architectural form of Pompeii's main urban sanctuary perfectly matched the regional tradition, which combined Etrusco-Italic architectural features with innovations of its own. It can therefore be read as a sign of adherence to the prevailing Campanian political network.

At the same time a totally different temple, now known as the 'Doric Temple', was erected on the so-called Foro Triangolare, a natural platform on the edge of the city facing the sea. It was built entirely of stone and had the form of a *peripteros*. However, this Greek temple concept was employed here in unusual proportions (2:3), which are characteristic of Etrusco-Italic temples. A second unusual feature is the uneven number of columns (seven) at the front and rear of the building. One of the very few existing parallels can be found in the immediate neighbourhood of Pompeii, at the Greek colony of Poseidonia. The so-called "Basilica" was erected there some decades before the Pompeian temple (shortly after 550 BC) with nine columns on front and rear (Mertens 1993).

The roof of the Doric temple also follows the Western Greek scheme, and includes some elements of Etruscan tradition. This temple was long considered an example of pure Greek architecture, but recently the opposite has been proposed, i.e. that it is of Etruscan origin (De Waele 2001). In my view, however, it is the first visual impression that is relevant here. This was *meant* to be that of a Greek temple and more specifically referred to Poseidonia.

Considering the complex eclectic combination of architectural elements in both of these strikingly different Archaic temples, it becomes highly improbable that they served to proclaim either a Greek or an Etruscan ethnic identity, as has so often been stated in the past. On the contrary, the common function of the two temples must have been to create and reflect a "Pompeian" urban identity and to place the city within the political network of the region by recalling the two main centres of power: Capua/Cuma in the north and Poseidonia in the south. They hereby transcended ethnic differences.

The religious context

To gain insight into religious beliefs and their role in creating identity, the cults practised in the two temples will now be considered.

The votive deposits of the temple of Apollo, which contained a high number of Greek *kraters*, indicate that the temple was dedicated to the Apollo of Delphi (De Caro 1986). This is confirmed by later inscriptions. The choice is particularly significant, because in the Archaic period the oracle of Delphi served as a diplomatic mediator between Etruscans and Greeks, whenever conflict threatened in the Tyrrhenian Sea (D'Agostino & Cerchiai 1998). Temples dedicated to the Delphic Apollo are to be found in many Etruscan seaports. Pompeii was underlining its role as an *emporion* by dedicating its principal sanctuary to a god worshipped by people who were ethnically different but who shared mutual economic interests.

The second Archaic cult also suggests that religious choices did not express ethnic identity. The Doric temple was probably dedicated to Herakles and Athena. The cult of the mythical founder of the city should be seen as another effective element in the creation of a specifically Pompeian identity.

The linguistic context

The final context takes language as a potential criterion for ethnic identity. The ten Archaic inscriptions from Pompeii are graffiti on ceramics that come from both domestic and votive settings (Christofani *et al* 1996). Most are in Etruscan and were found on bucchero sherds in the votive-deposit of the Apollo temple. They indicate that the lettered elite of 6th Century Pompeii wrote and read Etruscan. An inscription from a domestic context contains a complete Etruscan name, also known from Etruria itself. Because family names can be regarded as typical of a language and are not subject to short-term changes, it can be assumed that Etruscans comprised part of the Pompeian population (Zevi 1998).

Two inscriptions, which can be dated later than the rest, contain Oscan names but still use the Etruscan alphabet. This could indicate a shift within the Pompeian elite. In the 6th Century BC the introduction of Etruscan script to Campania accompanied the process of urbanisation, catalysed by the Etruscans, which integrated different ethnic groups (Cerchiai 1995: 102). The Oscan inscriptions may be the first signs of a Campanian ethnic identity, which emerged from the economic crisis around the middle of the 5th Century BC and is referred to as *ethnos Kampanon* in later sources.

Conclusion

An overview of the four contexts reveals that manifestations of ethnic identity are expressed in only the historic-mythological and the linguistic contexts, i.e. through historical and epigraphical sources. Bourdieu's theory would thus seem to be confirmed in this case.

However, these contexts suggest that the population of Archaic Pompeii was ethnically heterogeneous and was, at the very least, composed of indigenous Oscan-speaking people (whose single or plural ethnic origins remain unclear) and Etruscans. No signs of a stable Greek presence have yet been found.

It is striking that the references to ethnic identity remain indirect (e.g. names) and appear much less often than other forms of collective identity. Political or urban 'Pompeian' identity has clearly been shaped by aspects of culture and religion, expressed most notably in the sanctuaries. Evidence for the specific emergence of these forms of collective identity in Archaic Pompeii becomes even more plausible if viewed within the broader historical characteristics of the period, namely urbanisation and maritime trade networks. Sharp ethnic boundaries would only have impeded their development.

The emphasis placed on forms of identity which transcend ethnic differences proves that ethnicity was not the basis upon which the social and political organisation of the city depended. This must lead to the conclusion that ethnic identity was not of primary importance in Pompeii during the Archaic period.

Acknowledgements

This article presents the results of my master's thesis submitted in 2002 at the University of Amsterdam. I would like to thank the Nederlands Instituut Rome and the Prins Bernhard Cultuurfonds Foundation for their financial and institutional support.

Bibliography

Bourdieu P. 1977. *Outline of a theory of practice.* Cambridge.

Cerchiai L. 1995. *I Campani.* Milano.

Cristofani M., Pandolfini Angeletti M. & Coppola G. (eds.) 1996. *Corpus Inscriptionum Etruscarum, vol. 2.2 Inscriptiones et in Latium et in Campania repertae.* Roma.

D'Agostino B. & Cerchiai L. 1998. "Aspetti della funzione politica di Apollo in area tirrenica", in *I culti della Campania antica. Atti del convegno internazionale di studi in ricordo di Nazarena Valenza Mele Napoli 15.-17. Maggio 1995*. Roma: 119-128.

De Caro S. 1986. *Saggi nell' area del tempio di Apollo. Scavi stratigrafici de A. Maiuri nel 1931-32 e 1942-43*. Napoli.

Fulford F. & Wallace-Hadrill A. 1999. "Towards a history of Pre-Roman Pompeii: Excavations beneath the house of Amarantus (I.9.11-12), 1995-8" *Papers of the British School at Rome* 67: 37-144.

Hall J. 1997. *Ethnic identity in Greek antiquity*. Cambridge.

Hall J. 2002. *Hellenicity: Between ethnicity and culture*. Chicago.

Jones S. 1997. *The archaeology of ethnicity: Constructing identities in the past and present*. New York.

Maiuri A. 1973. *Alla ricerca di Pompei preromana*. Napoli.

Mau A. 1908. *Pompeji in Leben und Kunst*. Leipzig.

Mertens D. 1993. *Der alte Heratempel in Paestum und die archaische Baukunst in Unteritalien*. Mainz.

Renfrew C. 1996. "Prehistory and the identity of Europe, or, don't let's be beastly to the Hungarians", in P. Graves-Brown, S. Jones & C. Gamble (eds.) *Cultural identity and archaeology: The construction of European communities*. London: 125-137.

Smith A. 1986. *The ethnic origin of nations*. Oxford.

De Waele J.A.K.E. (ed.) 2001. *Il tempio dorico del Foro triangolare di Pompei*. Roma.

Zevi F. 1998. "I Greci, gli Etruschi, il Sele. Note sui culti arcaici di Pompei", in *I culti della Campania antica. Atti del convegno internazionale di s tudi in ricordo di Nazarena Valenza Mele Napoli 15.-17. Maggio 1995*. Roma: 1-25.

Monument Conservation in the Mediterranean: Issues and Aspects of Anastylosis

Kalliopi Vacharopoulou – University College London

Introduction

In this paper, the concept of anastylosis as a conservation practice applied to archaeological monuments (*Figure 31.1*) is explored. Anastylosis has been defined in the international conservation charters as *"the reassembly of original but fragmented parts of a monument"*. Its choice, planning and implementation raises various questions and issues, whilst the ongoing debate on its best practice gives an indication of its significance. This paper will particularly focus on issues and aspects of:

terminology: official definition, etymology, use in the international conservation vocabulary;
theory: driving forces, principles and standards guiding anastylosis interventions;
philosophy: philosophical underpinnings of the concept;
practice: technical problems and issues emerging from its application.

Figure 31.1: The Parthenon on the Acropolis of Athens during anastylosis and restoration works (2002)

Anastylosis and the international conservation charters

Since the beginning of the 20th Century, a series of charters and recommendations have been delineated by the international conservation community, aiming at developing codes of ethics to guide practical actions for cultural heritage protection. These charters form the basis of the current international legislation philosophy (Bell 1997: 1-5).

Anastylosis was first introduced in the *Athens Charter for the Restoration of Historic Monuments*, the recommendations of an international conference held in Athens in 1931. Nikolaos Balanos, a Greek engineer who led the re-erection of the monuments in the Athenian Acropolis, described his working method, which he called anastylosis, as *"reassembly of existing but dismembered parts of monuments"* (Dimacopoulos 1985: 16; Schmidt 1997: 43-44).

In 1964, the *Second International Congress of Architects and Technicians of Historic Monuments* met in Venice. The congress concluded in the *International Charter for the Conservation and Restoration of Monuments and Sites*, known otherwise as the *Venice Charter*, which formulated binding principles for monument conservation and restoration (Schmidt 1997: 44) and endorsed anastylosis as a justified form of architectural conservation.

The Venice Charter has been the most influential international conservation document since 1964 (Demas 1997: 152), and remains a useful reference. However, since its creation, its validity has been questioned numerous times (Jokilehto 1996: 72). Despite that, the International Council on Monuments and Sites (ICOMOS) conservation professionals have resisted efforts to change it, as they recognised a need for stability and shared approaches in doctrinal references (Stovel 2001).

Regardless, anastylosis professionals involved with classical monuments have expressed their doubts regarding the Charter's provisions for such intervention. They ascertain that it forms a brief, flexible yet general code of ethics. It cannot give answers to more specific problems restorers face regarding the treatment of classical monuments, as it was drafted by architects in charge of mainly medieval and modern monuments (Archaeological Association 1996: 5, 26, 30, 75, 79-80).

Terminological and etymological aspects of anastylosis

The term *anastylosis* derives from the Greek language. The debate on its etymology is engaging. According to Dimacopoulos (1985: 24), the noun derives from the verb *αναστηλώ*, that etymologically is formed by the prefix *ανά-*, which means 'up', 'over', and the verb *στηλώ*, originating from the noun *στήλη* (*stele* – any upright piece of stone). On the other hand, Sanpaolesi (cited in Patricio & Van Balen 1993: 87-88), claims that *stele* is referring to 'column'.

However, the term has somewhat different uses in various countries and regions. In Italy and Greece, it has acquired a meaning that implies archaeological restoration (Dimacopoulos 1985: 19). Despite the fact that the term has been transliterated in various European languages, in practice it is hardly ever used by conservation specialists and other experts in other parts of Europe (Feilden 1994: 252).

In general, for conservation professionals, *"the confused and confusing use of terms"* can instigate great difficulties (Bell 1997: 23-26). Although ICOMOS has made efforts for consensus around the key words, linguistic, cultural and practical differences among those involved have always complicated their coherence (Stovel 2001).

Theoretical aspects of anastylosis

The theoretical framework of anastylosis is based on standards and principles embracing architectural conservation and restoration. At the same time, some further principles have been introduced in the Greek archaeological restoration field, developed after years of experience in anastylosis of classical monuments.

Significant is the debate originating from the justification for such intervention. Survival of original material and knowledge of the original location of existing members (Patricio *et al* 1993: 88) can form determinative factors. Structural stability emerges as an imperative motive though measures are not widely adopted for protection against weathering and pollution (Jokilehto 1995: 71). In addition, destruction of monuments due to accidental or violent phenomena, such as earthquakes, appears as valid motivation for anastylosis (Feilden 1994: 252; Patricio 1996: 102). Morphological and aesthetic reasons also come forward, as well as the fact that anastylosis can facilitate archaeological and architectural studies of a monument. Reinstatement of a monument's form, restoration of its shape and preservation of historic memory through retention of the monument's material are substantial reasons also.

However, a remarkable issue is related to the values and significance of archaeological heritage. *Values* determined in a monument can be historic, artistic, scientific, cultural, social, economic. *Significance* is ascribed when *"the place signifies or symbolises something larger and more important than merely ruins of its architecture"* (De la Torre & MacLean 1997: 8). Recently, a focus on archaeological heritage values has been advocated as the starting point in any conservation and management decision. By assessing which values should be endorsed, most proper conservation decisions can be made, involving not only the intervention method, but also its extent and manner (Clark 2001: 12; Pye 2001: 60-61).

Values cannot be explored in isolation from concepts of *authenticity* and *integrity*. These concepts indicate originality and material completeness, as well as genuineness in terms of materials, workmanship and date (Pye 2001: 58), although recently the concept of 'original' as synonymous with 'authentic' has come under question. At the same time, driving forces in implementing anastylosis are acknowledged as determinative of the extent and aim of intervention.

Political legitimisation, in the sense of creating or reinforcing identities, has played a considerable role in anastylosis interventions, since monuments can inspire people in forming a common cultural or national entity (*Figure 31.2*). As Herbert (1995: 49) notes,

> *"what a group or a nation seeks to preserve and represent to others, allows us to understand something of what a particular imagined community thinks it is".*

Figure 31.2: The Parthenon on the Acropolis of Athens forms a national landmark, symbolising the peak and achievements of Greek classical culture (2002)

Furthermore, *tourism*, though contributing to materialising the economic potentials of heritage, is not considered a well-intentioned factor in heritage conservation decision-making. A most aggravating occurrence is the disturbing trend of placing high priority on economic values rather than on other values ascribed to monuments (De la Torre & MacLean 1997: 11). It has been alleged that anastylosis *"is carried out in order to satisfy the demands of mass tourism"* (Melucco Vaccaro 1996a: 204; Ibid. 1996b: 330).

Educational values are interrelated with improvement of legibility and interpretation of monuments subjected to anastylosis. *Improvement of legibility* results in making a building's spatial volumes more easily visualised (*Figure 31.3*), since dispersed architectural elements in a monument's surroundings are not easy to comprehend. A reassembled monument can be palpably recorded in the viewer's memory, providing informative didactic treatment (Sivan 1997: 52; Feilden 1994: 94, 252; Mertens 1995: 114; Dimacopoulos 1985: 18). Additionally, *interpretation* enables visitors to understand archaeology and architecture, whilst bringing forth political struggles, cultural fashions, technological skills, artistic expressions, religious beliefs and other aspects of human behaviour (Sivan 1997: 52).

Besides, a critical dispute between Greek anastylosis professionals on the so-called *didactic anastylosis* has encouraged further debate on the educational aims of anastylosis. On the one side it is sustained that in *didactic anastylosis* the educational aim becomes certain, though should not be self-evident, whilst what is projected is a characteristic that is undoubtedly included in every anastylosis (AA 1996: 43, 46). The contradictory argument comprises the idea that non-experts, not possessing abilities to recompose a difficult architectural entity from dispersed members, should be entitled to experience monuments, being interested in them and educated by them (AA 1996: 76-77).

Figure 31.3: The Parthenon on the Acropolis of Athens during anastylosis and restoration works (2002)

Philosophical approaches to anastylosis

Philosophical speculations developed with regard to anastylosis are indicative of the impact of philosophical thought.

The desired determination of method and limits in anastylosis upholds a philosophical speculation. *Method* as a term expresses the way of a technical procedure, whilst *limit* is a stand that includes mathematical and philosophical meaning. Therefore, the *"limits of anastylosis should be a matter of philosophical speculation"* while methodology is purely technical (AA: 19, 21). It is realised then that theoretical and technical questions have a deeper reasoning and cannot be viewed in isolation.

Philippot (1996: 362), referring to ruins as *cultural objects*, asserts that the ruins have *"their own emotional values and appeal to imagination",* which will be completely destroyed by an attempt to restore the ruin to its original state. Any kind of intervention in a monument

> *"can only refer to a knowledge of the lost object, and such a knowledge cannot be identified with the real object without faking"* (Philippot 1996: 362).

Furthermore, anastylosis has been characterised -by Schmidt- as *ruin architecture* (Jokilehto 1995: 70), because the destroyed monument will not be regained, but the result will be an artificial ruin, a new construction with old fragments. As Giraud (AA 1996: 71) characteristically asserts, a hybrid monument with hypothetical and soulless forms will emanate from such intervention.

Additionally, another philosophical underpinning is based on the notion that the values of classical art and architecture are embodied in the meaning of the entity, which is never placed before the meaning of its parts. These monuments, even when dismembered, maintain their life and value. Preservation of their entity is not necessary for someone to be pleased by their beauty and acknowledge their value. Even when their initial function is lost, aesthetic and historical values are included in the smallest fragments (AA 1996: 8-9). As preservation of historic memory depends primarily on preservation of authentic form, special consideration is placed over the notion of material authenticity, which becomes the ultimate aim (AA 1996: 79).

Moreover, it has been sustained that reproduction of same quality ancient monuments is practically impossible simply because the contemporary craftsmen *"will not achieve that pagan flame that was burning in the soul of craftsmen who sculpted the stone"* (AA 1996: 11). What is needed is restoration of the philosophical framework that created these monuments (AA 1996: 72).

Technical aspects of anastylosis

In anastylosis, every element can take up its original unique position and structural role (Mertens 1995: 115, 121). However, a great number of technical issues and varying approaches emerge during the course of works. They are extremely significant for efficient completion of a project.

The main technical issues identified can be summarised as being related to:

The type of buildings with relation to their architecture and structure, their dating and geographical location. Experts seem to consent that anastylosis can be applied to ruined monuments originally erected with use of regularly cut pieces of stone, granite or marble, connected to each other with little or no mortar or even with small metal joints (Dimacopoulos 1985: 18; Mertens 1995: 115; Philippot 1996b: 362; AA 1996: 26). Even after the monument's destruction, architectural elements allow for restitution of original shape (*Figures 31.4 and 31.5*). This is particularly obvious in Greek and Roman ancient buildings, which could be easily dismantled and rebuilt in another location (AA 1996: 26).

Figure 31.4: Surviving architectural members from the Parthenon on the Acropolis of Athens. Example of individuality of construction (2002)

Figure 31.5: Surviving architectural members from the Parthenon on the Acropolis of Athens. Example of individuality of construction (2002)

The surviving ancient material with regard to its amount and state of preservation (*Figure 31.6*), and the use of eroded and weathered fragments. Anastylosis with distorted and altered members that were lying in the ground for centuries, can deprive the monument of its artistic interest (AA 1996: 28; Philippot 1996: 362) and does not guarantee restoration of static characteristics and architectural forms (Mertens 1995: 119-120). It is expected that the best preserved material should be used, because otherwise the quantity of additional material will be increased instead of minimised (Dimacopoulos 1985: 18; AA 1996: 45). Concerning proportions, there has never been a specified percentage of original material that justifies the implementation of anastylosis.

Figure 31.6: The Parthenon on the Acropolis of Athens and the surviving original material found dispersed around the monument (2002)

Completion of missing parts and integrations with new material for what reasons, from what material and in what quantities (*Figure 31.7*). It is acknowledged that reintegrations should only occur for structural, not for morphological reasons (AA 1996: 4). They should be limited in parts where their presence will alleviate the re-assembly of ancient elements. They should be based upon respect for original material, archaeological evidence, original design and authentic documents (AA 1996: 14-15). Every kind of new material used presents advantages and disadvantages. Natural stone has aesthetically satisfying results but is difficult, lengthy and costly to procure. Artificial stone is fast and simple to work and guarantees optimal adhesion of old material to new. It is costly too and presents decay problems (Mertens 1995: 120-121). In terms of integrations distributed in the building's surfaces, experts believe that they should be applied with such distances so as to avoid creating surfaces of new material and in limited dimensions (AA 1996: 6).

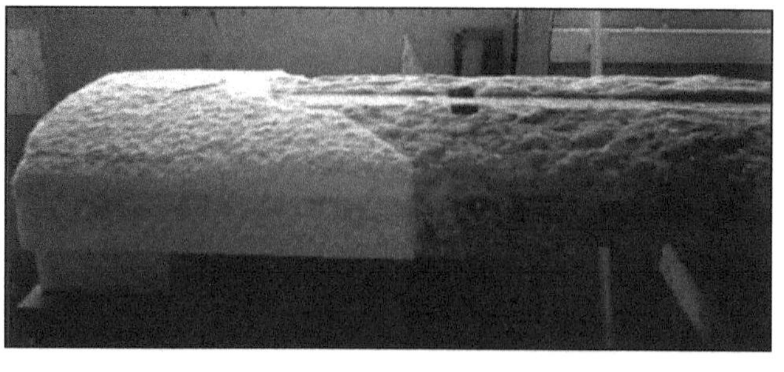

Figure 31.7: Original architectural element completed with new material. Propylaea on the Acropolis of Athens (2002)

The degree of differentiation or approximation of new to old material and how it should be achieved. The experts agree that new material should be clearly distinguished from the original whilst a harmonious co-existence should be accomplished (*Figure 31.8*). The intervention itself has to remain in the background whilst the unity of the whole should be preserved, with a continuity of form uninterrupted from imposed completions. (Mertens 1995: 115, 119; AA 1996: 15; Hueber cited in Mertens 1995: 119; AA 1996: 28). Solutions that have been used over time are: different composition material, different colour, different texture, abstraction of details, contemporary stamps or identifying marks on new members.

Figure 31.8: The Erectheion on the Acropolis of Athens after the anastylosis works. Example of the co-existence between new and original material (2002)

The design of a supporting system. The early projects show a tendency towards a complete construction that would resemble the ancient supporting system, as in the rebuilding of the Athenians' Treasury in Delphi. Later examples present extensive alterations of the construction system by inserting new supporting elements, mostly of reinforced concrete, as in the Marble Court at Sardis. In the latest examples, from the 1970s till today, a return to original supporting systems is achieved by replacing weak components with new elements of different material, as in the Library of Celsus in Ephesus (Starosta 1999: 84-88).

The joining of broken building blocks or of new ones and with what kind of materials. During the passage of time various materials have been used, whilst the results after a period of time have sometimes proved harmful for the monuments. Problems vary according to different original materials -marble, limestone, sandstone. For joining together broken fragments, epoxy glue had almost completely replaced cement, then stainless steel, followed by fibreglass armatures (Mertens 1995: 120). Lately titanium seems to be preferred (as in the anastylosis work on the Athenian Acropolis).

The use of techniques and working tools (*Figure 31.9*). Modern methods and techniques are cost-effective, easier to use and have faster results. Traditional methods and techniques ensure authenticity in workmanship and better results. Generally, there is a tendency to use all means of assistance and any device that was used in antiquity (Mertens 1995: 121).

Figure 31.9: Pantograph – a working tool that was used in antiquity (2002)

The people who undertake such projects. Anastylosis must be the work of a group of individuals with expertise in various fields -archaeologists, architects, civil engineers, art historians, chemists, physicians, conservation specialists and others (AA 1996: 65), together with skilled craftsmen (Michaelidis *et al* 1987: 260).

It is understood that anastylosis is an eminent method and a concept that has been applied to archaeological monuments-especially in the Mediterranean- since early on in architectural conservation history. Various issues are raised during its choice, planning and implementation. Although we should bear in mind that each monument is unique and demands a particular type of treatment, debates are present and more than engaging. All this prompts us to think more about possible ways in which the practice can be improved, so that the monuments of our archaeological heritage can be safeguarded for generations to come. My research is advancing in this direction, aiming to develop a methodological approach to be followed when employing anastylosis for the protection and preservation of ancient monuments.

Acknowledgements

I would like to express my gratitude to my supervisor Professor C.A. Price for kindly looking at a draft of this paper. I would also like to thank the Organising Committee of SOMA 2003 for giving me the opportunity to present this paper.

Bibliography

Archaeological Association (AA). 1996. "Οι Ελληνικές Αναστηλώσεις. Πρακτικά της Δεύτερης Συνεδρίας των Εταίρων" *Ο Μέντωρ: Δελτίο των Εταίρων της Εν Αθήναις Αρχαιολογικής Εταιρείας* 37 (Απρίλιος)

Bell D. 1997. *The Historic Scotland guide to international conservation charters. Historical Scotland Technical Advice Notes, no 8.* Edinburgh.

Clark K. 2001. *Informed conservation: Understanding historic buildings and their landscapes for conservation.* London.

De la Torre M & MacLean M. 1997. "The archaeological heritage in the Mediterranean region", in M. De la Torre (ed.) *The conservation of archaeological sites in the Mediterranean region. An international conference organised by the Getty Conservation Institute and the J. Paul Getty Museum. May 1995.* Los Angeles: 5-14.

Demas M. 1997. "Appendix A: Summary of charters dealing with the archaeological heritage", in M. De la Torre (ed.) *The conservation of archaeological sites in the Mediterranean region. An international conference organised by the Getty Conservation Institute and the J. Paul Getty Museum. May 1995.* Los Angeles: 151-153.

Dimacopoulos J. 1985. "Anastylosis and anasteloseis". *ICOMOS Information* 1: 16-25.

Feilden B.M. 1994. *Conservation of historic buildings.* Oxford.

Herbert D. T. (ed.) 1995. *Heritage, tourism and society.* London.

Jokilehto J. 1995. "Book review: 'Reconstruction of ancient ruins'". *Conservation and Management of Archaeological Sites* 1: 69-71.

Jokilehto J. 1996 "International standards, principles and charters of conservation", in St. Marks (ed.) *Concerning buildings. Studies in honour of Sir Bernard Feilden.* Oxford: 55-81.

Melucco Vaccaro A. 1996a. "Introduction to part III: The emergence of modern conservation theory", in N. Stanley-Price, Jr. M. Kirby Talley & A. Melucco Vaccaro (eds.) *Readings in conservation: Historical and philosophical issues in the conservation of cultural heritage.* Los Angeles: 202-211.

Melucco-Vaccaro A. 1996b. "Introduction to part VI: Reintegration of losses", in N. Stanley-Price, Jr. M. Kirby Talley & A. Melucco Vaccaro (eds.) *Readings in conservation: Historical and philosophical issues in the conservation of cultural heritage.* Los Angeles: 326-331.

Mertens D. 1995. "Planning and executing anastylosis of stone buildings", in N. P. Stanley-Price (ed.) *Conservation on archaeological excavations: With particular reference to the Mediterranean area.* Rome: 113-134

Michaelidis M., Filintra M. & Christofidou A. 1987. "Ο ρόλος του αρχιτέκτονα στην προστασία των μνημείων", in *Έργο και Λειτουργία μιας Υπηρεσίας για την Προστασία των Μνημείων Σήμερα. Υπουργείο Πολιτισμού.* Αθήνα: 255-266

Patricio T.C. 1996. "The nymphaeum at Sagalassos, Turkey: The anastylosis and the applicability", in *Archaeological remains, in situ preservation: Proceedings of the second ICOMOS International Committee on Archaeological Heritage Management international conference, Montreal, Quebec, Canada, October 11-15, 1994.* Ottawa: 101-109.

Patricio T. & Van Balen K. 1993. "The nymphaeum at Sagalassos and the anastylosis", in M. Waelkens & J. Poblome (eds.) *Sagalassos II: Report on the third excavation campaign of 1992. Acta Archaeologica Lovaniensia Monographiae 6.* Leuven: 87-91.

Philippot P. 1996. "Historic preservation: philosophy, criteria, guidelines", in N. Stanley-Price, Jr. M. Kirby Talley & A. Melucco Vaccaro (eds.) *Readings in conservation: Historical and philosophical issues in the conservation of cultural heritage.* Los Angeles: 358-363.

Pye E. 2001. *Caring for the past: Issues in conservation for archaeology and museums.* London.

Schmidt H. 1997. "Reconstruction of ancient buildings", in M. De la Torre (ed.) *The conservation of archaeological sites in the Mediterranean region. An international conference organised by the Getty Conservation Institute and the J. Paul Getty Museum. May 1995.* Los Angeles: 41-50.

Sivan R. 1997. "The presentation of archaeological sites", in M. De la Torre (ed.) *The conservation of archaeological sites in the Mediterranean region. An international conference organised by the Getty Conservation Institute and the J. Paul Getty Museum. May 1995.* Los Angeles: 51-59.

Starosta U. 1999. "Structural concepts of anastylosis". *Conservation and Management of Archaeological Sites* 3: 83-90.

Stovel H. 2001. *Notes on Authenticity*
http://www.unescobkk.org/culture/archives/herb_stovel.doc (accessed on 07/04/2003) Stovel H. 2001.

The skull cult of the Ancient Near East. Problems and new approaches.

Arne Wossink – Leiden University, Netherlands

Introduction

The mortuary behaviour of the Natufian, Pre-Pottery Neolithic A (PPNA), and B (PPNB) periods in the Ancient Near East (ca. 12500-6500 cal. BC) is characterized by what is referred to as the skull-cult. Its beginnings can be traced to the Natufian while it extends well into, and according to some even beyond the PPNB period. In terms of spatial distribution, evidence of the skull-cult has been found from modern southeastern Turkey to southern Jordan (see *Figure 32.1*). Since the excavation of the first plastered skulls by Kathleen Kenyon at Jericho (Tell es-Sultan) in 1953, this so-called skull cult has attracted considerable interest in the archaeology of the Ancient Near East (see Wossink 2003 for references). In general, the skull cult manifested itself through the decapitation of bodies, the separate burial of skulls, and specialised treatment through the plastering or painting of skulls. There are also conceptually connected artefacts such as stone masks, plastered statues, and decapitated figurines (cf. Bienert 1991). In the past, stimulating interpretations of the skull cult have been put forward, for example by Schmandt-Besserat (2002), and Kuijt (2000). For example, the latter author suggested that separated skulls functioned in a system to minimize social stress within communities. However, it is my opinion that any interpretation should take into account and reflect the fact that the periods under consideration were subject to considerable change.

This article seeks to situate a special class of artefacts, i.e. plastered skulls, in the context of contemporary mortuary behaviour, and tries to explain their occurrence as the result of certain developments that took place at that time. Firstly, I will present an overview of the mortuary behaviour of the Natufian, PPNA, and PPNB periods, which is necessary to understand the developments of the latter period, in which plastered skulls appeared. Secondly, I will discuss briefly the research that until now has been carried out on plastered skulls. Finally, I will argue why plastered skulls, or treated skulls in general, appeared only during the PPNB and not earlier, whereas the skull-cult had already been in existence for a long time.

Whilst I acknowledge that skull-removal and treatment continued to be practiced well after the end of the PPNB (see Meiklejohn *et al* 1992), it is my opinion that these occurrences should be studied in a framework other than that presented here. The evidence from Çatal Höyük in particular, with its separated skulls and paintings of vultures and headless bodies, is often employed in interpretations of the Levantine skull-cult. However, this evidence is separated both in time and space from the Levantine cluster of sites. This is equally the case for the plastered skulls from Köşk Höyük, so if it is used, it must be done with caution (cf. Wossink 2003).[1]

Population growth and social complexity

The developments that occurred during the Natufian, Pre-Pottery Neolithic A (PPNA), and Pre-Pottery Neolithic B (PPNB) periods can be summarized as follows (see Kuijt & Goring-Morris 2002 for a more general introduction). During these periods, a sedentary way of life, agriculture and animal husbandry were introduced. These developments were intimately linked with each other and it is useless to study one of them without recourse to the others. While these developments took place at different times in different regions, they were more or less complete by the end of the PPNB. Population growth combined with the adoption of a sedentary way of life was a crucial development from the Natufian onwards. During the PPNB especially, some sites reached an enormous size (Bienert 2001). A growing community crosses several thresholds that force the community to adopt new ways of reinforcing group cohesion. If the group grows so large that its members do not know each other personally anymore, public rituals may become necessary to ensure that the group members are still willing to make sacrifices that serve the group as a whole rather than just the household or the individual (Johnson & Earle 1987). The increasing complexity of settlement organization, and the appearance of communal buildings during the PPNB are two developments that can be seen in this light.

Untreated skulls

The number of untreated separated skulls exceeds by far the number of treated skulls. While treated skulls only occurred frequently during the PPNB, untreated skulls were present from the Natufian period onwards. Also, the majority of the treated skulls were found in modern Jordan and Israel, whereas untreated separated skulls were found as far north as the Taurus Mountains. The first separated skulls appeared during the Natufian period at sites such as Erq el Ahmar, and Hayonim. However, many Natufian burial grounds were found without a single skull removed, indicating that skull-removal was not a common practice during that period. In general, mortuary behaviour was dominated by local customs rather than regional traditions. This changed during the PPNA, when separated skulls were present in large quantities at Nemrik and Qermez Dere in the north down to Jericho in the south. They were buried both in houses

[1] This paper is an adaptation of my MA-thesis, which I finished shortly after SOMA 2003 was held, and presents some of its conclusions.

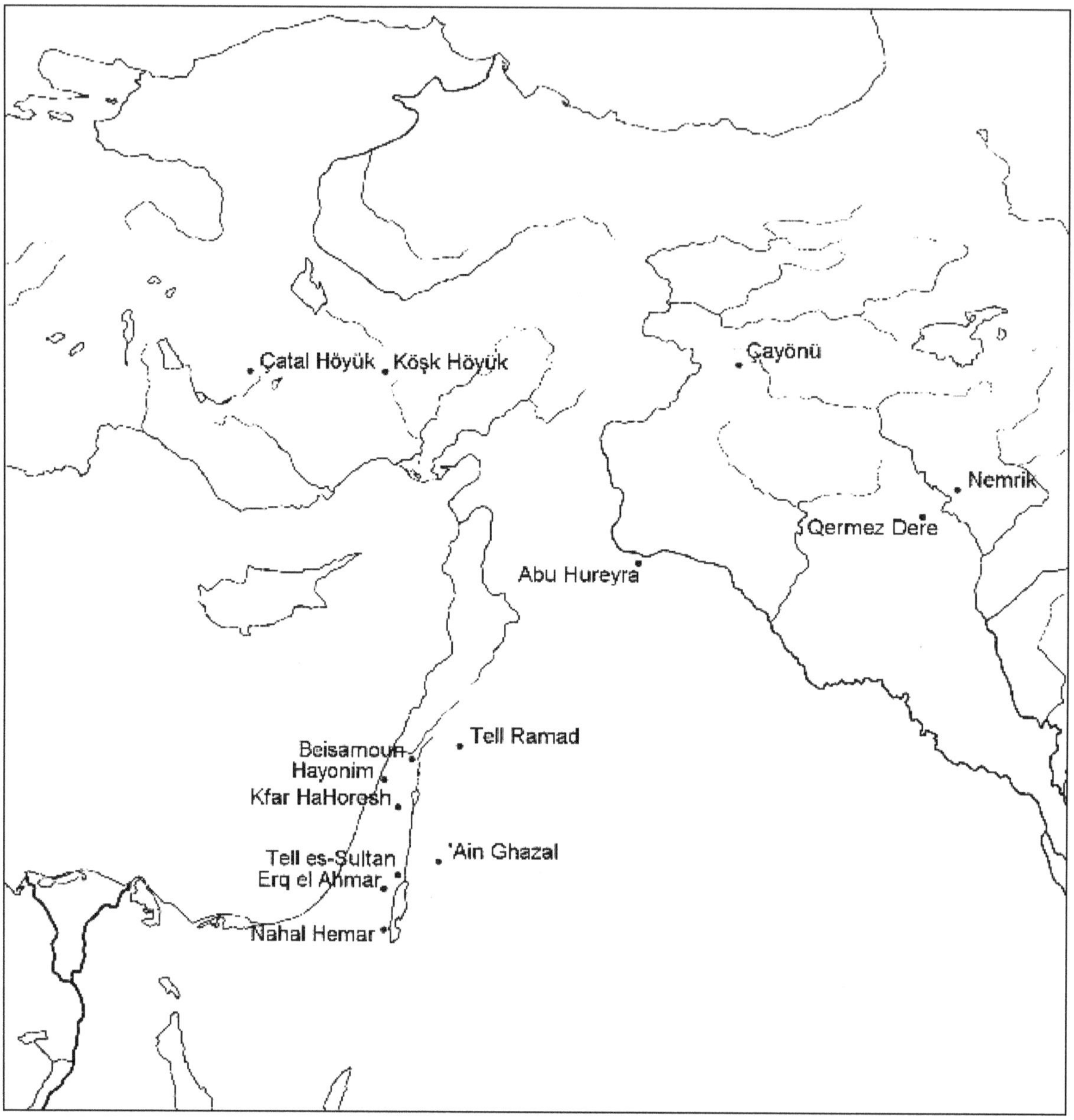

Figure 32.1: Map of the Near East with sites mentioned in the text.

and in public spaces and could be placed singly or in groups. This pattern was also observed during the PPNB.

During both periods, the majority of burials consisted of decapitated skeletons or separated skulls. Because of dubious and sometimes insufficient data, it is difficult to ascertain any patterning based on age or sex, though during the PPNA some sex-based selection may have been practiced, as female skeletons and skulls were predominantly found in private space, i.e. inside dwellings (Wossink 2003). As regards age, both children and adults could have their skulls removed and could also be placed together in skull-caches. Combined with the fact that PPNA and PPNB sites contain far too few burials to account for the site size (Bienert & Müller-Neuhof 2000), the possibility exists that on-site burials functioned primarily as the source for skulls (Wossink 2003).

Plastered skulls

Post mortem treated skulls were only found in the PPNA and PPNB periods, with Qermez Dere providing the oldest, although isolated, evidence in the form of anthropogenically-made holes in some of the skulls (Watkins *et al* 1991: 20). Some traces of paint were observed on skulls from Abu Hureyra, but the most common treatment remains plastering, which was practiced only during the PPNB. Plastered skulls where found at Beisamoun, Kfar HaHoresh, Tell Ramad, 'Ain Ghazal, and Jericho (see fig. 1). They were found both inside houses and in public spaces, and were sometimes grouped in caches of several skulls, and sometimes found singly. The study of plastered skulls can be divided roughly into three categories: firstly, the physical analysis of the construction and components of the plaster;

secondly, morphological studies of the skulls; and thirdly, studies of the artistic and cultural qualities of the skull treatment.

Except for the skulls from Beisamoun, all the other skulls were subjected to analysis to ascertain the composition of the plaster and the way it was applied to the skulls (e.g. Goren *et al* 2001; Bonogofsky 2002). Studies such as these proved that the composition of the plaster and the construction methods of a plastered skull differed from site to site. These conclusions are only strengthened by the skulls from Nahal Hemar, which were entirely different in both their execution and the material utilised (Yakar & Hershkovitz 1988). The plaster added to the skulls was of exceptional quality. Some care was taken to preserve the skulls, as one specimen from 'Ain Ghazal showed traces of repair carried out in antiquity (Goren *et al* 2001). At 'Ain Ghazal three plaster faces or masks were also found, that were once attached to skulls but apparently were removed with care and subsequently buried (Griffin *et al* 1998).

Studies of the morphological qualities of the skulls have resulted in quite different conclusions. The sex of the individuals from which the skulls were taken seems still to be a matter of debate. For example, according to Kurth and Röhrer-Ertl (1981), the skulls from Jericho belonged to young adult females, whereas Strouhal (1973) thought they came from old adult males. It was recently suggested that the plastered skulls from a number of sites including Beisamoun, Kfar HaHoresh and Jericho all belonged to young adult males (Hershkovitz Pers. Comm., cited by Goren *et al* 2001). Skull deformation, whether *in vivo* or *post mortem*, was uncommon. It has been suggested that skull deformation was used to give the skulls the looks of elderly people but the observed deformations are too small to have been noticed in antiquity (Zias 1993; Bonogofsky 2002). The general conclusion of these studies must be that selection of skulls was not based on sexual or morphological traits of the skulls. Age may have played a role in the selection process, but was obviously not the decisive factor, as children could also be subject to skull-removal.

The artistic qualities of the skulls have not received as much attention as the aspects already discussed above. Early articles focused on whether the plastered skulls represented true portraits or were merely idealized faces (Kenyon 1957; Narr 1982). Later, comparisons between the skulls from different sites and even from different skull caches at the same site became important (e.g. Schmandt-Besserat 2002). Studies such as these again emphasized the local differences, and even the unique characteristics of skull-caches of a single site.

In summary, these studies underline the importance of local traditions in the production of plastered skulls. In general, adult skulls were selected for treatment, but no other determinants were found to have affected this choice. Whether the skulls belonged to young or old adults remains a matter of discussion at the moment, but it is possible that the individual to whom the skull belonged did not matter at all in the selection process, i.e. whenever a skull was needed, one was dug up (cf. Schmandt-Besserat 2002; Wossink 2003).

Discussion

I will now turn to the central question of this paper and argue why the act of treating skulls, rather than just separating them, appeared only during the PPNB and not earlier. I have already outlined that during the PPNB, the size of some sites reached a threshold that forced its members to adopt new means to maintain group cohesion. Below, I will argue that plastered skulls, and the rituals in which they were involved, were one of these means.

The picture that emerges from the data presented above is one of an increasing importance attached to the removed skulls themselves, or to the act of removing them. From the Natufian onwards, the percentage of removed skulls and decapitated skeletons increased gradually and became a regional tradition, but probably not during the Natufian proper. This tradition established itself firmly during the PPNA, flourished during the PPNB, and diminished sharply or even disappeared afterwards. The PPNB not only featured the highest percentage of removed skulls, but also the subsequent plastering. Above, I have pointed out that neither untreated nor treated skulls seem to have been selected on the basis of sex, and also that age may have been an important but not the decisive factor in the selection process. On-site burials may have served the continual supply of skulls. The subsequent treatment of skulls can be seen as an elaboration of a pre-existing practice, rather than a new development. As such, its meaning must also be rooted in earlier practices prior to the PPNB.

During the PPNA, the practice of skull-removal was characterized by both intramural and extramural skull burials, which probably represent a difference between rituals conducted at the household and public level. Because skulls from different groups were used in public rituals, they reinforced membership of the community rather than membership of a specific group (Kuijt 2000). During the PPNB, the practice remained generally the same, but is expanded through the act of plastering skulls. However, it is important to note that plastered skulls were more often found outside any building or within a mortuary installation, whereas untreated skulls were mostly found buried in dwellings. The increased size of PPNB sites may have necessitated an elaboration of the communal rituals that ensured group cohesion. The presence of public buildings, caches of plastered statues, and stone masks at large sites such as 'Ain Ghazal points to such a development, and this elaboration may have led to the replacement of untreated skulls by plastered specimens. In the long term, the plastering of skulls may have contributed to increased hierarchical differences within the community, as has been suggested by Kuijt (2000). The fact that plastered skulls were created using local techniques, and that they were found at a non-residential site such as Kfar HaHoresh also points to their potential use in ritual spaces shared by different communities, during which the skulls may have represented the site and communities of their original manufacture (Wossink 2003). In this way, a generally accepted ritual system was created that allowed the fully sedentary communities of the PPNB to interact with each other and to face the social changes that had taken place from the Natufian onwards.[2]

Bibliography

Bienert H.-D. 1991. "Skull cult in the prehistoric Near East" *Journal of Prehistoric Religion* 5: 9-23.

Bienert H.-D. 2001. "The Pre-Pottery Neolithic B (PPNB) of Jordan: A first step towards proto-urbanism?" *Studies in the prehistory and archaeology of Jordan* VII. Amman: 107-119.

Bienert H.-D. & Müller-Neuhof, B. 2000. "Im Schutz der Ahnen? Bestattungssitten im präkeramischen Neolithikum Jordaniens" *Damaszener Mitteilungen* 12: 17-29.

Bonogofsky M. 2002. "Reassessing "dental avulsion" in Neolithic plastered skulls from the Levant through the use of Computed Tomography, direct observation, and photographs" *Journal of Archaeological Science* 29: 959-964.

Goren Y., Goring-Morris A.N., Segal I. 2001. "The technology of skull modelling in the Pre-Pottery Neolithic B (PPNB): Regional variability, the relation of technology and iconography and their archaeological implications" *Journal of Archaeological Science* 28: 671-690.

Griffin P. S., Grissom C.A., Rollefson G.O. 1998. "Three late eighth millennium plastered faces from 'Ain Ghazal, Jordan" *Paléorient* 24 (1): 59-70.

Johnson A.W. & Earle T. 1987. *The evolution of human societies. From foraging group to agrarian state.* Stanford.

Kenyon K. M. 1957. *Digging up Jericho*. London.

Kuijt I. 2000. "Keeping the peace. Ritual, skull caching, and community integration in the Levantine Neolithic", in I. Kuijt (ed.) *Life in Neolithic farming communities. Social organization, identity, and differentiation.* New York: 137-164.

Kuijt I. & Goring-Morris N. 2002. "Foraging, farming, and social complexity in the Pre-Pottery Neolithic of the Southern Levant: A review and synthesis" *Journal of World Prehistory* 16 (4): 361-440.

Kurth G. & Röhrer-Ertle O. 1981. "On the anthropology of the Mesolithic to Chalcolithic human remains from the Tell es-Sultan in Jericho, Jordan", in K.M. Kenyon & T.A. Holland (eds.) *Excavations at Jericho III. The architecture and stratigraphy of the tell.* London: 407-499.

Meiklejohn C.C., Agelarakis A., Akkermans P.A., Smith P.E.L. & Solecki R. 1992. "Artificial cranial deformation in the Proto-Neolithic and Neolithic Near East and its possible origin: Evidence from four sites" *Paléorient* 18 (2): 83-97.

Narr K.J. 1982. "Porträt oder Ahnenschädel?" *Boreas* 5: 16-26.

Schmandt-Besserat D. 2002. "From behind the mask: Plastered skulls from 'Ain Ghazal" *Origini* 24: 95-140.

Strouhal E. 1973. "Five plastered skulls from Pre-Pottery Neolithic B Jericho. Anthropological study." *Paléorient* 1 (2): 231-247.

Watkins T., Betts A., Dobney K., Nesbitt M., Gale R., Molleson, T. 1991. *Qermez Dere, Tel Afar: Interim report No. 2.* Edinburgh.

Wossink A. 2003. *The search for skulls. Mortuary behaviour of the Ancient Near East (12500-7000 cal BC).* Unpublished MA-thesis Leiden University.

Yakar R. & Hershkovitz I. 1988. "Nahal Hemar Cave. The modelled skulls" *Atiqot* 18: 59-63.

Zias J. 1993. "Artificial cranial deformation in the Proto-Neolithic and Neolithic Near East and its possible origin: A reply" *Paléorient* 19 (2): 109-110.

[2] The large sites of the Euphrates and Taurus Mountains regions, including Abu Hureyra and Çayönü, remain problematic because no plastered skulls were found. It is my intention to discuss the mortuary evidence of these sites in a separate article. As regards Çayönü, the fact that still almost no burials have been published, apart from those found in the Skull Building, makes the site difficult to use in any discussion on mortuary behaviour.

www.ingramcontent.com/pod-product-compliance
Lightning Source LLC
LaVergne TN
LVHW071212210926
848482LV00015B/785